621.4022 Heat conduction,
I61hc convection and radiation
1990

CMP

Acknowledgement is made to N.G. Gong et al for the use of Figure 1 on p. 162, which appears on the front cover of this book.

Advanced Computational Methods in Heat Transfer

Vol 1: Heat Conduction, Convection and Radiation

Proceedings of the First International Conference, held in Portsmouth, UK, 17-20 July 1990.

Editors: L.C. Wrobel
C.A. Brebbia
A.J. Nowak

Computational Mechanics Publications
Southampton Boston

Co-published with

Springer-Verlag
Berlin Heidelberg New York London Paris Tokyo

L.C. Wrobel
Wessex Institute of Technology,
Computational Mechanics Institute,
Ashurst Lodge, Ashurst
Southampton SO4 2AA, UK

C.A. Brebbia,
Wessex Institute of Technology,
Computational Mechanics Institute,
Ashurst Lodge, Ashurst
Southampton SO4 2AA, UK

A.J. Nowak
Institute of Thermal Technology
Technical University of Gliwice
44-101 Gliwice
Konarskiego 22, Poland

British Library Cataloguing in Publication Data

Advanced computational methods in heat transfer.
Vol 1. Heat conduction, convection and radiation.
1. Heat transfer. Analysis. Applications of computer systems
I. Wrobel, L.C. (Luiz Carlos) II. Brebbia, C.A. (Carlos
Alberto) *1938-* III. Nowak, A.J.
536.200285

ISBN 1-85312-085-5
ISBN 1-85312-105-3 Set

ISBN 1-85312-085-5 Computational Mechanics Publications, Southampton
ISBN 0-945824-68-8 Computational Mechanics Publications, Boston, USA
ISBN 3-540-52877-6 Springer-Verlag Heidelberg Berlin New York London Paris Tokyo
ISBN 0-387-52877-6 Springer-Verlag New York Heidelberg Berlin London Paris Tokyo

Set
ISBN 1-85312-105-3
ISBN 0-945824-97-1
ISBN 3-540-52879-2
ISBN 0-387-52879-2

Library of Congress Catalog Card Number 90 - 82732

Printed in Great Britain by Bell & Bain, Glasgow

CONTENTS

SECTION 1: HEAT CONDUCTION

PREFACE

Heat transfer problems in industry are usually of a very complex nature, involving the coupling of different types of mechanisms like conduction, convection and radiation, and non-linear features such as temperature-dependent thermophysical properties, phase change and other phenomena. The solution of these problems requires the application of sophisticated numerical methods and extensive use of computer resources.

The present book and its two companion volumes contain edited versions of the papers presented at the First International Conference on Advanced Computational Methods in Heat Transfer, held in Portsmouth, England, in July 1990. The main objective of the conference was to bring together scientists and engineers who are actively involved in developing numerical algorithms as well as in solving problems of industrial interest; to discuss the behaviour of such methods in these extreme conditions and to critically evaluate them by comparison with experiments or established benchmarks wherever possible. All papers have been reproduced directly from material submitted by the authors, and their content is a reflection of the authors' opinion and research work.

The editors would like to thank Prof. D.B. Spalding for his opening address, and all the distinguished scientists who accepted our invitation to deliver special lectures. We are also indebted to the secretarial staff of Computational Mechanics, in particular L. Newman, J.M. Croucher and J. Mackenzie, for the hard work which eventually led to a successful and fruitful meeting.

L.C. Wrobel, C.A. Brebbia, A.J. Nowak
Southampton, July 1990

SECTION 1: HEAT CONDUCTION

Transient Temperature Distribution in a Cylindrical Shell with a Time-Varying Incident Surface Heat Flux and Internal Heat Generation and Effects of Convective and Radiative Cooling

B. Sunden

Chalmers University of Technology, Department of Applied Thermodynamics and Fluid Mechanics, S-41296 Göteborg, Sweden

ABSTRACT

The transient temperature distribution in a cylindrical shell is calculated numerically by a finite difference method. The front surface of the shell is exposed to a time-varying incident surface heat flux. Within the shell, heat is generated uniformly due to an electrical current. The shell surface is cooled by combined convection and thermal radiation. The governing equations and the boundary conditons are cast into a non-dimensional form and the influence of leading physical parameters is analysed thoroughly. The algebraic equations are solved by using the TDMA algorithm. A mixture of implicit and explicit formulations is used and the condition of thermal radiation at the shell surface is linearized. Details of the numerical method as well as some results are provided.

INTRODUCTION

Transient heat conduction in a solid with convective and/or radiative cooling at the surfaces of the solid occurs frequently in practice. Several papers were published in the past but also during recent years some papers have appeared.

This paper deals with a numerical investigation of the transient temperature distribution in a cylindrical shell which is exposed to a time-varying incident heat flux such as a nuclear thermal pulse. Within the shell, heat is generated uniformly due to an electrical current. The outer surface of the shell is cooled by combined convection and thermal radiation. The heat conduction is taken as two-dimensional in space and the inner surface of the shell is insulated. The main objectives of this paper are to present and discuss the numerical solution method and to provide some results. The paper is an extension and complement to a previous paper by the author ref. [1].

Heat transfer problems related to the one to be presented are found for space vehicles and satellites exposed to solar radiation, see refs. [2-9].

In handbooks, e.g. ref. [10], several examples of transient conduction with various surface conditions are available but they are not applicable to the case which is considered in this paper.

If the cylindrical shell is only heated internally by e.g. an electrical current and if cooling by convection occurs, a steady-state solution can be found.

The present problem with an incident time-varying surface heat flux, internal heat generation and cooling by combined convection and thermal radiation has not been treated in the past. Thus the paper will provide new and additional knowledge.

PROBLEM UNDER CONSIDERATION

Consider a circular cylindrical shell as shown in Fig. 1.

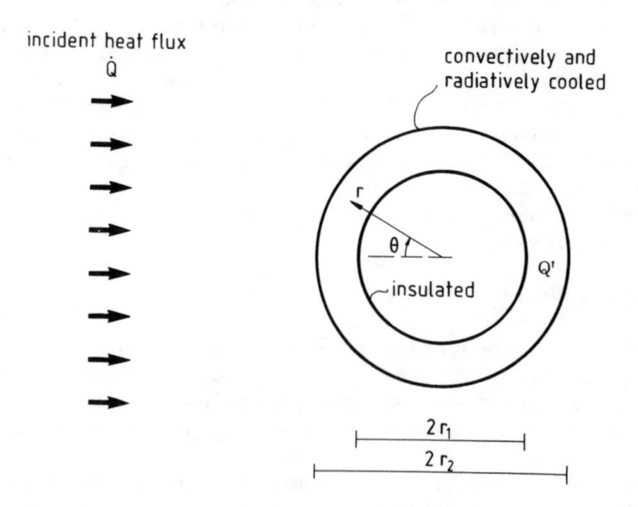

Figure 1. Problem under consideration.

Initially the solid material has a temperature distribution corresponding to a uniform internal heat source (Q' W/m^3) and convective cooling at the surface. Suddenly its outer front surface is exposed to a high incident radiative heat flux which is varying in time as depicted in Fig. 2. The incident surface heat flux can in this case be regarded as an optically parallel beam and thus at every position on the front surface, the heat flux has to be multiplied by the cosine of the angle between the surface normal and the direction of the incident heat flux. The whole outer surface of the solid exchanges heat with the surrounding environment by convection and radiation. The inner surface is in direct contact with a centrebody which has such a low thermal conductivity that the inner surface can be regarded as thermally insulated at every instant of time. The conduction in the solid is assumed to be two-dimensional and the coordinate system is placed as shown in Fig. 1. The solid is opaque to thermal radiation and the physical properties are uniform and independent of temperature. The ambient air is transparent to thermal radiation and its temperature T_{o2} is independent of time. The background radiation temperature is constant and denoted by T_{o1}.

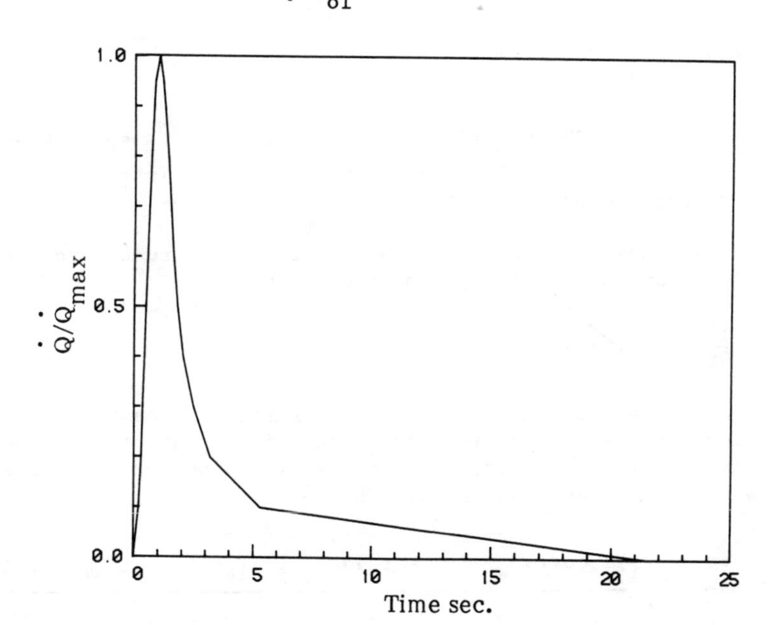

Figure 2. Incident heat flux vs time.

BASIC EQUATIONS AND BOUNDARY CONDITIONS

The two-dimensional form of the heat conduction equation in cylindrical polar coordinates yields

$$\frac{\partial T}{\partial t} = a_1\{\frac{\partial^2 T}{\partial r^2} + \frac{1}{r}\frac{\partial T}{\partial r} + \frac{1}{r^2}\frac{\partial^2 T}{\partial \theta^2}\} + \frac{Q'}{\rho c} \tag{1}$$

where a_1 is the thermal diffusivity of the solid shell. This equation is cast into a non-dimensional form by introducing the following variables:

$$T' = (T - T_{ref})/T_{ref}, \quad r' = (r - r_1)/(r_2 - r_1),$$
$$t' = a_1 t/(r_2 - r_1)^2, \quad \varphi = \theta/\pi \tag{2}$$

One then has

$$\frac{\partial T'}{\partial t'} = \frac{\partial^2 T'}{\partial r'^2} + \frac{1}{r' + c_1}\frac{\partial T'}{\partial r'} + \frac{1}{\pi^2 (r' + c_1)^2}\frac{\partial^2 T'}{\partial \varphi^2} + \frac{Q'}{kT_{ref}}(r_2 - r_1)^2 \tag{3}$$

where $c_1 = r_1/(r_2 - r_1)$.

A non-dimensional group, $Q'(r_2 - r_1)^2/kT_{ref}$, appears in equation (3). This will hereafter be called the internal heat flux number and denoted by HNI.

The inner surface of the shell is assumed to be insulated and the boundary condition to be

$$\frac{\partial T'}{\partial r'}(0,\varphi,t') = 0 \tag{4}$$

For the front part of the outer surface, a heat balance between the absorbed incident radiative heat flux, the thermal radiation loss, the convective cooling and the outward heat conduction yields:

$$\frac{(r_2 - r_1)\alpha\dot{Q}\cos(\gamma)}{kT_{ref}} - \frac{(r_2 - r_1)\sigma\epsilon T_{ref}^3((T_s' + 1)^4 - (T_{o1}' + 1)^4)}{k} - \frac{(r_2 - r_1)h}{k}(T_s' - T_{o2}') - \frac{\partial T'}{\partial r'} = 0 \tag{5}$$

where γ is the angle between the surface normal and the direction of the incident parallel beam of radiant energy, T'_{o1} and T'_{o2} the dimensionless temperatures of the background radiation and the surrounding air, respectively, and T'_s is the dimensionless surface temperature $T'(1,\varphi,t')$. k is the thermal conductivity of the solid and h the convective heat transfer coefficient.

For the rear part of the outer surface, equation (5) is also valid but with $\dot{Q} = 0$.

In equation (5), three non-dimensional groups appear. They are identified as: the Biot number $Bi = h(r_2 - r_1)/k$, the radiation number $M = \sigma\epsilon T^3_{ref}(r_2 - r_1)/k$ and the external heat flux number $\alpha\dot{Q}(r_2 - r_1)/kT_{ref}$. Since the latter varies with time, it is more convenient to use the maximum value of $\dot{Q}$ to represent the incident heat flux. Thus the notation HN is introduced as $HN = \alpha\dot{Q}_{max}(r_2 - r_1)/kT_{ref}$.

The importance of the radiative and convective cooling can be studied by applying various values of the radiation number M and the Biot number Bi while the influence of the strength of the incident heat flux can be simulated by using various values of HN. The influence of the internal heat source can be studied by varying the internal heat flux number HNI. However, when generalizing the results, one has to observe that the variation of the incident heat flux is given as a function of real time. This implies that for solids having different thermal diffusivities the order of magnitude of the Fourier number ($t' = Fo$) may differ considerably. Also, since the thermal conductivity k appears in the dimensionless numbers Bi, M, HN, HNI, the values of these numbers differ for good and poor conducting materials at the same magnitude of the heat fluxes and convective heat transfer coefficient.

STEADY-STATE SOLUTION WITH INTERNAL HEAT GENERATION

If the cylindrical shell is only heated by a uniformly distributed heat source (Q' W/m^3) and cooled by convection, a steady temperature distribution is found. An analytical solution is easily obtained from equation (3) if only radial conduction is considered. For the special case of $r_1 = 0$ the solution will be identical to that one found in most textbooks, e.g. ref. [11]. The solution for the cylindrical shell reads

$$T' - T'_{o2} = HNI \cdot \{ \frac{1}{4} (\frac{r_2^2}{(r_2 - r_1)^2} - (r' + \frac{r_1}{r_2 - r_1})^2) +$$

$$\frac{r_1^2}{2(r_2 - r_1)^2} \cdot (\ln \frac{r_2 - r_1}{r_2} + \ln(r' + \frac{r_1}{r_2 - r_1})) +$$

$$\frac{1}{Bi} \cdot \frac{r_2 + r_1}{2r_2} \} \tag{6}$$

NUMERICAL SOLUTION PROCEDURE

Equation (3) is solved numerically by using finite difference approximations. The symmetry is taken into account and the solution domain is bounded by $\varphi = 0$ and 1. In the space coordinates (r', φ) central second-order approximations are employed while for the discretization in time a first-order approximation is used. The derivatives in equations (4) and (5) are calculated by second-order forward and backward finite difference approximations, respectively.

With the finite difference approximations introduced, equation (3) can be written (i, circumferential direction; j, radial direction; n, time level)

$$a_{ij} \cdot T'^{n+1}_{i,j} = b_{ij} \cdot T'^{n+1}_{i,j+1} + c_{ij} \cdot T'^{n+1}_{i,j-1} + d_{ij} \tag{7}$$

where

$$a_{ij} = \frac{1}{\Delta t'} + \frac{2}{\Delta r'^2} + \frac{2}{R'^2 \pi^2 \Delta\varphi^2} \tag{8}$$

$$b_{ij} = \frac{1}{\Delta r'^2} + \frac{1}{2R' \Delta r'} \tag{9}$$

$$c_{ij} = \frac{1}{\Delta r'^2} - \frac{1}{2R' \Delta r'} \tag{10}$$

$$d_{ij} = \frac{T'^{n}_{i,j}}{\Delta t'} + \frac{T'^{n}_{i+1,j} + T'^{n+1}_{i-1,j}}{\pi^2 R'^2 \Delta\varphi^2} + HNI \tag{11}$$

In equations (8) - (11) $R' = r' + c_1$. The step sizes $\Delta r'$ and $\Delta\varphi$ are constants but the proper sizes have to be found by test calculations. From equations (7) and (11) it is obvious that a fully implicit formulation is used in the radial direction while in the circumferential direction the formulation is explicit.

Equation (5) is solved for the dimensionless surface temperature T'_s. However, the term $(T'_s + 1)^4$ is linearized according to

$$(T'_s + 1)^4 = (T'^{n+1}_s + 1)\cdot(T'^{n}_s + 1)^3 \tag{12}$$

The derivative $\partial T'/\partial r'$ in equation (5) is approximated as:

$$\frac{\partial T'}{\partial r'}(1,\varphi) = \frac{3T'^{n+1}_s(\varphi) - 4T'^{n}(1-\Delta r',\varphi) + T'^{n}(1-2\Delta r',\varphi)}{2\Delta r'} \tag{13}$$

The TDMA algorithm is employed in the solution procedure. With the linearization (12), a non-iterative solution procedure is achieved and thus a rapid solution is obtained at each time level. The linearization (12) was not used in ref. [1].

Sample calculations

The appropriate number of grid points in the space plane was determined by test calculations of the steady-state problem with a uniformly distributed heat source in the shell. The analytical solution was given by equation (6). For the particular size of the shell (r_1 = 5.9 mm, r_2 = 7.3 mm), it was found that 141 grid points in the radial direction and 32 grid points in the circumferential direction gave sufficient accuracy. For another shell size the necessary number of grid points may differ.

The time step $\Delta t'$ was determined by considering the time scale of the shell material and the time dependence of the incident radiative heat flux. For a shell material having a thermal diffusivity $a_1 = 6.6\cdot 10^{-8}$ m^2/s, a dimensionless time step of $3.4\cdot 10^{-6}$ (real time $\Delta t = 10^{-4}$ s) was less than necessary. T_{o1} and T_{o2} was set equal to T_{ref}.

RESULTS AND DISCUSSION

Figure 3 shows the initial temperature profile. It is valid for HNI = 0.4625 and Bi = 5.

In Fig. 4, the surface temperature $T'_s(1,0,t')$ is shown for HN = 8.8, Bi =5, M = 0.15 and HNI = 0.4625. As the front surface is exposed to the incident radiative heat flux the dimensionless surface temperature starts to increase rapidly, reaches a maximum and then begins to decrease gradually. The maximum temperature is greatly affected by the Biot number, the radiation number and the magnitude of the incident heat flux (HN). This was already shown in [1].

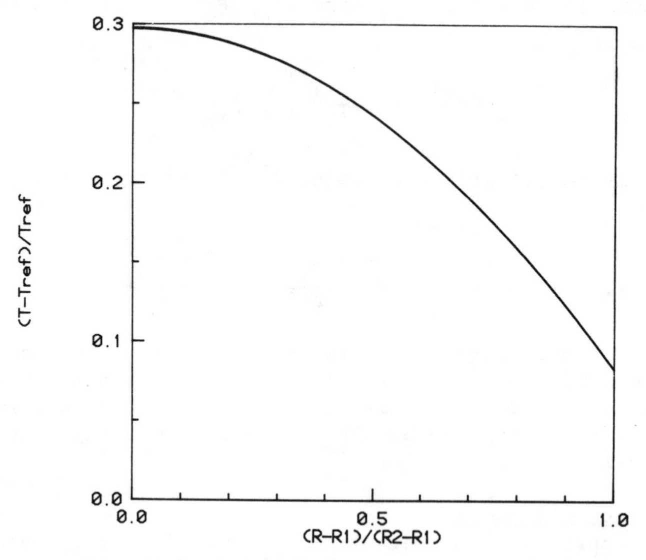

Figure 3. Initial temperature profile.

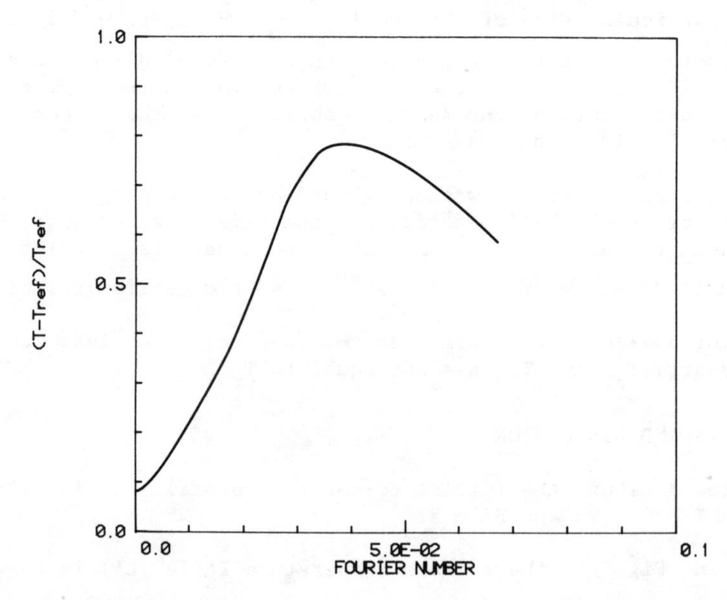

Figure 4. Surface temperature at $\varphi = 0$ vs dimensionless time.

Figure 5 provides temperature distributions within the shell. T'(r',0) is given at various instants of time. As is obvious, at first only a thin layer close to the surface is affected but gradually more and more of the shell material is affected by the strong incident heat flux. The temperature profile then differs considerably from the initial one. As time proceeds and the incident heat flux has decayed totally, the temperature distribution asymptotically approaches the initial profile.

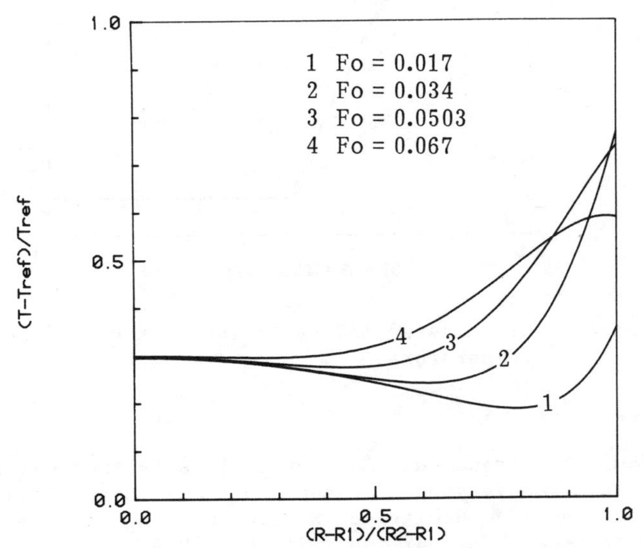

Figure 5. Radial temperature distributions at $\varphi = 0$.

In Fig. 6 the surface temperature $T_s'(1,\varphi)$ is depicted. The rear part of the shell surface remains at its initial temperature since it is not hitted by the incident heat flux and due to the fact that the shell material is conducting heat poorly.

CONCLUSIONS

A numerical study of the transient temperature distribution in a cylindrical shell exposed to a time-varying incident surface heat flux and with internal heat generation has been carried out.

A numerical method based on finite difference approximations and a mixture of implicit and explicit formulations was developed and found to work well.

Some interesting results with realistic values of the leading physical parameters were presented.

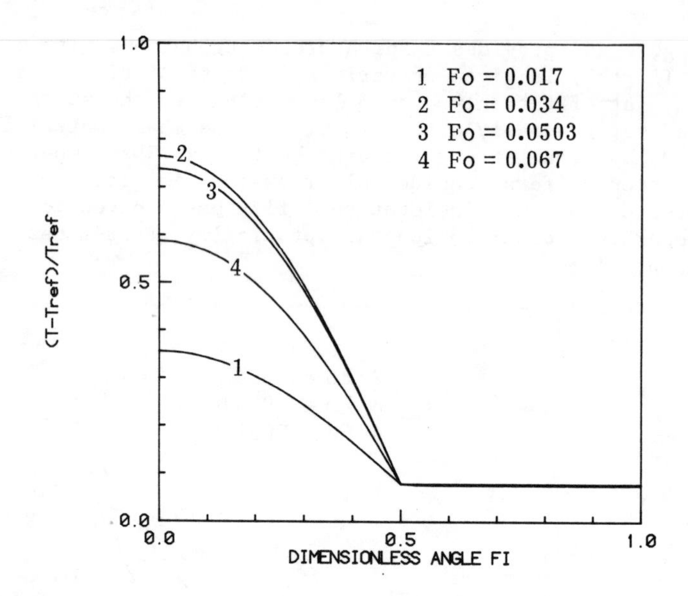

Figure 6. Circumferential variation of the surface temperature.

REFERENCES

1. Sunden, B. Transient Conduction in a Cylindrical Shell with a Time-Varying Incident Surface Heat Flux and Convective and Radiative Surface Cooling, Int. J. Heat Mass Transfer, Vol. 32, pp. 575-584, 1989.

2. Nichols, L.D. Surface Temperature Distribution on Thin-Walled Bodies Subjected to Solar Radiation in Interplanetary Space, NACA TN D-584, 1961.

3. Charnes, A. and Raynor, S. Solar Heating of a Rotating Cylindrical Space Vehicle, ARS J., Vol. 30, pp. 479-484, 1960.

4. Olmstead, W.E. and Raynor, S. Solar Heating of a Rotating Spherical Space Vehicle, Int. J. Heat Mass Transfer, Vol. 5, pp. 1165-1177, 1962.

5. Olmstead, W.E., Peralta, L.A. and Raynor, S. Transient Radiation Heating of a Rotating Cylindrical Shell, AIAA J., Vol. 1, pp. 2166-2168, 1963.

6. Prelewicz, D.A. and Kennedy, L.A. Radiant Heating of a Rotating Thick-Walled Spherical Satellite, AIAA J., Vol. 5, pp. 179-181, 1967.

7. Iqbal, M. and Aggarwala, B.D. Solar Heating of a Long Circular Cylinder with Semigray Surface Properties, J. Spacecraft, Vol. 5, pp. 1229-1231, 1968.

8. Iqbal, M. and Aggarwala, B.D. Radiant Heating of a Solid Spherical Satellite, AIAA J., Vol. 7, pp. 784-786, 1969.

9. Sikka, S., Iqbal, M. and Aggarwala, B.D. Temperature Distribution and Curvature Produced in Long Solid Cylinders in Space, J. Spacecraft, Vol. 6, pp. 911-916, 1969.

10. Schneider, P.J. Conduction. Section 3, Handbook of Heat Transfer, (Ed. Rohsenow, W.M. and Hartnett, J.P.), McGraw-Hill, New York, 1973.

11. Incropera, F.P. and Dewitt, D.P. Fundamentals of Heat and Mass Transfer, 2nd Edn, J. Wiley & Sons, New York, 1985.

An Iterative Graphical Mapping Technique for the Prediction of Temperatures Arising from Essentially Two-Dimensional Heat Flow

D.J. Dean

Atomic Weapons Research Establishment,
Aldermaston, Berkshire, England

ABSTRACT

An iterative graphical method may be used to produce a temperature map of a two-dimensional region containing several heat sources; it is an extension [1a] of the graphical method used in the design of refrigerators as early as 1929 [2] and more recently of high frequency antennae [3].

1. INTRODUCTION

The graphical method is a versatile technique, far more than just a means of temperature evaluation. Besides providing quick solutions having errors of only a few percent, the technique inevitably leads to an ability to visualise heat flow itself.

This paper first describes the original method as applied to a long strip heat source mounted on the surface of a substrate whose other face is bonded to an isothermal heat sink; and second, the method's extension to multiple heat source problems.

2 BASICS OF THE GRAPHICAL METHOD

The graphical method consists of drawing a thermal map by plotting a series of isotherms and flow lines. An isotherm is defined as a line of constant temperature, that is a line of zero temperature gradient. A flow line is a line across which there is no net flow of thermal energy at any point; hence there is no temperature gradient normal to flow lines. Isotherms must therefore intersect flow lines at right angles. The fundamental basis for the graphical method is the principle of thermal resistance per square.

Consider a square sheet of thickness 't', edge length 'a', and thermal conductivity of 'k'. If heat flows normal to one edge at the rate of '$\dot{Q}$', the temperature drop across the square will be:

$$\theta = \frac{\dot{Q}a}{akt} = \frac{\dot{Q}}{kt}$$

That is, the temperature difference is independent of square dimension. The temperature drop per unit rate of heat flow is defined as being the thermal resistance per square.

In the graphical technique the sequence of construction involves the sketching of isotherms and lines of heat flow such that the figures formed by the intersection of the two sets of lines will approximate to curvilinear squares (Figure 1). Curvilinear squares have many properties in common with rectilinear squares, in particular that of having a thermal resistivity independent of size. Isotherms and flow lines are re-drawn alternately until the majority of shapes formed approximate to curvilinear squares, and those few deformed 'squares' are well scattered in both position and direction. The graphical method is not an exact analytical technique but an iterative method controlled visually rather than numerically. The temperature distribution is obtained by counting isotherms and is described in section 2.2.

2.1 Illustration of the Method as Originally Applied

Consider a long strip heat source mounted upon a thin rear cooled substrate. In the region approximately midway along the source the heat flow will be essentially two-dimensional and in a plane that has the lengthwise direction of the strip as a normal.

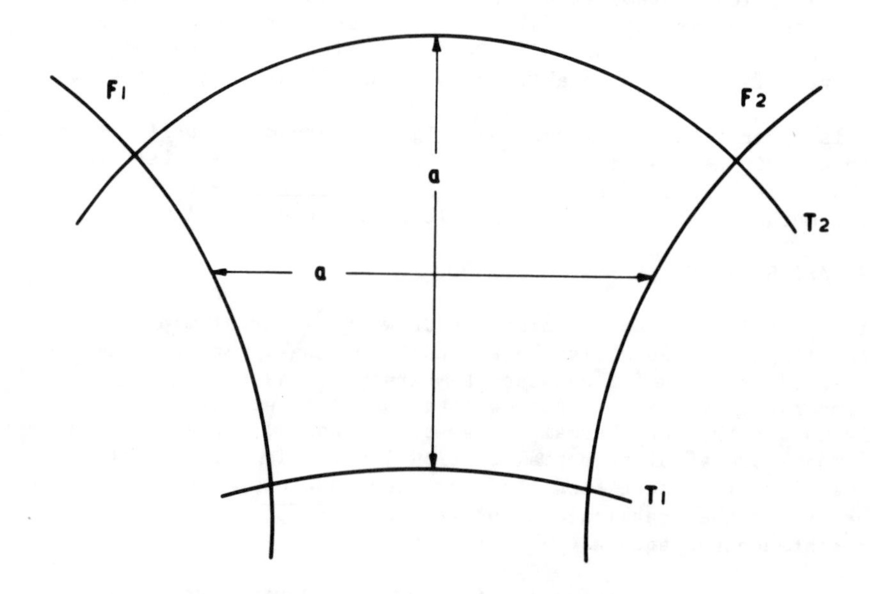

FIGURE 1 Example of a Curvilinear Square

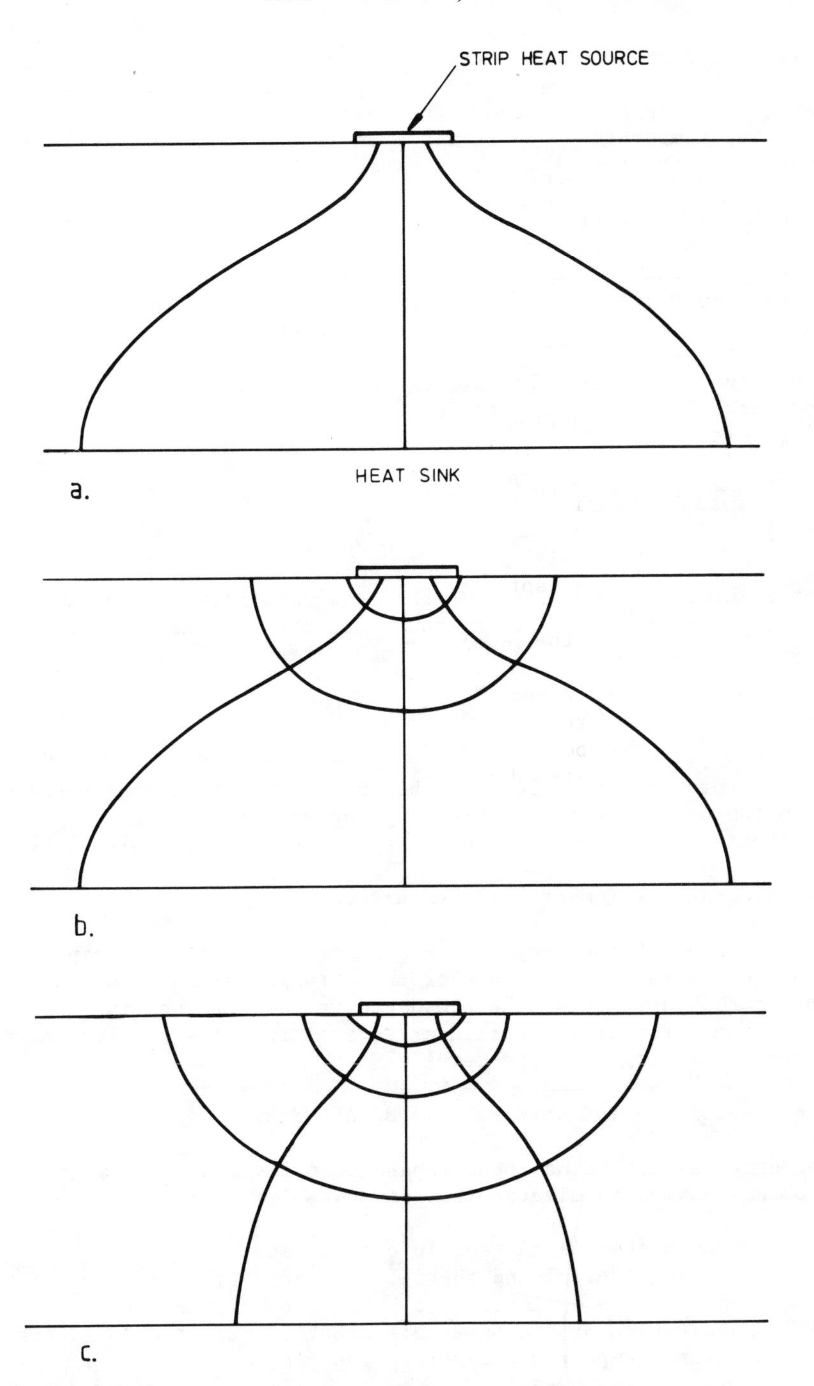

FIGURE 2 Graphical Analysis of a Long Strip Source

Assume four individual heat sources go together to make up the strip source; the flow paths will spread from the strip in a manner similar to that shown in figure 2a. The heat sink is at a constant temperature and is hence an isotherm. The flow lines must therefore meet the heat sink at right angles.

Once the initial flow lines have been illustrated, the next step is to attempt to construct curvilinear squares using these flow lines and adding isotherms as in figure 2b. As the heat sink is already an isotherm it is better, in all but the most simple cases, to draw the isotherms starting with those closest to the heat sink. The flow lines are then re-drawn to give better curvilinear squares. The sequence of modifying flow lines and contours alternately is continued until a temperature map similar to Figure 2c results.

2.2 Temperature Evaluation

The temperature interval between successive isotherms is based upon the establishment of a basic unit of heat dissipation '$\dot{Q}$' for each flow path. Consider the flow path between the two flow lines F_1 and F_2 in figure 1; the temperature interval between successive isotherms has already been shown to be:

$$\Delta T = \frac{\dot{Q}}{kt} .$$

Temperatures can therefore be obtained by counting the isotherms from the heat sink and multiplying the number by ΔT. Temperatures between successive isotherms are interpolated by eye.

2.3 Accuracy of Temperature Evaluation

The accuracy of the temperature evaluation depends principally upon the overall correctness of the curvilinear squares, which is visually assessed. The human eye is rather sensitive to deviations from true curvilinear squares and this results in the technique giving accurate results.

3 EXPANSIONS OF THE ORIGINAL GRAPHICAL METHOD

The original method has been expanded to cover a wide range of problems common in electronics, for example:

a) Air- and paste-filled gaps in contact areas [1b].
b) Changes in material and thermal conductivity [1c].
c) Voidage and the optimising of substrate thickness [1d].
d) The evaluation of the power distribution required to maintain a constant temperature distribution [1e].

4 APPLICATION OF METHOD TO MULTIPLE HEAT SOURCE PROBLEMS

In microelectronics thermal design requires the prediction of the temperatures that arise on edge-cooled circuit boards containing multiple heat sources. The original graphical method had to be extended to cope with the effects of interaction between different heat sources.

To assist with the application of the extended graphical method a viscous fluid analogy to heat flow has been developed. This analogy also helps in the description of the problems that prevent the direct application of the original method to multiple source situations.

4.1 The Viscous Fluid Analogy

The viscous fluid analogy is a means of obtaining a very good three-dimensional image of two-dimensional heat flow. A typical electronic circuit board having several heat sources cooled by conduction to an edge-mounted heat sink is analogous to a table representing the circuit board with holes connected to fluid pumps representing the heat sources. These fluid pumps produce viscous (treacle-like) volcanoes. One edge of the table is left clear to represent the heat sink while the others are bounded by barriers to represent the insulated edges. When the fluid pumps are switched on a representation of the temperature profile results, initially of the transient performance and then of the steady state once equalibrium is reached.

As a result of the interaction between the different heat sources the original graphical method had to be extended to cope with the ensuing difficulties. How these interactions affect the completion of thermal maps and how to overcome these problems can be visualised and described using the viscous fluid analogy. A further problem is that multiple heat sources cannot always be reduced to a practicable number of individual sources each generating the same quantity of heat. This problem can also be tackled by considering the viscous fluid analogy.

It is helpful to note that although the viscous fluid analogy was developed to simplify instruction in graphical methods, the graphical method could also be described as a means of quantifying the heat flow visualised in the analogy.

4.2 Techniques for Extending the Graphical Method to Multiple Heat Sources

In order that multiple heat source problems can be tackled, flow path distribution and interaction have to be considered.

4.2.1 Lack of Uniformity in the Heat Generation of Different Sources

The temperature contours in one flow path have to be continuous with those in adjacent paths. For this to be so each path must contain the same heat flow. This would not constitute a problem if each heat source had a power level equal to a small multiple of a common value. However, the power dissipations of individual heat sources are not usually in simple small ratios to each other.

To maintain equal heat flows in each path some compromise has to be made in the assumed heat source distribution. Once the user has gained experience in the technique of solving problems in which the heat sources have power levels in simple proportions, less rigidly defined problems may be tackled.

The establishment of uniform flow path heat content will involve the lumping together of some heat sources and the borrowing of some heat input from one to give to another. Consideration of the viscous fluid analogy will indicate those areas where such action will not have a significant effect or where the changes counterbalance each other. As the thermal map develops it may be appropriate to modify the redistribution of heat generation. This is all part of the iterative procedure and is judged from visualisation of the viscous fluid analogy, which has in turn been influenced by the current thermal map.

4.2.2 Interaction Between the Heat Flow from Different Sources

In the original method only an isotherm could meet a flow line at right angles. Consider the viscous fluid analogy for two heat sources in a line normal to the heat sink. The line joining the two sources will appear as a ridge, peaking at the two viscous fluid volcanoes. The fluid height falls away on both sides of this ridge and there can be no net heat flow across such a ridge. The ridge is therefore a flow line. At the lowest point on the ridge between the two volcanoes valleys will start to go down both sides. Ignoring momentum, which heat flow does not possess, the viscous fluid will not cross the valley floor as this would require it to flow uphill. The base of the valley where a stream would exist is therefore also a flow line.

Physically what is happening is that the fluid from one volcano has to divide to flow around the back-flow from the second volcano. The ridge between these two sources is a crest flow line and the respective fluid flows meet at the lowest point with equal opposing pressure. The flows then proceed together down the valley, maintaining equal and opposite pressure along a trough flow line. A trough flow line always separates the heat flow from different heat sources or is at an insulating boundary.

The crest and trough flow lines cross at right angles and the point at which they meet is a saddle point. Previously it was claimed that any line meeting a flow line at right angles was a temperature contour, that is a line of zero temperature gradient. However, a saddle point is also a point of zero temperature gradient and the conflict is accommodated by establishing a temperature plateau at this point.

Similar temperature plateaux exist where the flows from two adjacent heat sources have to divide when they meet the backward flow from another heat source. Another plateau occurs where the backward flow from a heat source reaches an insulated boundary.

4.2.3 The Incorporation of Malleable Membranes in the Viscous Fluid Analogy

The graphical method is an iterative method, and as with all iterative methods a good initial choice of solution will speed up the evaluation. Wrong decisions are eliminated during the evaluation but this sometimes means that the same ground has to be covered twice. On its own the viscous fluid analogy greatly speeds up the drawing of flow lines. Visualisation can be further improved by using an additional analogy for the flow lines themselves.

There is no net flow across a flow line, so in the viscous fluid analogy a barrier to fluid flow would automatically be a flow line itself. In order that such a barrier does not affect the final fluid contours it must not impede the actual fluid flow. The barrier that is visualised is a perfectly malleable membrane which: automatically takes up a position where the fluid pressures on both sides are equal, and can be deformed in any way with no effect on the fluid flow.

The visualisation of malleable membranes greatly helps in deciding where the flow paths are likely to occur. However, its most powerful use is in preventing the commonest error that occurs when using the graphical method with more than one heat source: as the thermal map is being developed new users frequently include some flow paths which do not contain any heat flow. This error is not restricted to the graphical method, as from time to time papers are published which are based on finite element and finite difference evaluations yet show temperature gradients without any heat flow.

Unless a fluid is contained and has no route of escape it is not possible for two membranes to be kept apart without fluid flowing between them. Similarly it is not possible to separate two heat flow lines without heat being generated and flowing between them.

4.3 Solution of Multiple Heat Source Problems

The temperature map is most rapidly constructed by alternately using the flow lines to generate the temperature contours and the most recent contours to generate a new set of flow lines. As in the original applications of the method this procedure is repeated until a visually acceptable picture has been obtained. In order that the early lines can be easily discarded the most effective technique is to use pads of tracing paper.

A scale drawing of the problem layout is made on card and this is placed under the rear sheet of tracing paper in the pad. Flow lines and contours are then drawn alternately, each on separate sheets, moving the card with the layout up through the pad.

4.3.1 Initial Stage of Thermal Map Construction

Initial and guiding flow lines are drawn to follow the temperature crests and troughs. It has already been established that a temperature ridge occurs between two heat sources with a plateau between them. These ridges are the first lines to be drawn and are described as a crest flow lines. Other temperature crests occur between the sources and the heat sink. The initial positions for the crests and troughs are established by visualising the situation using the viscous fluid analogy. As the influence of other heat sources can mask these features temperature crests and troughs are defined as follows:

A temperature crest is:
    The shortest heat flow line from a heat source to a heat sink or temperature plateau. (This does not necessarily mean the shortest distance, since other sources can cause the flow line to curve.)

A temperature trough is:
    The dividing line between the heat flowing from one source and that flowing from another source.

By definition no flow can take place into a perfect insulator, so any such boundary flow must be parallel to the surface. An insulated boundary is hence taken as one of the chosen flow lines. As no flow can take place across a line of symmetry within the board, a straight insulated boundary could be replaced by a mirror image circuit [1f]. The flow line at an insulated boundary is therefore a trough.

4.3.2 Equalisation of the Power Level of the Flow Paths

Section 4.2.1 described how uniform flow path content could be established by borrowing some heat content from one source to lend to another and by lumping some heat sources together. Once temperature crests and troughs are established two flow paths for each individual or lumped power source exist. Additional flow lines are then added such that if each heat generating element is considered as a number of equal discrete heat sources, then each unit source generates its own flow path. All sources must, therefore, be subdivided into multiples of the basic unit of dissipation. Alternatively, it is sometimes more convenient to achieve this equality of flow in each path by removing some of the crest flow lines which go directly to a heat sink.

The decisions on distribution of generated heat are made by considering the viscous fluid analogy. In section 4.1 the graphical method was described as the quantification of the viscous fluid analogy. As iterations proceed the thermal map may change the visualisation of the viscous analogy. This can result in the modification of the initial choice of power level distribution. This is all part of the iterative method and enables early decisions to be made without fear.

4.3.3 Establishing the Initial Flow Paths

Every flow path has to have an input and an output and each starts at a heat source and ends at a heat sink. As a stepping stone to making the temperature prediction the individual routes and widths of the flow paths are identified. Initial decisions as to the form and position of the individual flow paths are corrected as the iteration proceeds. Users will find, however, that some of the early stages of the iteration can be avoided by considering the viscous fluid analogy and including some initial plateaux at the start. Provided that each flow path starts at a heat source and finishes at a heat sink, and does not cross another the initial choice of paths will not adversely affect the final result.

The initial flow paths are therefore drawn so that each reaches a heat sink. Some will have a lot of space while others will have to be squeezed together to go between two heat sources. At this stage it is sometimes clear that some of the groups of flow lines from a source or pair of sources will divide around another heat source. Normally it is now that the first temperature plateaux are established and a new set of starting flow lines drawn.

4.3.4 The First Set of Temperature Contours

The first isotherm is drawn so that it, the heat sink, and the flow lines generate figures that appear likely to have the property of fixed thermal resistance per square. Similarly the subsequent isotherms are then drawn using themselves and the previous isotherm to produce more curvilinear squares. This is repeated until contours have approached all of the heat sources.

4.3.5 Examination of Isotherms

Consider the height contours on the viscous fluid analogy:

a) Contours on the crest flow line will be concave when viewed from the heat source.
b) Contours on the trough flow lines will be concave when viewed from the heat sink.
c) With small viscous fluid generators a definite 'conical' volcano is created. In this case the contour closest to each heat source must encircle it. Such a contour may be too close to the previous contour to generate curvilinear squares. If this is the case it will disappear with future iterations. The circular contours are included at this stage to clarify the existence of a saddle point and facilitate the correct choice of flow paths.

To ensure that the existence of a saddle point is not missed on the thermal map, temperature plateaux are clearly marked where required. The establishment of a temperature plateau must occur with at least one crest flow line which can be considered as two malleable membranes coherent with each other and without any flow between them. At the plateau the membranes separate and become two trough flow lines. Also at the plateau a double flow line, either a crest or trough, from behind the downstream heat source separates to collapse alongside the two trough flow lines.

At the latter stages of the iteration some temperature plateaux in the vicinity of a large heat source can become absorbed in the area of the source but the separation of the heat flow still occurs.

4.3.6 Completing the Temperature Diagram

Using the indications obtained as in section 4.3.5 redraw the flow lines on the next page of tracing paper. Separating the flow lines reduces the number of contours and concentrating them increases the numbers. Such changes should not be too dramatic until the user is experienced in the method. These new flow lines are then used to guide the drawing of the next set of isotherms. Repeat this operation until a satisfactory scattered distribution of deformed curvilinear squares results.

As part of this operation flow paths can frequently be moved to follow paths going round several other heat sources. If the decision is wrong it will become apparent at a later stage and can then be corrected.

In order to make a better assessment where the flow paths are very wide it is often easier to visualise the curvilinear squares if the flow is subdivided. Such a subdivision can be limited to just one flow path.

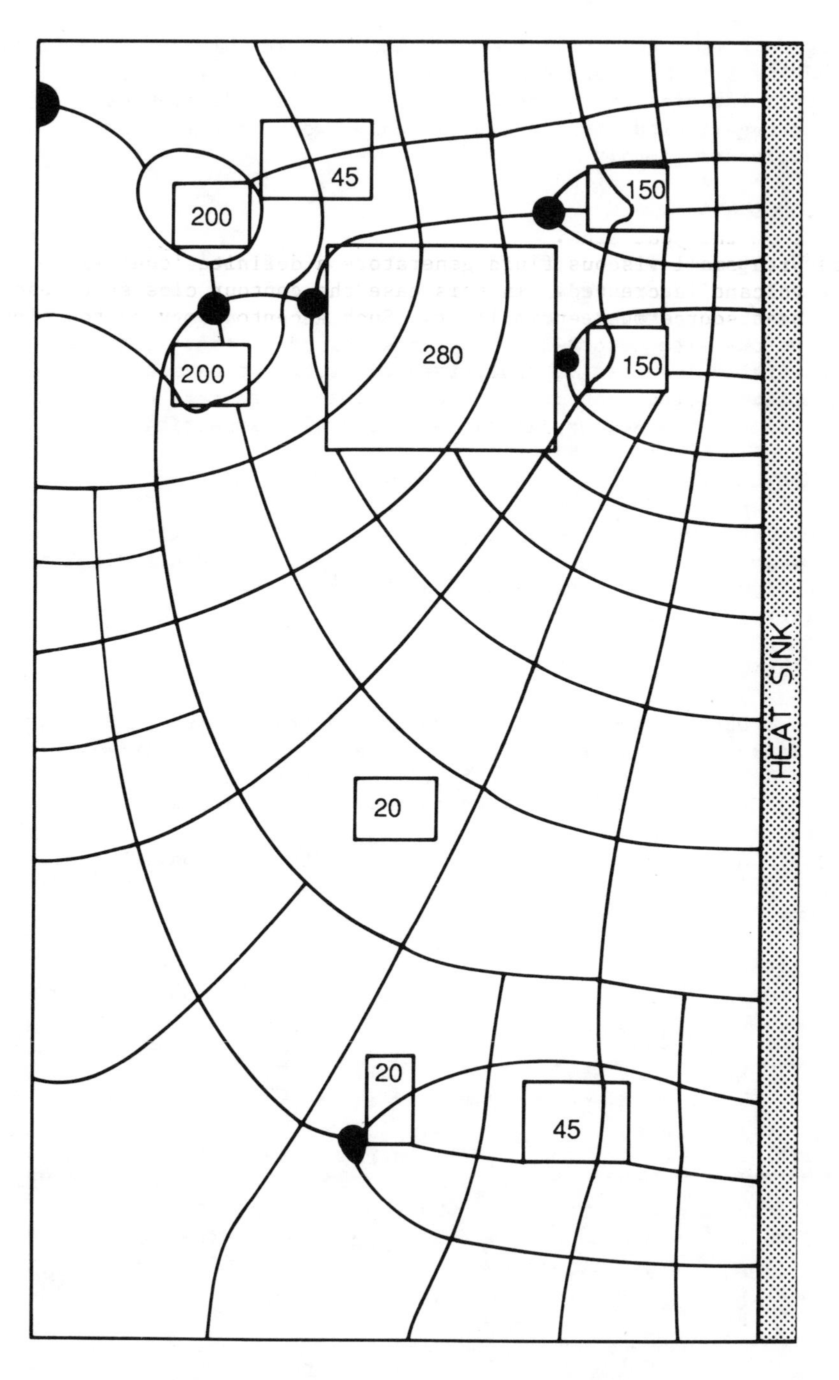

FIGURE 3 Graphical Solution of Example Problem

Figure 3 is an example of one such completed temperature diagram showing all the features described in this paper. This particular diagram was completed in less than 30 minutes by a course participant who had drawn only two simple teaching examples previously.

5 CONCLUSION

The graphical method is probably the most versatile technique in the thermal designer's repertoire, as it is much more than just a means of temperature prediction. Besides providing quick solutions with errors of only a few per cent, the technique inevitably leads to an understanding of heat flow itself. The graphical method has great potential when used as a qualitative technique with new applications continually appearing.

ACKNOWLEDGEMENT

The work reported in this paper is part of an on-going development of the graphical method. Some of the descriptions and figures are therefore used with the permission of Electrochemical Publications Ltd.

REFERENCES

1) D J Dean, 'The Thermal Design of Electronic Circuit Boards and Packages' (Electrochemical Publications, London, 1985)
a) Chapter 3, b) Section 5.5.1, c) Section 5.2.2,
d) Section 5.6.3.1, e) Section 3.5, f) Section 2.3.3.

2) J H Awbery and F H Schofield, Proc. 5th Int. Cong. Refgn. (1929).

3) H P Williams, 'Antenna Theory and Design', Vol. 2, Pitman (1966).

Analytical Solutions to Parabolic Multi-Dimensional Diffusion Problems within Irregularly Shaped Domains

J.B. Aparecido(*), R.M. Cotta

Programa de Engenharia Mecânica, COPPE/UFRJ, Universidade Federal do Rio de Janeiro, Cidade Universitária, Cx. Postal 68503, Rio de Janeiro, RJ 21945, Brasil

(*) Permanent address: Depto. de Engenharia Mecânica, C.P. 31, 15378 UNESP, Ilha Solteira, SP, Brasil

ABSTRACT

The formalism in the generalized integral transform technique is utilized to provide analytical solution to transient multidimensional diffusion problems within arbitrarily shaped regions that do not regularly fit into the orthogonal coordinate system chosen to describe the domain. The integral transformation process then yields an infinite system of coupled ordinary differential equations with constant coefficients, that can be readily handled through the appropriate eigensystem analysis.

An application is considered, for illustration purposes, related to internal forced convection problems within polygonal ducts under laminar flow conditions.

INTRODUCTION

The analytical solution of heat and mass diffusion problems has been recently reviewed and unified [1], when formal solutions were made available for a number of reasonably general classes of problems, through the use of the classical ntegral transform technique. The analysis in [1], however, is only applicable to regular domains described by orthogonal coordinate systems that allow for separation of variables in the related auxiliary eigenvalue problem.

An analytical approach was then developed [2] for the solution of elliptic multidimensional diffusion problems within irregular domains, based on extension of the so-called generalized integral transform technique [3-6] for a priori non-transformable convection-diffusion problems, and applied to the analysis of fully developed laminar flow within various irregularly shaped ducts [2,7-8]. In parallel, the analytical solution of problems with nonseparable Sturm-Liouville type systems was also accomplished by extending the ideas in the generalized approach [9], allowing for the solution of parabolic multidimensional problems within regular domains. The present work is aimed at combining the developments in [2,9] to provide an interesting alternative to classical purely numerical approaches in solving transient multidimensional diffusion problems within irregularly shaped regions. The analysis is then illustrated by considering thermally developing flow inside right triangular ducts under prescribed uniform wall temperature boundary condition.

ANALYSIS

For the sake of simplicity, and without loss of generality, our starting point is the parabolic version of the two-dimensional problem considered in [2], obtained by adding a transient term combined with a general nonseparable coefficient [9], and written as:

$$u(x,y)\,\frac{\partial T(x,y,t)}{\partial t} + \left[w(x)\,L_y + L\right]T(x,y,t) = P(x,y,t),$$

$$\text{for } t > 0\ ,\ y_o < y < y_1\ ,\ x_o(y) < x < x_1(y) \qquad (1.a)$$

subjected to the initial and boundary conditions

$$T(x,y,0) = g(x,y) \qquad (1.b)$$

$$B_{x,k}\,T(x,y,t) = 0\ ,\ \text{in } x = x_k(y)\ ,\ k = 0,1 \qquad (1.c,d)$$

$$B_{y,k}\,T(x,y,t) = f_k(x,y,t)\ ,\ \text{in } y = y_k\ ,\ k = 0,1 \qquad (1.e,f)$$

where the operators above are defined as [2]

$$L_y \equiv -\,a(y)\,\frac{\partial}{\partial y}\left[b(y)\,\frac{\partial}{\partial y}\right] \qquad (2.a)$$

$$L \equiv -\,\frac{\partial}{\partial x}\left[K(x)\,\frac{\partial}{\partial x}\right] + d(x) \qquad (2.b)$$

$$B_{x,k} \equiv \alpha_k - \left(-1\right)^k \beta_k \frac{\partial}{\partial x} \tag{2.c}$$

$$B_{y,k} \equiv \delta_k - \left(-1\right)^k \gamma_k \frac{\partial}{\partial y} \tag{2.d}$$

By combining the formalism in [2,9], the appropriate auxiliary problem for integral transformation in the x-direction is taken as

$$\mu^2(y)\ w(x)\ \psi(x,y) = L\ \psi(x,y) \quad , \quad \text{in } x_o(y) < x < x_1(y) \tag{3.a}$$

with boundary conditions

$$B_{x,k}\ \psi(x,y) = 0 \quad , \quad \text{at} \quad x = x_k(y) \quad , \quad k = 0,1 \tag{3.b,c}$$

The solution of this y-dependent eigenvalue problem of the Sturm-Liouville type is assumed to be known at this point, through application of well-established algorithms [1], providing as many eigenvalues, μ_i's, and eigenfunctions, ψ_i's, as needed. Problem (3) above allows definition of the following integral transform pair:

$$\bar{T}_i(y,t) = \int_{x_o(y)}^{x_1(y)} w(x)\ \mathbb{K}_i(x,y)\ T(x,y,t)\ dx \ , \ \text{transform} \tag{4.a}$$

$$T(x,y,t) = \sum_{i=1}^{\infty} \mathbb{K}_i(x,y)\ \bar{T}_i(y,t) \ , \ \text{inversion} \tag{4.b}$$

where the normalized eigenfunction is defined as

$$\mathbb{K}_i(x,y) = \frac{\psi_i(x,y)}{N_i^{1/2}(y)} \tag{4.c}$$

and the normalization integral given by

$$N_i(y) = \int_{x_o(y)}^{x_1(y)} w(x)\ \psi_i^2(x,y)\ dx \tag{4.d}$$

Integral transformation of eq. (1.a) is now attempted through application of the operator $\int_{x_o(y)}^{x_1(y)} \mathbb{K}_i(x,y)\, dx$, to yield:

$$\sum_{j=1}^{\infty} C_{ij}(y) \frac{\partial \overline{T}_j(y,t)}{\partial t} + \mu_i^2(y)\, \overline{T}_i(y,t) + L_y\, \overline{T}_i(y,t) -$$

$$- 2a(y)b(y) \sum_{j=1}^{\infty} A_{ij}(y) \frac{\partial \overline{T}_j(y,t)}{\partial y} - a(y) \sum_{j=1}^{\infty} \left[\frac{db(y)}{dy} A_{ij}(y) + \right.$$

$$\left. + b(y)\, B^{+}_{ij}(y) \right] \overline{T}_j(y,t) = \overline{P}_i(y,t) \qquad (5.a)$$

where,

$$C_{ij}(y) = \int_{x_o(y)}^{x_1(y)} u(x,y)\, \mathbb{K}_i(x,y)\, \mathbb{K}_j(x,y)\, dx \qquad (5.b)$$

$$A_{ij}(y) = \int_{x_o(y)}^{x_1(y)} w(x)\, \mathbb{K}_i(x,y) \frac{\partial\, \mathbb{K}_j(x,y)}{\partial y}\, dx \qquad (5.c)$$

$$B^{+}_{ij}(y) = \int_{x_o(y)}^{x_1(y)} w(x)\, \mathbb{K}_i(x,y) \frac{\partial^2 \mathbb{K}_j(x,y)}{\partial y^2}\, dx \qquad (5.d)$$

$$\overline{P}_i(y,t) = \int_{x_o(y)}^{x_1(y)} \mathbb{K}_i(x,y)\, P(x,y,t)\, dx \qquad (5.e)$$

Similarly, the initial and boundary conditions, eqs. (1.b) and (1.e,f), can be integral transformed in the x-direction with the operator $\int_{x_o(y)}^{x_1(y)} w(x)\ \mathbb{K}_i(x,y)\ dx$, to provide, respectively:

$$\bar{T}_i(y,0) = \bar{g}_i(y) \tag{6.a}$$

$$B_{y,k}\ \bar{T}_i(y,t) = \bar{F}_{i,k}(y,t) \quad , \quad \text{at} \quad y = y_k \quad , \quad k = 0,1 \tag{6.b,c}$$

where,

$$\bar{g}_i(y) = \int_{x_o(y)}^{x_1(y)} w(x)\ \mathbb{K}_i(x,y)\ T(x,y,0)\ dx \tag{6.d}$$

$$\bar{F}_{i,k}(y_k,t) = \bar{f}_{i,k}(y_k,t) + \left(-1\right)^k \gamma_k \sum_{j=1}^{\infty} A_{ij}(y_k)\ \bar{T}_j(y_k,t) \tag{6.e}$$

$$\bar{f}_{i,k}(y_k,t) = \int_{x_o(y)}^{x_1(y)} w(x)\ \mathbb{K}_i(x,y_k)\ f_k(x,y_k,t)\ dx \tag{6.f}$$

Eq. (5.a) has now to be integral transformed in the y-direction and the related auxiliary problem is chosen as:

$$a(y)\ \lambda^2\ \phi(y) = L_y\ \phi(y) \quad , \quad y_o < y < y_1 \tag{7.a}$$

with boundary conditions

$$B_{y,k}\ \phi(y) = 0 \quad , \quad \text{at} \quad y = y_k \quad , \quad k = 0,1 \tag{7.b,c}$$

which allows definition of the following integral transform pair:

$$\tilde{\bar{T}}_{im}(t) = \int_{y_o}^{y_1} Z_m(y)\ \bar{T}_i(y,t)\ dy \quad , \quad \text{transform} \tag{8.a}$$

$$\tilde{\bar{T}}_i(y,t) = \sum_{m=1}^{\infty} Z_m(y)\, \tilde{\bar{T}}_{im}(t) \quad , \quad \text{inversion} \tag{8.b}$$

where the normalized eigenfunction is given by

$$Z_m(y) = \frac{\phi_m(y)}{M_m^{1/2}} \tag{8.c}$$

and the norm

$$M_m = \int_{y_o}^{y_1} \phi_m^2(y)\, dy \tag{8.d}$$

Eq. (5.a) is now operated on with $\int_{y_o}^{y_1} \frac{Z_m(y)}{a(y)}\, dy$, to yield the fully transformed problem:

$$\sum_{n=1}^{\infty} \sum_{j=1}^{\infty} D_{ijmn} \frac{d\, \tilde{\bar{T}}_{jn}(t)}{dt} + \sum_{n=1}^{\infty} \sum_{j=1}^{\infty} H_{ijmn}\, \tilde{\bar{T}}_{jn}(t) = \tilde{\bar{h}}_{im}(t),$$

$$i,m = 1,2,... \tag{9.a}$$

where,

$$D_{ijmn} = \int_{y_o}^{y_1} a^{-1}(y)\, C_{ij}(y)\, Z_m(y)\, Z_n(y)\, dy \tag{9.b}$$

$$H_{ijmn} = E_{iimn}\, \delta_{ij} - F_{ijmn} - G_{ijmn} + \lambda_m^2 \delta_{ij} \delta_{mn} - R_{ijmn} \tag{9.c}$$

$$E_{iimn} = \int_{y_o}^{y_1} a^{-1}(y)\, \mu_i^2(y)\, Z_m(y)\, Z_n(y)\, dy \tag{9.d}$$

$$F_{ijmn} = 2 \int_{y_o}^{y_1} b(y)\, A_{ij}(y)\, Z_m(y) \frac{\partial\, Z_n(y)}{\partial y}\, dy \tag{9.e}$$

$$G_{ijmn} = \int_{y_o}^{y_1} \left[\frac{db(y)}{dy} A_{ij}(y) + b(y) B^{+}_{ij}(y) \right] Z_m(y) Z_n(y) dy \quad (9.f)$$

$$R_{ijmn} = \sum_{k=0}^{1} (-1)^k \gamma_k A_{ij}(y_k) Z_n(y_k) \left[\frac{Z_m(y_k) + (-1)^k b(y_k) \dfrac{\partial Z_m(y_k)}{\partial y}}{\delta_k + \gamma_k} \right] \quad (9.g)$$

and,

$$\tilde{\overline{h}}_{im}(t) = \tilde{\overline{P}}_{im}(t) + \sum_{k=0}^{1} \overline{f}_{i,k}(y_k,t) \left[\frac{Z_m(y_k) + (-1)^k b(y_k) \dfrac{\partial Z_m(y_k)}{\partial y}}{\delta_k + \gamma_k} \right] \quad (9.h)$$

$$\tilde{\overline{P}}_{im}(t) = \int_{y_o}^{y_1} a^{-1}(y) Z_m(y) \overline{P}_i(y,t) dy \quad (9.i)$$

with the transformed initial conditions

$$\tilde{\overline{T}}_{im}(0) = \tilde{\overline{g}}_{im} \equiv \int_{y_o}^{y_1} Z_m(y) \overline{g}_i(y) dy \quad (9.j)$$

Eqs. (9) above form an infinite system of first order ordinary differential equations with constant coefficients, which is to be truncated for the purpose of obtaining numerical results, as follows:

$$\sum_{n=1}^{N^*} \sum_{j=1}^{N} D_{ijmn} \frac{d\,\tilde{\bar{T}}_{jn}(t)}{dt} +$$

$$+ \sum_{n=1}^{N^*} \sum_{j=1}^{N} H_{ijmn}\, \tilde{\bar{T}}_{jn}(t) = \tilde{\bar{h}}_{im}(t) \quad , \quad t > 0 \qquad (10.a)$$

$$\tilde{\bar{T}}_{im}(0) = \tilde{\bar{g}}_{im} \quad , \quad i = 1,2,\ldots,N \quad , \quad m = 1,2,\ldots,N^* \qquad (10.b)$$

In matrix form, eqs. (10) are given by

$$\mathit{D}\, \underset{\sim}{y}'(t) + \mathit{H}\, \underset{\sim}{y}(t) = \underset{\sim}{h}(t) \qquad (11.a)$$

$$\underset{\sim}{y}(0) = \underset{\sim}{g} \qquad (11.b)$$

where the various arrays of order $N^*.N$ are obtained from

$$\underset{\sim}{y}(t) = \left\{ \tilde{\bar{T}}_{1,1}(t),\ldots,\tilde{\bar{T}}_{1,N^*}(t),\ldots,\tilde{\bar{T}}_{N,1}(t),\ldots,\tilde{\bar{T}}_{N,N^*}(t) \right\}^T \qquad (11.c)$$

$$\underset{\sim}{h}(t) = \left\{ \tilde{\bar{h}}_{1,1},\tilde{\bar{h}}_{1,2},\ldots,\tilde{\bar{h}}_{1,N^*},\ldots,\tilde{\bar{h}}_{N,1},\tilde{\bar{h}}_{N,2},\ldots,\tilde{\bar{h}}_{N,N^*} \right\}^T \qquad (11.d)$$

$$\underset{\sim}{g} = \left\{ \tilde{\bar{g}}_{1,1},\tilde{\bar{g}}_{1,2},\ldots,\tilde{\bar{g}}_{1,N^*},\ldots,\tilde{\bar{g}}_{N,1},\tilde{\bar{g}}_{N,2},\ldots,\tilde{\bar{g}}_{N,N^*} \right\}^T \qquad (11.e)$$

$$\mathit{D} = \left\{ d_{k,\ell} \right\} \quad , \quad k,\ell, = 1,2,\ldots,(N.N^*) \qquad (11.f)$$

where, $d_{k,\ell} = D_{ijmn}$, with

$$i = \text{int}\left[(k-1)/N^*\right] + 1 \quad ; \quad j = \text{int}\left[(\ell-1)/N^*\right] + 1$$

$$m = k - (i-1)N^{*} \quad ; \quad n = \ell - (j-1)N^{*}$$

and,

$$H = \left\{ h_{k,\ell} \right\} \quad , \quad k,\ell = 1,2,\ldots,(N.N^{*}) \qquad (11.g)$$

where, $h_{k,\ell} = H_{ijmn}$, with i,j,m, and n defined as above.

System (11) can be accurately and automatically solved by making use of well-established scientific subroutines packages, such as in the IMSL library, either through the related matrix eigensystem analysis or through direct numerical integration of the initial value problem (11). Once the transformed potentials have been numerically evaluated, the inversion formula is recalled to provide the original potential:

$$T(x,y,t) = \sum_{m=1}^{N^{*}} \sum_{i=1}^{N} \mathbb{K}_i(x,y)\, Z_m(y)\, \tilde{\bar{T}}_{im}(t) \qquad (12)$$

with sufficiently large values of N^{*} and N for the desired convergence.

APPLICATION AND DISCUSSION

In order to illustrate the approach here advanced, an application is considered dealing with thermally developing laminar flow inside a right triangular duct, subjected to an uniform wall temperature distribution [10], which has the following problem formulation in dimensionless form:

$$U(x,y) \frac{\partial T(x,y,z)}{\partial z} = \frac{\partial^2 T}{\partial x^2} + \frac{\partial^2 T}{\partial y^2} \quad , \quad 0 < y < \beta \ , \ 0 < x < x_1(y) \ ,$$

$$z > 0 \qquad (13.a)$$

with inlet and boundary conditions

$$T(x,y,0) = 1 \quad , \quad 0 \le y \le \beta \quad , \ 0 \le x \le x_1(y) \qquad (13.b)$$

$$T(0,y,z) = 0 \quad ; \quad T(x_1(y),y,z) = 0 \quad , \quad z > 0 \qquad (13.c,d)$$

$$T(x,0,z) = 0 \quad ; \quad T(x,\beta,z) = 0 \quad , \quad z > 0 \qquad (13.e,f)$$

where the irregular boundary is given in terms of the dimensionless sides of the triangle

$$x_1(y) = \alpha \left(1 - \frac{y}{\beta}\right) \tag{13.g}$$

and the axial coordinate z plays the role of the t-variable in the preceding analysis. The decoupled velocity field for hydrodinamically fully developed flow, $U(x,y)$, has been previously obtained [2] in explicit form.

Quantities of practical interest can be determined once the dimensionless temperature distribution along the duct has been obtained, such as:

- Dimensionless bulk temperature

$$T_{av}(z) = \frac{1}{A_c} \int_{A_c} U(x,y)\ T(x,y,z)\ dA_c \tag{14.a}$$

- Local Nusselt number

$$Nu(z) = \frac{h(z)\ D_h}{K} = -\frac{1}{4\ T_{av}(z)} \frac{d\ T_{av}(z)}{dz} \tag{14.b}$$

- Average Nusselt number

$$Nu_{av}(z) = \frac{1}{z} \int_0^z Nu(z)\ dz = -\frac{1}{4z} \ell n\ T_{av}(z) \tag{14.c}$$

Problem (13) above was solved with $N^* = N \le 20$, which proved to be sufficiently large for the desired convergence in the range of interest for the dimensionless axial coordinate, $10^{-4} \le Z \le 10^{0}$. Table I presents results for the limiting Nusselt number ($Nu(\infty)$), which is attained at the fully developed region ($z \longrightarrow \infty$), for different values of the triangle apex angle, θ, as defined in [10]. The results are in excellent agreement with those from reference [10] where Galerkin functions were employed. It should be noted that convergence for the limiting Nusselt number is achieved with just a few terms in the expansion, being faster for smaller values of θ, when the regular rectangular geometry is approached.

TABLE I - Comparison of Nu(∞) from complete solution, eq. 14.b, and numerical results from [10], for different apex angle of right triangular duct (θ)

θ	Nu(∞) - PRESENT	Nu(∞) - REF. [10]
80°	1.706	1.70
75°	1.894	1.90
70°	2.045	2.05
65°	2.162	2.16
60°	2.249	2.25
55°	2.309	2.31
50°	2.345	2.35
45°	2.357	2.36

REFERENCES

1. Mikhailov, M.D. and M.N. Ozisik, "Unified Analysis and Solutions of Heat and Mass Diffusion", John Wiley, New York, 1984.

2. Aparecido, J.B., R.M. Cotta, and M.N. Ozisik, "Analytical Solutions to Two-dimensional Diffusion Type Problems in Irregular Geometries", J. Franklin Inst., V. 326, 3:421-434, 1989.

3. Ozisik, M.N. and R.L. Murray, "On the Solution of Linear Diffusion Problems with Variable Boundary Condition Parameters", J. Heat Transfer, V. 96, 48-51, 1974.

4. Cotta, R.M. and M.N. Ozisik, "Diffusion Problems with General Time-Dependent Coefficients", Braz. J. Mech. Sciences, RBCM, V. 9, 4:269-292, 1987.

5. Cotta, R.M. and M.N. Ozisik, "Laminar Forced Convection in Ducts with Periodic Variation of Inlet Temperature", Int. J. Heat & Mass Transfer, V. 29, 10:1495-1501, 1986.

6. Cotta, R.M., "Hybrid Numerical-Analytical Approach to Nonlinear Diffusion Problems", Num. Heat Transfer, Part B, V. 17, pp. XXX-XXX, 1990.

7. Aparecido, J.B. and R.M. Cotta, "Fully Developed Laminar Flow in Trapezoidal Ducts", 9th Brazilian Congress of Mechanical Engineering, IX COBEM, V. 1, 25-28, Florianópolis, Brasil, 1987.

8. Aparecido, J.B. and R.M. Cotta, "Laminar Flow Inside Hexagonal Ducts", Computational Mech., V. 6, pp. 93-100, 1990.

9. Aparecido, J.B. and R.M. Cotta, "Thermally Developing Laminar Flow Inside Rectangular Ducts", Int. J. Heat & Mass Transfer, in press.

10. Lakshminarayanan, R. and A. Haji-Sheik, "A Generalized Close-Form Solution to Laminar Thermal Entrance Problems", Proc. of the 8th Int. Heat Transfer Conf., Eds. C.L. Tien, V.P. Carey, and J.K. Ferrel, San Francisco, USA, V. 3, pp. 871-876, July 1986.

Alternative Approach to the Integral Transform Solution of Non-Homogeneous Diffusion Problems

F. Scofano Neto(*), R.M. Cotta, M.D. Mikhailov(**)
Programa de Engenharia Mecânica, COPPE/UFRJ, Universidade Federal do Rio de Janeiro, Cidade Universitária, Cx. Postal 68503, Rio de Janeiro, RJ 21945, Brasil
Permanent addresses: (*) IME, Instituto Militar de Engenharia, Rio de Janeiro, Brasil
(**) WMEI 'LENIN', Institute of Applied Mathematics and Informatics, Sofia, Bulgaria

ABSTRACT

An alternative approach is proposed to the analytical treatment of linear nonhomogeneous diffusion problems through the classical integral transform technique. Based on direct integration of the original partial differential equation and manipulation of related boundary conditions, this approach provides alternative infinite sums for the potential and its flux with enhanced convergence characteristics, specially for locations in the vicinity of the nonhomogeneous boundaries.

Within the context of an application, comparisons are made with the direct application of the related inversion formula.

INTRODUCTION

The exact analysis of heat and mass diffusion problems through application of the integral transform technique has been unified and systematically presented for seven different classes of problems [1]. The formal general expressions obtained are not in a convenient form for computational purposes when the problem formulation involves nonhomogeneous source terms, since the eigenfunction expansion might become slowly converging at points close to the boundaries, where the original boundary equation is not satisfied by the auxiliary eigenfunctions. To alleviate this difficulty, alternative approaches based on integration by parts of the formal solutions [2] and on splitting-up of the original system into simpler problems [1,3] were proposed. In particular, the splitting-up procedure of [1,3], provides the fastest converging expressions for the class of linear problems considered, by separating the solution of the general problem into a set of simpler problems containing a homogeneous transient problem and a set of steady-state problems for which separate solutions can be obtained. However, the available expressions are limited to time-dependent source functions represented by exponentials and q-order polynomials in t, and the analytical involvement makes it a not so practical approach for the more general situations of arbitrary source functions and multidimensional problems. Motivated by the development of automatic solvers through the recently advanced and more flexible generalized integral transform technique [4,5], an alternative and quite straightforward approach is here proposed for the computational enhancement of analytical solutions for nonhomogeneous diffusion problems, based on integration of the original partial differential equation over the whole region and appropriate manipulation of the boundary conditions, as now demonstrated.

ANALYSIS

To be concise, the following transient one-dimensional diffusion problem with nonhomogeneous boundary and equation source functions is considered

$$w(x)\frac{\partial T(x,t)}{\partial t} = \frac{\partial}{\partial x}\left[K(x)\frac{\partial T(x,t)}{\partial x}\right] - d(x)T(x,t) + P(x,t) \; ,$$

$$\text{in } x_o < x < x_1 \; , \; t > 0 \tag{1.a}$$

with initial and boundary conditions

$$T(x,0) = f(x) \quad , \quad x_o \le x \le x_1 \tag{1.b}$$

$$\alpha_o T(x_o,t) - \beta_o K(x_o)\left.\frac{\partial T}{\partial x}\right|_{x=x_o} = \phi_o(t) \quad , \quad t > 0 \tag{1.c}$$

$$\alpha_1 T(x_1,t) + \beta_1 K(x_1)\left.\frac{\partial T}{\partial x}\right|_{x=x_1} = \phi_1(t) \quad , \quad t > 0 \tag{1.d}$$

Problem (1) above belongs to the so-called class I problems of reference [1], for which the exact formal solution is readily obtained through application of the classical integral transform technique, as follows. The appropriate auxiliary eigenvalue problem is taken as

$$\frac{d}{dx}\left[K(x)\frac{d\psi_i(x)}{dx}\right] + \left[\mu_i^2 w(x) - d(x)\right]\psi_i(x) = 0 \quad , \quad \text{in } x_o < x < x_1 \qquad (2.a)$$

with boundary conditions

$$\alpha_k \psi_i(x_k) + (-1)^{k+1}\beta_k K(x_k)\left.\frac{d\psi_i}{dx}\right|_{x=x_k} = 0 \ , \ k=0,1 \qquad (2.b,c)$$

which allows the establishment of the integral transform pair with a symmetric kernel

$$\bar{T}_i(t) = \frac{1}{N_i^{1/2}}\int_{x_o}^{x_1} w(x)\,\psi_i(x)\,T(x,t)\,dx \ , \ \text{transform} \qquad (3.a)$$

$$T(x,t) = \sum_{i=1}^{\infty}\frac{1}{N_i^{1/2}}\psi_i(x)\,\bar{T}_i(t) \qquad , \ \text{inversion} \qquad (3.b)$$

Eq. (1.a) is operated on with $\int_{x_o}^{x_1}\frac{\psi_i(x)}{N_i^{1/2}}dx$, the inversion formula above is employed and the boundary conditions, eqs. (1.c,d) and (2.b,c) are manipulated to yield the decoupled system of ordinary differential equations for the transformed potentials

$$\frac{d\bar{T}_i(t)}{dt} + \mu_i^2\bar{T}_i(t) = \bar{g}_i(t) \quad , \quad t > 0 \quad , \quad i=1,2,\ldots \qquad (4.a)$$

where the contribution from the nonhomogeneous terms is given as

$$\bar{g}_i(t) = \frac{1}{N_i^{1/2}} \left[\sum_{k=0}^{1} \phi_k(t)\, \Omega_i(x_k) + \int_{x_o}^{x_1} P(x,t)\, \psi_i(x)\, dx \right] \quad (4.b)$$

and,

$$\Omega_i(x_k) = \frac{\psi_i(x_k) + \left(-1\right)^k K(x_k)\, d\psi_i(x_k)/dx}{\alpha_k + \beta_k} \quad (4.c)$$

while the transformed initial condition becomes

$$\bar{T}_i(0) = \bar{f}_i \equiv \frac{1}{N_i^{1/2}} \int_{x_o}^{x_1} w(x)\, \psi_i(x)\, f(x)\, dx \quad (4.d)$$

System (4) is readily solved and the inversion formula recalled to provide the formal exact solution

$$T(x,t) = \sum_{i=1}^{\infty} \frac{1}{N_i^{1/2}} \psi_i(x)\, e^{-\mu_i^2 t} \left[\bar{f}_i + \int_0^t \bar{g}_i(t') e^{\mu_i^2 t'} dt' \right] \quad (5)$$

The infinite summation represented by eq. (5) is not uniformly convergent for nonhomogeneous boundary conditions and can be particularly slowly convergent for regions in the vicinity of the nonhomogeneous boundaries. To alleviate this difficulty, alternative expressions are here proposed for the potential, $T(x,t)$, and its flux, both at the boundaries and within the medium, that provide convergence rates enhancement for computational purposes. In fact, the present approach has the interesting feature of making use of the transformed potentials already obtained, without requiring the solution of additional problems. Instead, an integral balance equation over any portion of the region is employed, together with the direct use of the boundary equations, as now demonstrated.

Eq. (1.a) is operated on with $\int_{x_o}^{x_1}$ dx, to yield, after use is made of the inversion formula (3.b):

$$K(x_1)\left.\frac{\partial T}{\partial x}\right|_{x_1} - K(x_o)\left.\frac{\partial T}{\partial x}\right|_{x_o} = \sum_{i=1}^{\infty}\left\{\bar{f}_i^* \, \bar{g}_i(t) + \right.$$

$$\left. + \bar{T}_i(t)\left[\bar{h}_i - \bar{f}_i^* \mu_i^2\right]\right\} - g(t) \qquad (6.a)$$

where,

$$\bar{f}_i^* = \frac{1}{N_i^{1/2}}\int_{x_o}^{x_1} w(x)\,\psi_i(x)\,dx \qquad (6.b)$$

$$\bar{h}_i = \frac{1}{N_i^{1/2}}\int_{x_o}^{x_1} d(x)\,\psi_i(x)\,dx \qquad (6.c)$$

$$g(t) = \int_{x_o}^{x_1} P(x,t)\,dx \qquad (6.d)$$

Eq. (6.a) is a working expression that relates the fluxes at the two boundaries to known quantities readily available. In fact, if one of the fluxes is explicitly known, either from a second kind boundary condition or from a homogeneous condition when the inverse formula can be directly applied, then eq. (6.a) suffices to provide the flux at the other boundary and, consequently, the potential is obtained from the boundary equation for the more general 3rd kind condition. The flux within the medium is evaluated, for instances, by operating on eq. (1.a) with $\int_{x_o}^{x}$ dx, to yield:

$$K(x)\left.\frac{\partial T}{\partial x}\right|_{x} = K(x_o)\left.\frac{\partial T}{\partial x}\right|_{x_o} + \sum_{i=1}^{\infty}\left\{\bar{f}_i^*(x)\,\bar{g}_i(t) + \right.$$

$$+ \bar{T}_i(t) \left[\bar{h}_i(x) - \bar{f}^*_i(x) \, \mu_i^2 \right] \Bigg\} - g(x,t) \qquad (7.a)$$

where,

$$\bar{f}^*_i(x) = \frac{1}{N_i^{1/2}} \int_{x_o}^{x} w(x) \, \psi_i(x) \, dx \qquad (7.b)$$

$$\bar{h}_i(x) = \frac{1}{N_i^{1/2}} \int_{x_o}^{x} d(x) \, \psi_i(x) \, dx \qquad (7.c)$$

$$g(x,t) = \int_{x_o}^{x} P(x,t) \, dx \qquad (7.d)$$

Or, alternatively, by operating with $\int_x^{x_1} dx$:

$$K(x) \left. \frac{\partial T}{\partial x} \right|_x = K(x_1) \left. \frac{\partial T}{\partial x} \right|_{x_1} - \sum_{i=1}^{\infty} \Bigg\{ \hat{f}^*_i(x) \, \bar{g}_i(t) +$$

$$+ \bar{T}_i(t) \left[\hat{h}_i(x) - \hat{f}^*_i(x) \, \mu_i^2 \right] \Bigg\} + \hat{g}(x,t) \qquad (8.a)$$

where,

$$\hat{f}^*_i(x) = \frac{1}{N_i^{1/2}} \int_{x}^{x_1} W(x) \, \psi_i(x) \, dx = \bar{f}_i(x_1) - \bar{f}_i(x) \qquad (8.b)$$

$$\hat{h}_i(x) = \frac{1}{N_i^{1/2}} \int_{x}^{x_1} d(x) \, \psi_i(x) \, dx = \bar{h}_i(x_1) - \bar{h}_i(x) \qquad (8.c)$$

$$\hat{g}(x,t) = \int_x^{x_1} P(x,t)\, dx = g(t) - g(x,t) \qquad (8.d)$$

The potential within the medium is determined, for example, from integration of eq. (7.a) from x to x_1 to furnish:

$$T(x,t) = T(x_1,t) - \hat{K}(x)\, K(x_o) \left.\frac{\partial T}{\partial x}\right|_{x_o} -$$

$$- \sum_{i=1}^{\infty} \left\{ \hat{\bar{f}}{}^*_i(x)\, \bar{\bar{g}}_i(t) + \bar{T}_i(t) \left[\hat{\bar{h}}{}^*_i(x) - \hat{\bar{f}}{}^*_i(x)\, \mu_i^2 \right] \right\} + \hat{g}(x,t) \qquad (9.a)$$

where,

$$\hat{K}(x) = \int_x^{x_1} \frac{1}{K(x)}\, dx \qquad (9.b)$$

$$\hat{\bar{f}}{}^*_i(x) = \int_x^{x_1} \frac{\bar{f}^*_i(x)}{K(x)}\, dx \qquad (9.c)$$

$$\hat{\bar{h}}{}^*_i(x) = \int_x^{x_1} \frac{\bar{h}_i(x)}{K(x)}\, dx \qquad (9.d)$$

$$\hat{g}(x,t) = \int_x^{x_1} \frac{g(x,t)}{K(x)}\, dx \qquad (9.e)$$

It suffices now to show how the most general situation of nonhomogeneous third kind boundary conditions on both surfaces is handled by the present approach. The quantities at the boundaries, potentials and fluxes, have to be explicitly obtained to allow for the evaluations within the medium through eqs. (7.a), (8.a) and (9.a). The two boundary conditions, eqs. (1.c,d), together with eq. (6.a) and eq.

(9.a) evaluated at $x=x_o$, provide four simultaneous equations for the unknowns $T(x_o,t)$, $T(x_1,t)$, $K(x_o) \left.\frac{\partial T}{\partial x}\right|_{x_o}$, and $K(x_1) \left.\frac{\partial T}{\partial x}\right|_{x_1}$, which are readily solved to yield the final expressions:

$$K(x_o) \left.\frac{\partial T}{\partial x}\right|_{x_o} = \frac{1}{\frac{\beta_o}{\alpha_o} + \frac{\beta_1}{\alpha_1} + \hat{K}(x_o)} \left[B(t) - \frac{\beta_1}{\alpha_1} A(t) + \frac{\phi_1(t)}{\alpha_1} - \frac{\phi_o(t)}{\alpha_o} \right] \tag{10.a}$$

$$K(x_1) \left.\frac{\partial T}{\partial x}\right|_{x_1} = K(x_o) \left.\frac{\partial T}{\partial x}\right|_{x_o} + A(t) \tag{10.b}$$

$$T(x_o,t) = \frac{\phi_o(t)}{\alpha_o} + \frac{\beta_o/\alpha_o}{\frac{\beta_o}{\alpha_o} + \frac{\beta_1}{\alpha_1} + \hat{K}(x_o)} \left[B(t) - \frac{\beta_1}{\alpha_1} A(t) + \frac{\phi_1(t)}{\alpha_1} - \frac{\phi_o(t)}{\alpha_o} \right] \tag{10.c}$$

$$T(x_1,t) = \frac{\phi_1(t)}{\alpha_1} - \frac{\beta_1/\alpha_1}{\frac{\beta_o}{\alpha_o} + \frac{\beta_1}{\alpha_1} + \hat{K}(x_o)} \left[B(t) - \frac{\beta_1}{\alpha_1} A(t) + \frac{\phi_1(t)}{\alpha_1} - \frac{\phi_o(t)}{\alpha_o} \right] - \frac{\beta_1 A(t)}{\alpha_1} \tag{10.d}$$

where,

$$A(t) = \sum_{i=1}^{\infty} \left\{ \bar{f}_i^*(x_1)\, \bar{g}_i(t) + \bar{T}_i(t) \left[\bar{h}_i(x_1) - \right.\right.$$

$$\left.\left. - \bar{f}_i^*(x_1)\, \mu_i^2 \right] \right\} - g(t) \qquad (10.e)$$

$$B(t) = - \sum_{i=1}^{\infty} \left\{ \hat{\bar{f}}_i^*(x_o)\, \bar{g}_i(t) + \bar{T}_i(t) \left[\hat{\bar{h}}_i^*(x_o) - \right.\right.$$

$$\left.\left. - \hat{\bar{f}}_i^*(x_o)\, \mu_i^2 \right] \right\} + \hat{g}(x_o,t) \qquad (10.f)$$

and the alternative expressions for both potential and flux within the medium can then be directly employed.

APPLICATION AND DISCUSSION

An application related to transient heat conduction in a slab is now considered so as to illustrate the convergence enhancement achievable by the present procedure, according to the following formulation:

$$\frac{\partial T(x,t)}{\partial t} = \frac{\partial^2 T(x,t)}{\partial x^2} \quad , \quad 0 < x < 1 \quad , \quad t > 0 \qquad (11.a)$$

with initial and boundary conditions

$$T(x,0) = 1 \quad , \quad 0 \le x \le 1 \qquad (11.b)$$

$$\left. \frac{\partial T(x,t)}{\partial x} \right|_{x=0} = 0 \; ; \; \left. \frac{\partial T(x,t)}{\partial x} \right|_{x=1} + Bi\,T(1,t) = e^{-t} \; , \; t>0 \qquad (11.c,d)$$

Following the formalism just described, the final alternative series for both potential and flux are written as:

$$T(x,t) = \frac{e^{-t}}{Bi} + \sum_{i=1}^{\infty} \frac{1}{N_i^{1/2}} \cos \mu_i x \left[\bar{T}_i(t) - \frac{\cos \mu_i . e^{-t}}{N_i^{1/2} \mu_i^2} \right] \qquad (12.a)$$

$$\frac{\partial T(x,t)}{\partial x} = - \sum_{i=1}^{\infty} \frac{1}{N_i^{1/2}} \mu_i \sin \mu_i x \left[\bar{T}_i(t) - \frac{\cos \mu_i . e^{-t}}{N_i^{1/2} \mu_i^2} \right] \qquad (12.b)$$

where the eigenvalues are obtained from the transcendental equation

$$\mu_i \tan \mu_i = Bi \qquad (12.c)$$

and the norm is given by

$$N_i = \frac{\mu_i^2 + Bi\ (Bi + 1)}{2\ (\mu_i^2 + Bi^2)} \qquad (12.d)$$

In Table I we present some numerical results for potential and flux from eqs. (12.a,b) above, at different spatial positions and t = 0.01. Also shown are the results from the formal solution, eq. (5), obtained from direct application of the inverse formula. Clearly, the convergence rates are markedly improved, specially at the nonhomogeneous boundary where convergence isn't in fact achieved by the formal solution, most noticiably for the flux computation. The approach here advanced is quite straightforward and convenient for computer solvers implementation, since only a few additional evaluations are required. Its extension to multidimensional and more involved problems shall follow.

TABLE I - Convergence of potential, T(x,t), and flux, K(x) $\left.\frac{\partial T}{\partial x}\right|_{x}$, at different x positions (t = 0.01; Bi = 0.1)

NUMBER OF TERMS	POTENTIAL x = 0	POTENTIAL x = 0.8	POTENTIAL x = 1.0	FLUX x = 0.8	FLUX x = 1.0
1	0.8655 (*) 1.025 (**)	1.144 0.9933	1.299 0.9757	0.6921 - 0.0785	0.8601 - 0.0976
2	1.024 0.9884	1.014 1.023	1.141 1.012	0.4075 - 0.0128	0.8759 - 0.1012
3	0.9937 1.008	1.004 1.030	1.111 1.032	0.2285 - 0.1051	0.8789 - 0.1032
4	1.002 0.9941	1.006 1.025	1.103 1.046	0.1552 0.2317	0.8798 - 0.1046
5	0.9995 1.004	1.008 1.017	1.101 1.056	0.1382 0.3080	0.8800 - 0.1056
10	1.000 0.9989	1.009 1.007	1.100 1.079	0.1400 0.0473	0.8801 - 0.1079
∞	1.000	1.009	1.100	0.1400	0.8801

(*) Present approach

(**) Inverse formula, eq. (5)

REFERENCES

1. Mikhailov, M.D. and M.N. Özisik, "Unified Analysis and Solutions of Heat and Mass Diffusion", John Wiley, New York, 1984.

2. Özisik, M.N., "Heat Conduction", John Wiley, New York, 1980.

3. Mikhailov, M.D., "Splitting Up of Heat Conduction Problems", Letters Heat & Mass Transf., V. 4, pp. 163-166, 1977.

4. Cotta, R.M. and M.N. Özisik, "Diffusion Problems with General Time-Dependent Coefficients", Braz. J. Mech. Sciences, RBCM, V. 9, no. 4, pp. 269-292, 1987.

5. Cotta, R.M., "Hybrid Numerical-Analytical Approach to Diffusion-Convection Problems", Proc. of the XV National Summer School (Invited Lecture), Institute of Applied Mathematics & Computer Science, Bulgaria, August 1989.

New Load Dependent Methods for Modal Solution of Transient Heat Conduction

F.C. dos Santos, A.L.G.A. Coutinho, L. Landau

Laboratory for Computer Methods in Engineering, Department of Civil Engineering, COPPE/Federal University of Rio de Janeiro, Caixa Postal 68506, Rio de Janeiro, RJ 21945, Brasil

ABSTRACT

In this work, new methods based on direct iteration for generation of Ritz or Lanczos load dependent vectors for the modal solution of transient heat conduction are presented.

INTRODUCTION

The finite element solution of linear transient heat conduction requires the integration of coupled semi-discrete first-order ordinary differential equations of the form

$$\underset{\sim}{C}\,\dot{\underset{\sim}{u}} + \underset{\sim}{K}\,\underset{\sim}{u} = \underset{\sim}{f}(t) \tag{1}$$

where $\underset{\sim}{C}$ and $\underset{\sim}{K}$ are, respectively, the heat capacity and conductivity matrices; $\underset{\sim}{f}(t)$ is the heat supply vector; $\underset{\sim}{u}$ is the vector of nodal temperatures and the dot denotes time differentiation. All vectors and matrices are n-dimensional, where n is the total number of unknown temperatures. Traditionally,the solution of (1) is obtained using a direct step-by-step method [1]. Recently, it has been shown that a modal solution technique where a load dependent basis generated by Lanczos algorithm [2,3] was more cost-effective than the usual direct step-by-step procedure. Further, if the time variation of the heat supply vector is piecewise linear, it was also shown in Ref.[3] that the modal solutions are always numerically stable and accurate. In the early approaches, the load dependent basis was constructed by Lanczos-type algorithms utilizing inverse iteration. These type of algorithms were borrowed from structural dynamics [4,5], where the response is usually dominated by the lower modes, which are approximated first. However, it is well known that in solution of Eq. (1) the higher modes are the most important. Therefore, we suggest in this work, the utilization of Ritz or Lanczos algorithms utilizing direct iteration instead of inverse iteration. To control the quality of the approximation furnished by these algorithms, we include the two cut-off criteria proposed in Ref. [5]. The first one is the usual load participation factor, where the second is a cut-off criteria based on the heat capacity matrix representation. The remainder of this paper is organized as follows. In the next section we briefly review the general reduction techniques applied to trasient heat conduction. Then, we present the Ritz and Lanczos algorithms utilized in this work, with the cut-off criteria to generate a basis with the maximum accuracy. It follows the numerical results obtained in the analysis of a linear strip with convection boundary conditions. The final section addresses the main features of the proposed approach.

REDUCTION METHOD FOR TRANSIENT HEAT CONDUCTION

The reduction method replaces the set of equations(1), governing transient heat conduction, by a system of equations with considerably fewer unknowns, through the following co-ordinate transformation:

$$\underset{\sim}{u} = \underset{\sim}{\Phi}\underset{\sim}{x} \tag{2}$$

where $\underset{\sim}{x}$ is a generalized temperature vector in the new co-ordinates and $\underset{\sim}{\Phi} = [\underset{\sim}{\Phi}_1, \underset{\sim}{\Phi}_2, \ldots, \underset{\sim}{\Phi}_m]$, m<n is the co-ordinate transformation matrix. Thus, the transient response $\underset{\sim}{u}(t)$ will be approximated by a linear combination of the generalized temperatures in the new co-ordinates $\underset{\sim}{x}(t)$. The approximate response will depend entirely on the quality of the co-ordinate transformation matrix, since m<n. Applying this co-ordinate transformation and its time derivatives to equation (1), one arrives at

$$\bar{\underset{\sim}{C}}\dot{\underset{\sim}{x}} + \bar{\underset{\sim}{K}}\underset{\sim}{x} = \bar{\underset{\sim}{f}}(t) \tag{3}$$

in which the generalized heat capacity and conductivity matrices, and heat supply vector, are given by

$$\bar{\underset{\sim}{C}} = \underset{\sim}{\Phi}^T\underset{\sim}{C}\underset{\sim}{\Phi} \tag{4}$$

$$\bar{\underset{\sim}{K}} = \underset{\sim}{\Phi}^T\underset{\sim}{K}\underset{\sim}{\Phi} \tag{5}$$

$$\bar{\underset{\sim}{f}}(t) = \underset{\sim}{\Phi}^T\underset{\sim}{f}(t) \tag{6}$$

The classical reduction method utilizes, as the co-ordinate transformation matrix, the modal matrix, whose columns are the first m eigenvectors of the eigenproblem related to equation(1),

$$\underset{\sim}{K}\underset{\sim}{\Phi} = \underset{\sim}{C}\underset{\sim}{\Phi}\underset{\sim}{\Lambda}_m \tag{7}$$

where $\underset{\sim}{\Lambda}_m$ is a diagonal matrix listing the eigenvalues. The modal matrix satisfies the following relations:

$$\underset{\sim}{\Phi}^T\underset{\sim}{K}\underset{\sim}{\Phi} = \underset{\sim}{\Lambda}_m \tag{8}$$

$$\underset{\sim}{\Phi}^T\underset{\sim}{C}\underset{\sim}{\Phi} = \underset{\sim}{I}_m \tag{9}$$

where $\underset{\sim}{I}_m$ is the identity matrix of order m. Pre-multiplying equation (1) by $\underset{\sim}{\Phi}^T$, applying the co-ordinate transformation (2) and relations (8),(9), the reduced uncoupled transient linear heat conduction equations can be written as

$$\dot{\underset{\sim}{x}} + \underset{\sim}{\Lambda}_m\underset{\sim}{x} = \underset{\sim}{\Phi}^T\underset{\sim}{f}(t) \tag{10}$$

The solution of this system of equations can be obtained, if the heat supply vector time variation is piecewise linear, through the exact integration of each uncoupled equation. This solution scheme does not present any numerical instability, and the quality of the results will depend entirely on the co-ordinate transformation matrix.

RITZ AND LACZOS VECTORS PLUS CUT-OFF CRITERIA FOR MODE SUPERPOSITION ANALYSIS

The algorithms for generation of Ritz and Lanczos vectors based on the utilization of direct iteration constructs an orthonormal basis for the Krylov space.

$$\kappa(\underset{\sim}{f}) = \text{span}\ [\underset{\sim}{f},\ \underset{\sim}{C}^{-1}\underset{\sim}{K}\underset{\sim}{f},\ (\underset{\sim}{C}^{-1}\underset{\sim}{K})^2\underset{\sim}{f},\ldots,(\underset{\sim}{C}^{-1}\underset{\sim}{K})^m\underset{\sim}{f}] \tag{11}$$

Where the starting vector for the Krylov sequence is the amplitudes of the heat supply vector $\underset{\sim}{f}$. Therefore the essential part is the consideration of the spatial distribution of the loading vector to generate a load dependent basis. The algorithm to generate $\underset{\sim}{C}$-orthogonal Ritz vectors is as follows:

1. Solve $\underset{\sim}{C}\,\underset{\sim}{x}_1^* = \underset{\sim}{f}$

2. $\underset{\sim}{C}$ - normalize $\quad \underset{\sim}{x}_1 = \underset{\sim}{x}_1^* \,/\, ||\underset{\sim}{x}_1^*||_{\underset{\sim}{C}}$

3. Calculate additional vectors $(i = 2,\ldots,m)$

3.1 Solve $\underset{\sim}{C}\,\underset{\sim}{x}_i^* = \underset{\sim}{K}\,\underset{\sim}{x}_{i-1}$ (direct iteration)

3.2 $\underset{\sim}{C}$ - orthogonalize:

for $j = 1,\ldots, i-1$

$$c_j = \underset{\sim}{x}_j^T \underset{\sim}{C}\,\underset{\sim}{x}_i^*$$

$$\underset{\sim}{x}^{(j+1)} = \underset{\sim}{x}_i^{(j)} - c_j \underset{\sim}{x}_j$$

end loop.

$\underset{\sim}{C}$ - normalize $\underset{\sim}{x}_i = \underset{\sim}{x}_i^{(i)} / ||\underset{\sim}{x}_i^{(i)}||_{\underset{\sim}{C}}$

4. Compute $\bar{\underset{\sim}{K}} = \underset{\sim}{X}\underset{\sim}{K}\underset{\sim}{X}$, where $\underset{\sim}{X} = [\underset{\sim}{x}_1, \underset{\sim}{x}_1, \ldots, \underset{\sim}{x}_m]$

5. Solve $\underset{\sim}{K}\,\underset{\sim}{Z} = \underset{\sim}{Z}\,\underset{\sim}{\Lambda}$

6. Compute $\underset{\sim}{\Phi} = \underset{\sim}{X}\,\underset{\sim}{Z}$, $\underset{\sim}{\Phi} = [\underset{\sim}{\Phi}_1, \underset{\sim}{\Phi}_2, \ldots, \underset{\sim}{\Phi}_m]$.

In the above sequence of operations $||\cdot||_{\underset{\sim}{C}}$ is a norm defined by

$$||\underset{\sim}{x}||_{\underset{\sim}{C}} = (\underset{\sim}{x}^T \underset{\sim}{C}\,\underset{\sim}{x})^{1/2} \tag{12}$$

This algorithm is particularly efficient if we note that when $\underset{\sim}{C}$ is diagonal the solution of the systems of equations in steps 1 and 3.1 are trivial. The $\underset{\sim}{C}$ - orthogonalization step utilizes the modified Gram - Schimdt orthogonalization to keep the Ritz vectors orthogonal. However, if the number of vectors to be computed is relatively great, then would be preferable to use the Lanczos type recurrence, where the orthogonality is only imposed between two successive vectors. Therefore, the step 3.2 of the previous algorithm should be modified to:

Calculate additional vectors $(i=2,\ldots,m)$ with Lanczos iteration:

$$\alpha = \underset{\sim}{x}_{i-1}^T \underset{\sim}{C}\,\underset{\sim}{x}_i^*$$

$$\underset{\sim}{x}_i = \underset{\sim}{x}_i^* - \alpha\,\underset{\sim}{x}_{i-1}$$

$$\bar{\underset{\sim}{x}}_i = \underset{\sim}{x}_i - \beta_i\,\underset{\sim}{x}_{i-1}$$

$$\beta_{i+1} = ||\bar{\underset{\sim}{x}}_i||_{\underset{\sim}{C}}$$

$$x_i = \bar{x}_i \; / \; \beta_{i+1}$$

The number of basis vectors to reach a desired accuracy can be determined in the Ritz and Lanczos algorithms using the two cut-off criteria introduced in Ref.[5]. The first one is related to the participation of the spatial distribution of the heat supply vector f. The load vector of Eq(1) can be expressed as,

$$f = X \alpha \tag{13}$$

where $X = [x_1, x_2, \ldots, x_m]$ are the Ritz or Lanczos vectors and α are the generalized coordinates of the heat supply vector. The matrix X is the transfer operator from the f-coordinates to the α-coordinates. Since the transfer operator is C-orthonormal, the components of α can be expressed as

$$\alpha_i = f^T C \, x_i \tag{14}$$

and thus, the approximation of the heat supply vector is expressed by

$$f_m = \sum_{i=1}^{m} x_i f^T C \, x_i \tag{15}$$

Therefore, the error in the spatial distribution of the external loading is,

$$e_m = f - f_m \tag{16}$$

and the first stopping criteria can be derived from the above relation as

$$\varepsilon_m = \left| f^T e_m \right| \; / \; f^T f \tag{17}$$

The second cut-off criteria is related to the representation of the heat capacity matrix, which can be expressed by the relation,

$$\zeta_m = \sum_{i=1}^{m} (\bar{C}^T x_i)^2 \; / \; \sum_{i=1}^{m} \bar{C} \tag{18}$$

where $\bar{C}$ is a vector whose entries are computed as,

$$\bar{\mathbf{C}}_i = \sum_{j=1}^{n} \mathbf{C}_{ij}$$

If $\mathbf{C}$ is diagonal, $\bar{\mathbf{C}}_i = C_{ii}$. Also, it should be noted that ζ_m converges monotonically to 1 as $m \to n$. The values of ε_m and ζ_m can be computed and accumulated as soon as a new Ritz or Lanczos vector is evaluated, and if their values are below some user supplied tolerances, the generation process is stopped.

NUMERICAL APPLICATION

The numerical application selected was a linear strip with convection boundary conditions analysed first in Ref[6]. The domain is $0 \leq x \leq 2.5$m, and the material properties are: thermal conductivity, $k = 1.4$ W/m/ºC, thermal capacity, $\rho C = 1000$ W/m²/ºC. The convection coefficient, α_c, is 21 W/m²/ºC and the external temperatures in $x = 0$ is $T_a = 40$ºC and $x = 2.5$ is $T_a = 0$ºC. The initial temperature of the body is ºC and the time step is 4.0 seconds. The problem was discretized by ten equally spaced bilinear isoparametric finite elements considering a lumped heat capacity matrix. In Tables 1a and 1b the relative errors for

TIME (SEC)	ANALYTICAL SOLUTION (ºC)	RITZ AND LANCZOS ALGORITHMS DIRECT ITERATION	
		m = 2	m = 4,6,8,11
4.0	24.9207	- 27.37	- 27.37
8.0	27.7313	- 5.88	- 5.88
12.0	29.4729	0.77	0.77
16.0	30.6537	2.43	2.43
20.0	31.5126	2.82	2.82
24.0	32.1719	2.57	2.57
28.0	32.6961	2.15	2.15
32.0	33.1275	2.03	2.03
36.0	33.4901	1.22	1.52
40.0	33.8006	0.89	1.48
44.0	34.0703	0.67	1.26
48.0	34.3075	0.27	1.14
52.0	34.5183	0.24	1.11
56.0	34.7072	- 0.02	0.84
60.0	34.8778	- 0.22	0.92

Table 1.a - Solution Data in $x = 0$ for a Linear Strip with Convection Boundary Conditions

TIME (SEC)	RITZ AND LANCZOS ALGORITHMS INVERSE ITERATION		
	m = 2	m = 4	m = 6,8,11
4.0	- 58.27	- 28.97	- 27.37
8.0	- 35.45	- 6.60	- 5.88
12.0	- 21.28	0.77	0.77
16.0	- 11.59	3.09	2.43
20.0	- 5.43	3.45	2.82
24.0	- 1.16	3.20	2.57
28.0	1.54	2.76	2.15
32.0	3.54	2.33	2.03
36.0	4.51	1.82	1.52
40.0	5.03	1.48	1.48
44.0	5.37	1.26	1.26
48.0	5.52	1.14	1.14
52.0	5.45	0.82	1.11
56.0	5.45	0.84	0.84
60.0	5.22	0.64	0.92

Table 1.b - Solution Data for x = 0 for a Linear Strip with Convection Boundary Conditions

the temperatures in x = 0 computed utilizing the Ritz and Lanczos algorithms considering direct iteration, are compared with the standard Ritz an Lanczos procedures, which are based on inverse iteration. In this Table are included the analytical solution of this problem given also in Ref.[6].

The magnitude of the errors for almost all the responses in Table 1a and 1b indicate a good agreement with the analitycal solution. The only exception is the solution for m = 2 in Table 1b. Further, the solutions with direct iteration converges quickly than its inverse iteration counterparts, which stabilize only for $m \geqq 6$. It should be also stressed that, for Lanczos algorithms with inverse iteration considering m = 11, the computed solution presented some numerical instabilities related to the loss of orthogonality among Lanczos vectors. Tables 2 and 3 show, for m = 1 to 11, the computed values for ε_m and ζ_m respectively for the direct and inverse iteration versions of the Ritz procedure.

The results for direct iteration provide an excellent load representation for all m. This is directly associated to the characteristics of direct iteration, where the higher frequencies are approximated first. In constrast, the inverse iteration approach needs more than 6 vetores to have less than 5% of error in the load representation. Also, in both Tables the ζ_m values converge monotonically to 1.0 as we increase m. However, for this problem the heat capacity matrix representation appears to be less important than the error in the spatial load distibution. For m= 2 in Table 2, we have ζ_m = 0.15, but

m	ε_m	ζ_m
1	0.2475E-15	0.05
2	0.2475E-15	0.15
3	0.2475E-15	0.25
4	0.2475E-15	0.35
5	0.2475E-15	0.45
6	0.2475E-15	0.55
7	0.2475E-15	0.65
8	0.2475E-15	0.75
9	0.2475E-15	0.85
10	0.2475E-15	0.95
11	0.2475E-15	1.00

Table 2 - Cut-off Parameters for Ritz Procedure with Direct Iteration

m	ε_m	ζ_m
1	0.8318	0.6989
2	0.5896	0.8714
3	0.5283	0.8861
4	0.5204	0.9365
5	0.3433	0.9396
6	0.1125	0.9565
7	0.04632	0.9602
8	0.03880	0.9652
9	0.03644	0.9682
10	0.03584	0.9702
11	0.03557	0.9710

Table 3 - Cut-off Parameters for Ritz Procedure with Inverse Iteration

the results for the transient response in Table 1a show a good agreement with the analytical solution. Finally, the computed values of ε_m, ζ_m for the Lanczos procedures are equal to those presented in Tables 2 and 3. The only exception is the values for m = 11 in Table 3, that are meaningless, since the orthogonality between the Lanczos vectors were lost.

CONCLUSIONS

The utilization of Ritz and Lanczos procedures based on direct iteration to generate a suitable basis for the modal solution of transient heat conduction offers the following advantages against the usual procedures based on inverse iteration:

(a) Provides a better representation of the spatial distribution of the heat supply vector.

(b) If the heat capacity matrix is lumped, the algorithms involve less floating point operations than their inverse iteration counterparts, thus being more computationally attractive for the solution of large problems.

REFERENCES

1. HUGHES, T.J.R., The Finite Element method; Linear Static and Transient Finite Element Analysis, Prentice-Hall Englewood Cliffs, NJ. 1987.
2. COUTINHO, A.L.G.A., Landau, L., Wrobel, L.C. and Ebecken, N.F.F., Modal Solution of Transient Heat Conduction Utilizing Lanczos Algorithm, Int. Journal for Numerical Methods in Engineering, Vol. 28, pp.13-27, 1989.
3. NOUR-OMID, B., Lanczos Method for Heat Conduction Analysis, Int. Journal for Numerical Methods in Engineering, Vol. 24, pp. 251-262, 1987.
4. COUTINHO. A.L.G.A., Landau, L., Lima, E.C.P. and Ebecken, N.F.F., The Application of the Lanczos Mode Superposition Method in Dynamic Analysis of Offshore Structures, Computer & Structures, Vol. 25, pp. 615-625, 1987.
5. JOO, K.I., Wilson, E.and Leger, P., Ritz Vectors and Generation Criteria for Mode Superposition Analysis, Earthquake Engineering and Structural Dynamics, Vol.18, pp. 149-167, 1989.
6. OWEN, D.R.J. and Damjanic, F., Reduced Numerical Integration in Thermal Transient Finite Elements Analysis, Computer & Structures, Vol. 17, pp. 261-176, 1983.

Finite Element Analysis of Heat Transfer in a Solid Coupled with Cooling Pipes

N.G. Gong, G.R. Li(*), R.W. Lewis(**)

Department of Civil Engineering, University of Nottingham, Nottingham, NG7 2RD, England

(*) Department of Hydraulic Engineering, Chengdu University of Science & Technology, Chengdu, China

(**) Department of Civil Engineering, University College of Swansea, Swansea, SA2 8PP, Wales

ABSTRACT

A finite element method is used in this paper to simulate the heat transfer in a solid body cooled by embedded cooling pipes. The heat relationship between the cooling fluid and the solid body is discussed. An one-dimensional pipe element has been used to simulate the cooling pipes. It couples with a two-dimensional or three-dimensional finite element mesh. The thermal boundary conditions of the solid body cooled are based on the energy balance between the fluid and the solid body. An iterative solution procedure is adopted to seek the convergent solution of the energy equilibrium. The method shows a good convergence conditionally. Numerical results are compared with the theoretical results and show that a reasonable accuracy has been achieved.

INTRODUCTION

The purpose in this paper is to consider the transient heat transfer in a solid with embedded cooling pipes. This problem can be found in many industry applications. For example, to remove the extra hydration heat and to control the temperature of a massive concrete structures in a dam construction, cooling pipes are embedded. In this situation, not only should the temperature of the cooling fluid be considered, but also the thermal state of the concrete has to be concerned by engineers. It is not difficult to obtain the thermal state in the concrete using a thermal analysis method in the solid body whose boundary conditions are known. However, these boundary conditions include the heat flux at the interface of the concrete and the cooling pipes. The difficult of this problem is that the heat flux is dependent on both the cooling fluid temperature and the solid temperature at the same time. This coupled heat transfer problem has been studied by many researchers, and many numerical methods have been developed. Those previous studies were summarized by Gane *et al* [1]. A finite element model was developed by Gong *et al* [2]. A 2-D calculation plane was coupled with a string of 1-D curved cooling pipe elements, as shown in Fig. 1. In this paper, further development of the model has been presented and some effects of the calculation parameters has been discussed.

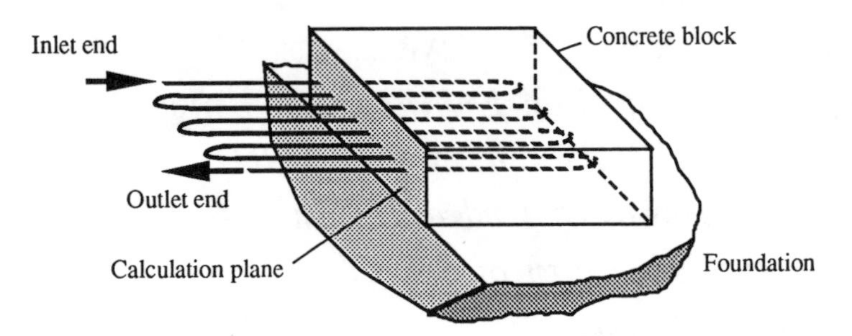

Fig. 1 Solid body is cooled by embedded cooling pipes

THE HEAT RELATIONSHIP BETWEEN SOLID AND FLUID

When cooling fluid passes an hot solid body through an embedded cooling pipe, it is warmed by the solid body continuously. The change of fluid temperature depends on the thermal state of the solid and itself at the same time. It is assumed that the fluid temperature is not changed suddenly along the cooling pipe axis. Therefore the axis heat flux within the fluid is negligible, and the direction of heat flux between the fluid and the solid body is along the normal of the pipe wall. In the pipe, the fluid flow is supposed to be fully developed pipe flow if the pipe is not bending rapidly and the concerned position of the pipe is far from the inlet end. The typical profiles of the flow velocity and the temperature are shown in Fig. 2. The distribution of the fluid temperature T_f at any cross section of the pipe is axis-symmetric so that

$$\frac{\partial T_f}{\partial \theta} = 0 , \qquad r \leq R \tag{1}$$

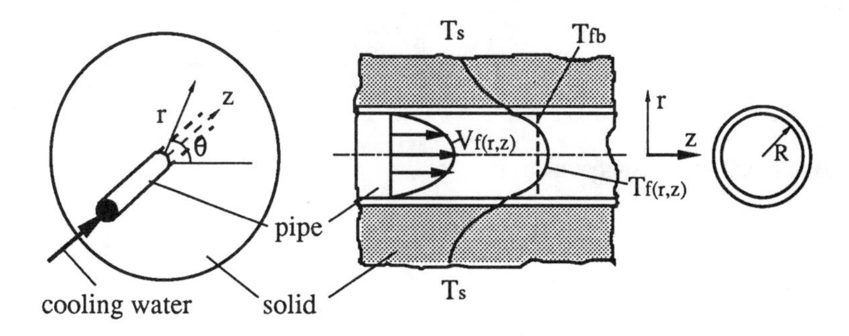

Fig. 2 Fluid temperature in the pipe

Based on an energy balance, at the interface between the solid and the fluid, the absorbed heat in the cooling fluid is equal to the heat loss in the solid within a pipe length of dz.

$$c_f \rho_f q_f \; dT_{fb} = kR \left(\int_0^{2\pi} \frac{\partial T_s}{\partial n} d\theta \; dz + \frac{\partial}{\partial z} \int_0^{2\pi} \frac{\partial T_s}{\partial n} d\theta \; dz \right) \qquad (2)$$

where dT_{fb} is the increment of bulk temperature of the fluid, c_f,ρ_f and q_f are the fluid specific heat, density and the rate of fluid flow. k is the thermal conductivity of the solid. R is the radius of the cooling pipe and $\frac{\partial T_s}{\partial n}$ is the solid temperature gradient at the interface - pipe wall. Within a pipe length L, the fluid temperature increment should be

$$\Delta T_{fb} = \frac{kR}{c_f \rho_f q_f} \left(\int_0^L \int_0^{2\pi} \frac{\partial T_s}{\partial n} d\theta \; dz + \int_0^L \frac{\partial}{\partial z} \int_0^{2\pi} \frac{\partial T_s}{\partial n} d\theta \; dz \right) \qquad (3)$$

The temperature at pipe wall, T_w, is governed by a convective boundary condition. Therefore, in a unit pipe length, the absorbed heat flux q_f in the fluid is

$$q_f = 2\pi R h \, (T_w - T_{fb}) \qquad (4)$$

where h is the heat transfer coefficient of the fluid.

On the other hand, the heat loss of the cooled solid body in the same area of the wall surface can be expressed by the solid temperature gradient at the surface.

$$q_s = -kR\int_0^{2\pi}\frac{\partial T_s}{\partial n}d\theta \tag{5}$$

Based on the energy balance

$$q_s = -q_f \tag{6}$$

Thus

$$k\int_0^{2\pi}\frac{\partial T_s}{\partial n}d\theta = 2\pi h\,(T_w - T_{fb}) \tag{7}$$

Substituting Eq. (7) into Eq. (2), the pipe wall temperature increment along the pipe from point 0 to point L can be obtained by

$$\Delta T_w\Big|_0^L = \frac{kR}{c_f\rho_f q_f}\left(\int_0^L\int_0^{2\pi}\frac{\partial T_s}{\partial n}d\theta\,dz + \int_0^L\frac{\partial}{\partial z}\int_0^{2\pi}\frac{\partial T_s}{\partial n}d\theta\,dz\right) +$$

$$\frac{k}{2\pi h}\int_0^{2\pi}\left(\frac{\partial T_s}{\partial n}_L - \frac{\partial T_s}{\partial n}_0\right)d\theta \tag{8}$$

Eq. (8) is the boundary condition of the following governing Eq. in a solid body:

$$k\,\nabla^2 T_s - \rho\,c\frac{\partial T_s}{\partial t} = -Q(x,y,z,t) \tag{9}$$

where ρ and c are the density and specific heat of solid respectively and $Q(x,y,z,t)$ is the rate of heat generated in the solid body.

To solve Eq. (9) with boundary condition of Eq. (8) is extremely difficult using a mathematic method. In this paper, a numerical method has been used to solve this problem in terms of a pipe element which links with a two-dimensional or three-dimensional finite element model.

PIPE ELEMENT AND SOLUTION PROCEDURE

In a three-dimensional domain, the cooling pipe is discretized by a string of pipe elements along the pipe axis. At each pipe node, there is a reference plane that is perpendicular to the pipe axis in the solid body. This can be seen in Fig. 3. In a situation of the bending pipe coils, a unified solid reference plane could be employed.

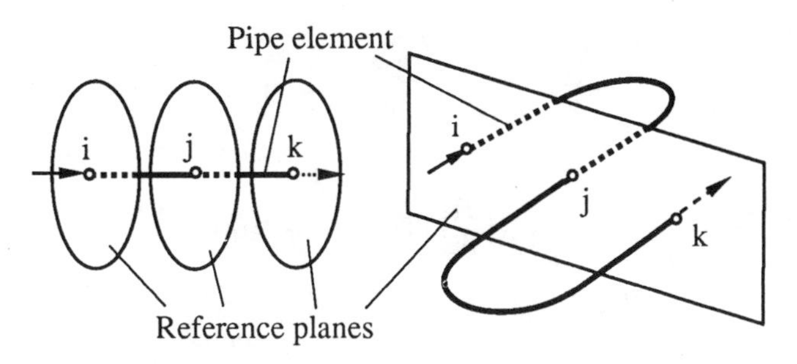

Fig. 3 Pipe element and reference plane

A three-dimensional problem can be reduced to a two-dimensional problem provided that the heat flux along the cooling pipe in the solid can be neglected. In this case, all the reference planes become to the two-dimensional finite element computed planes.

Along the pipe axis z, the distribution of the solid temperature gradient $\frac{\partial T_s}{\partial n}$ is defined as

$$\frac{\partial T_s}{\partial n} = \sum_i N_i \frac{\partial T_s}{\partial n}_i \tag{10}$$

$\frac{\partial T_s}{\partial n}_i$ is the nodal gradient value at node i. N_i are the shape functions. For a quadratic pipe element, they are

$$N_i = \frac{1}{2}\left[\frac{(z - L)^2}{L^2} - \frac{z - L}{L}\right]$$

$$N_j = 1 - \frac{(z - L)^2}{L^2} \tag{11}$$

$$N_k = \frac{1}{2}\left[\frac{(z - L)^2}{L^2} + \frac{z - L}{L}\right]$$

Using Eq.(10), the Eq. (8) can be evaluated in an element. The increments of pipe wall temperature from node i to node j and k are

$$\Delta T_w|_i^j = \frac{kRL}{c_f \rho_f q_f} \left[\left(\frac{5}{12} - \frac{1}{L}\right) \int_0^{2\pi} \frac{\partial T}{\partial n_i} d\theta + \left(\frac{2}{3} + \frac{1}{L}\right) \int_0^{2\pi} \frac{\partial T}{\partial n_j} d\theta \right) -$$

$$\frac{1}{12} \int_0^{2\pi} \frac{\partial T}{\partial n_k} d\theta) \Big] + \frac{k}{2\pi rh} \int_0^{2\pi} \left(\frac{\partial T}{\partial n_j} - \frac{\partial T}{\partial n_i} \right) d\theta \tag{12}$$

$$\Delta T_w|_i^k = \frac{kRL}{c_f \rho_f q_f} \left[\left(\frac{1}{3} - \frac{1}{L}\right) \int_0^{2\pi} \frac{\partial T_s}{\partial n_i} d\theta + \frac{4}{3} \int_0^{2\pi} \frac{\partial T_s}{\partial n_j} d\theta \right. +$$

$$\left. \left(\frac{1}{3} + \frac{1}{L}\right) \int_0^{2\pi} \frac{\partial T_s}{\partial n_k} d\theta \right] + \frac{k}{2\pi h} \int_0^{2\pi} \left(\frac{\partial T_s}{\partial n_k} - \frac{\partial T_s}{\partial n_i} \right) d\theta \tag{13}$$

If a linear pipe element is used, the increment of pipe wall temperature from node i to node j must be

$$\Delta T_w|_i^j = \frac{kRL}{2c_f \rho_f q_f} \left[\left(1 - \frac{1}{L}\right) \int_0^{2\pi} \frac{\partial T_s}{\partial n_i} d\theta + \left(1 + \frac{1}{L}\right) \int_0^{2\pi} \frac{\partial T_s}{\partial n_j} d\theta \right] +$$

$$\frac{k}{2\pi h} \int_0^{2\pi} \left(\frac{\partial T_s}{\partial n_j} - \frac{\partial T_s}{\partial n_i} \right) d\theta \tag{14}$$

The initial fluid temperature $T_{f,ini}$ at the inlet end of the cooling pipe is known usually. It is supposed that the pipe wall temperature at the inlet end is the same as $T_{f,ini}$. Therefore, the distribution of the pipe wall temperature along the pipe can be obtained by adding the increments of temperature that are calculated by Eq. (12) and Eq. (13) or Eq. (14).

$$T_w = T_{w,ini} + \sum_i \Delta T_{w,i} \tag{15}$$

Eq. (15) is the boundary condition at the solid-fluid interface in the finite element model. $T_{w,ini}$ is known initially but all the $\Delta T_{w,i}$ are dominated by the solid temperature gradients at the pipe walls. Using numerical differentiations and integrations in Eq. (12), (13) or (14), temperature increments can be expressed with matrices.

$$\Delta \mathbf{T}_W = \mathbf{P}\ \mathbf{T}_S \tag{16}$$

$\Delta \mathbf{T}_W$ is the nodal pipe wall temperature increment vectors. **P** is an heat control matrix. It is very sparse because only a small part of the total solid nodal temperatures are used to calculate the temperature gradients at the pipe nodes. $\mathbf{T}_S$ is all the nodal temperatures in the solid body. It should be noted that $\mathbf{T}_S$ includes the pipe wall nodal temperature $\mathbf{T}_W$.

The global finite element linear equation for the solid body at a time-step is

$$\mathbf{K}\ \mathbf{T}_S = \mathbf{F} \tag{17}$$

with specific boundary values described by Eq.(15). It is difficult to solve it directly with Eq. (15). Nevertheless, a simple iteration procedure can be established here using Eq. (15), (16) and (17).

$$\begin{gathered} \Delta \mathbf{T}_W{}^{k+1} = \mathbf{P}\ \mathbf{T}_S{}^{k} \\ \mathbf{T}_W{}^{k+1} = \mathbf{T}_{W,ini} + \Sigma\, \Delta \mathbf{T}_W{}^{k+1} \\ \mathbf{K}\ \mathbf{T}_S{}^{k+1} = \mathbf{F} \end{gathered} \tag{18}$$

The convergent criteria is that

$$\left|\, \| \mathbf{T}_W{}^{k+1} \| - \| \mathbf{T}_W{}^{k} \| \,\right| \leq \varepsilon \tag{19}$$

where ε is a positive small number, say, 0.0001° C.

It is clear that this iteration procedure is to modify the boundary values $\mathbf{T}_W$ in each iteration while the global coefficient matrix K is no change.

CONVERGENCE AND NUMERICAL RESULTS

In the iteration procedure (18), the direct solution of Eq.(17) is a normal initial-boundary value problem in finite element method. It has a unique solution if the boundary value T_W is specified. For an any initial value of $\mathbf{T}_W$, it is obvious that the iteration procedure (18) is convergent if the temperature increments $\Delta \mathbf{T}_W$ is convergent. Therefore, the convergent condition

$$\| \mathbf{P} \| < 1 \tag{20}$$

should be satisfied.

The norm of matrix **P** is defined by Eqs.(12-14) and for a quadratic pipe element it may be calculated as

$$\| \mathbf{P} \| = \left(\frac{4\pi kRL}{c_f \rho_f q_f} + \frac{k}{h} \right) \frac{1}{d} \tag{21}$$

where d is the radial size of the solid element near the pipe wall.

A cooled solid hollow cylinder in Fig. 4 has been computed using this method. The outer surface of the cylinder is thermal insulated and the outer diameter is 2.5 metres. The diameter of the cooling pipe is 0.025 metres. The total length of the cylinder is 200 meters. k= 10 W/m-°C. c=1.5 kJ/kg-°C. ρ=2450 kg/m^3. The initial fluid temperature at the inlet end of the pipe is 10°C. The initial temperature of the cylinder is 20°C. The rate of cooling fluid flow is 1,000 litres per hour. 5 quadratic identical pipe elements are adopted.

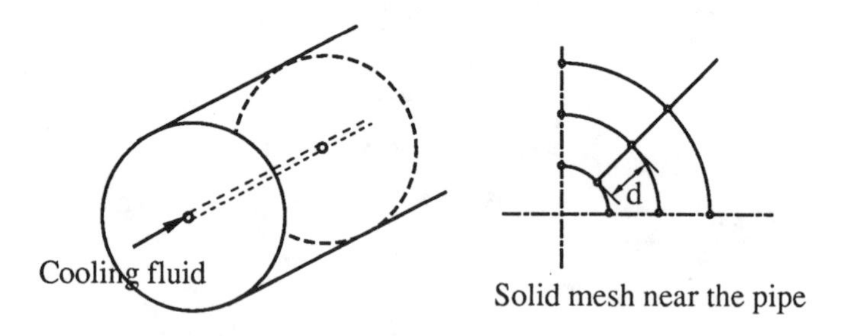

Fig. 4 Solid hollow cylinder and mesh near the pipe

It is difficult to decide the value of the fluid heat transfer coefficient h in use. In fact, it is affected by many factors and it is not easy to measure. It could be defined as infinite so that the pipe wall temperature is equal to the bulk fluid temperature. The fluid temperatures at pipe wall after 1 day cooling (in the case k/h=0) is shown in Fig. 5. The distributions of the pipe wall temperature are changed while the ratio of k/h becomes greater. When k/h is greater than 0.01, the procedure is divergent due to $\|\mathbf{p}\|$ is greater than 1. As $\|\mathbf{p}\|$ less than 1, the convergent solutions (ε<0.0001 C°) are obtained with 10-20 iterations.

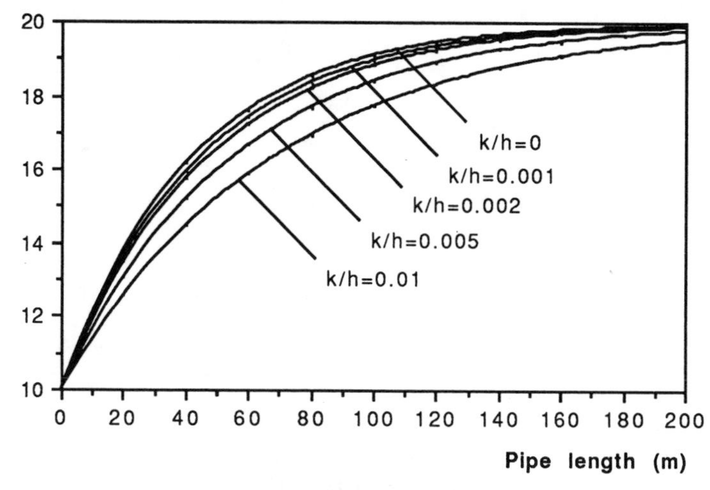

Fig. 5 Effect of k/h

In Eq.(21), the most variable parameter is the element size d at the pipe wall. It depends on the solid finite element mesh used. The numerical results have been affected by it even though the iterative procedure is convergent. It is obvious that a small value of d could lead to the divergence of the iteration. On the other hand, the accuracy of the solid temperature gradients $\frac{\partial T_s}{\partial n}$ at the pipe wall depends on the fineness of the solid mesh as well. Therefore, it is essential to use an good solid element type, a suitable element size and an accurate calculation method for the temperature gradients. The results in Fig. 6 are the fluid temperature distributions at 1 day. They are based on a two-dimensional bi-linear finite element for the solid temperatures and the simplest calculation for the temperature gradients. It can be seen that the effect of the mesh size d is significant. When the ratio of d/R is less than 0.5, the procedure is divergent since the norm of **p** exceeds 1.

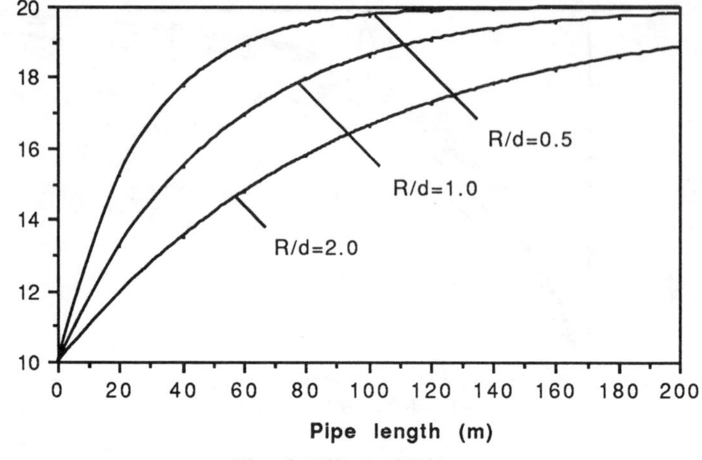

Fig. 6 Effect of R/d

It should be pointed out that the convergence is not affected by the length of the adopted pipe elements. This has been verified by the numerical results shown in Fig. 7. Two sets of results are almost same.

Fluid temperature at pipe wall (C)

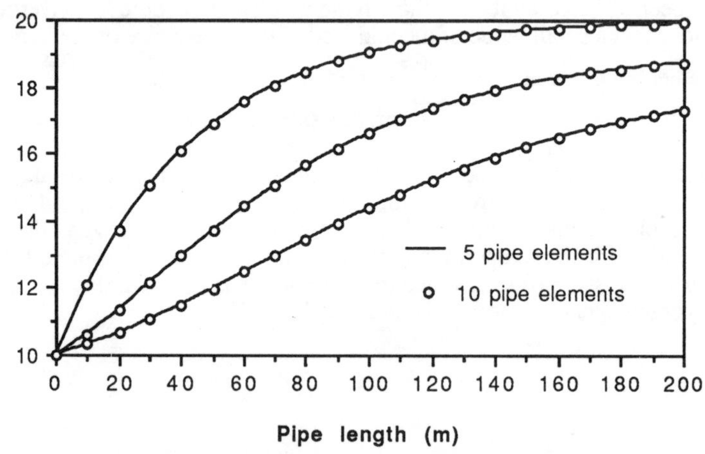

Pipe length (m)

Fig. 7 Effect of the number of pipe elements used

Finally, to compare with the theoretical results of McClellan[3], the numerical results using this method and the theoretical results are shown in Fig. 8. The size of d is 1 cm and the ratio of k/h is 0.0068 in this case. The numerical results show that a reasonable accuracy has been achieved.

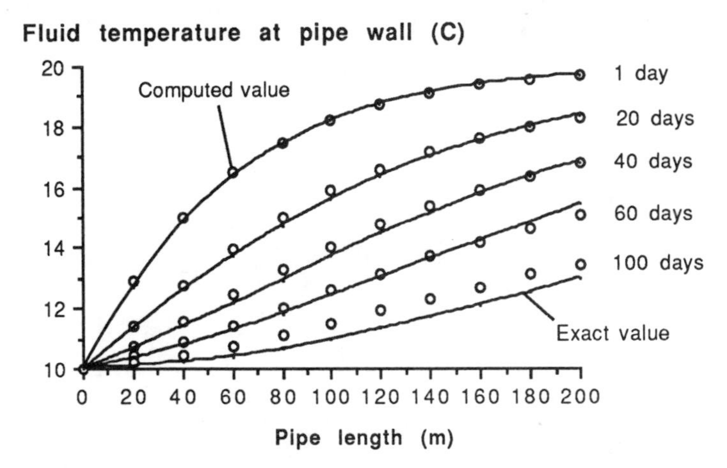

Fig. 8 Comparisons of the results

CONCLUSIONS

The coupled heat transfer problem of a solid cooled by embedded cooling pipes has been successfully simulated by the pipe element used in this paper in conjunction with a 2-d or 3-d finite element method for heat transfer. The iterative procedure for the modifying the specific temperature boundary conditions at the cooling pipe wall converges quickly provided that the solid mesh size d is in a suitable range. It is expected that the effect of the solid mesh size d could be reduce if an higher order solid element and a more accurate method for temperature gradients are adopted.

ACKOWNLEGEMENT

The authors would like to thank Dr. M G Coutie and Dr. B S Choo of the University of Nottingham for their help.

REFERENCES

1. Gane, C. R., Oliver, A. J., Soulsby, D. R. and Stephenson, P. L., Numerical solution of coupled conduction-convection problems using lumped-parameter methods, Numerical methods in heat transfer, Vol. II, Edited by R. W. Lewis *et al* , 227-274, John Wiley & Sons Ltd, 1983
2. Gong, N. G., Li, G. R. and Lewis, R. W., The finite element analysis of the cooling of concrete with embedded water pipes, Numerical methods in thermal problems, Part 2, Vol. VI, Edited by R. W. Lewis and K. Morgan, 1622-1632, Pineridge press, Swansea, U.K., 1989
3. McClellan, L.N., Cooling of Concrete Dams, Boulder Canyon Project Final Reports, Part 7, U.S. Bureau of Reclamation, 1949.

Transputer Solution of the Three-Dimensional Non-Linear Heat Conduction Equation

C.C. Wong, Y.K. Chan, F. Tam(*), Y. Wong

Department of Electronic Engineering, City Polytechnic of Hong Kong, Kowloon Tong, Kowloon, Hong Kong

(*) Now with the Department of Computing, Staffordshire Polytechnic, England

ABSTRACT

Based on the method of transmission-line modelling (TLM) an approved algorithm for three-dimensional, nonlinear heat conduction problems is described. The algorithm retains the explicit nature of the method and at the same time allows highly nonlinear problems to be handled efficiently. Examples are solved in both a VAX super-mini-computer and a transputer system with four T800s. Numerical experiments show that the transputer system is a suitable environment for the implementation of the TLM method.

1 INTRODUCTION

The application of the method of transmission-line modelling (TLM) for solving the diffusion equation was first described by Johns [1]. Subsequently the method has been applied to various problems [2,3]. In all previous publications only two-dimensional problems were considered. In this paper, the method is generalized to include three-dimensional nonlinear cases. The method is essentially an explicit time-stepping algorithm which is unconditionally stable [1]. Due to its explicit nature, the algorithm is particularly suitable for parallel processing. Chan et al [4] have recoded this method for two-dimensional problems in a parallel environment (using four NS floating-point coprocessors) and resonable results have been obtained. In this paper, the parallel enviroment comprises an AT and four T800 transputers. The bottle neck found in [4] is virtually eliminated. Numerical experiments show that a properly configured parallel transputer system can easily outperform a super-mini-computer, such as the VAX 8820.

2 THE TRANSMISSION-LINE MODEL AND SOLUTION

In a conduction region where radiation and convection are negligible, the governing equations are

$$\underline{q} = -k_1 \nabla \phi \qquad (1)$$

$$\nabla . \underline{q} = -k_2 \frac{\partial \phi}{\partial t} \qquad (2)$$

where q, ϕ are the stream (e.g., rate of heat flow) and the potential (e.g., temperature) functions respectively; k_1 and k_2 are constants or functions of ϕ. Combining the above equations, the diffusion equation can be obtained

$$\nabla^2 \phi = \frac{k_2}{k_1} \frac{\partial \phi}{\partial t} \qquad (3)$$

Consider now a field region which is discretized by a three-dimensional transmission-line matrix of fig.1 with all the line parameters (l, c and r) having the appropriate units in per unit length. It can be shown that

$$\begin{aligned} \frac{\partial v}{\partial x} &= -l \frac{\partial i_x}{\partial t} - r.i_x \\ \frac{\partial v}{\partial y} &= -l \frac{\partial i_y}{\partial t} - r.i_y \\ \frac{\partial v}{\partial z} &= -l \frac{\partial i_z}{\partial t} - r.i_z \end{aligned} \qquad (4)$$

and if the effect of the capacitance is assumed lumped at the node

$$\frac{\partial i_x}{\partial x} + \frac{\partial i_y}{\partial y} + \frac{\partial i_z}{\partial z} = -3c \frac{\partial v}{\partial t} \qquad (5)$$

Comparing eqs(1,2) and (4,5), the following analogy can be obtained
$v = \phi$; $\underline{i} = \underline{q}$; $1/r = k_1$; $c = k_2/3$ and the terms with l represent the error (see also eq(6)).
Furthermore, from eqs(4,5)

$$\nabla^2 v = 3lc \frac{\partial^2 v}{\partial^2 t} + 3rc \frac{\partial v}{\partial t} \qquad (6)$$

Provided that the first term of the RHS can be made negligibly small the diffusion equation can be approximated. Johns gave a detail account of this method for one- and two-dimensional problems in reference [1]. The method has been extended to include nonlinear media [3] (also in two dimensions) where any increase of $c(= k_2/3)$ can be modelled by additional shunt stubs (c_s) at the appropriate nodes. The disadvantage of this approach is that c may become relatively small such that the time-step must be kept within an upper limit in order to reduce the error due to l [3,5]. This will affect the efficiency of the algorithm and the potential of the TLM cannot be fully exploited. The aforesaid limitation can be removed as described in the following paragraph.

Since k_2 is assumed nonlinear and is a function of ϕ only, eqs(4,5) can be rewritten as (for convenience, terms with l have been droped)

$$\begin{aligned}\frac{\partial v}{\partial x} &= -\frac{r}{k_2}(i_x k_2)\\ \frac{\partial v}{\partial y} &= -\frac{r}{k_2}(i_y k_2)\\ \frac{\partial v}{\partial z} &= -\frac{r}{k_2}(i_z k_2)\end{aligned} \tag{7}$$

and

$$\frac{\partial i_x k_2}{\partial x} + \frac{\partial i_y k_2}{\partial y} + \frac{\partial i_z k_2}{\partial z} = -3ck_2\frac{\partial v}{\partial t} \tag{8}$$

The new analogy between the transmission-line network and the heat conduction problem becomes

$v = \phi$; $k_2\underline{i} = \underline{q}$; $1/r = k_1/k_2$ and $c = 1/3$

With this rearrangement, c is kept constant and the resistance of the line is rendered nonlinear. Indeed any suitable values of c can be assigned with r appropriately scaled. Solution of the transmission-line network can be easily obtained through the Thevenin's equivalent circuit which is shown in fig.2. A typical node voltage at node (L,M,N) is

$$V(L,M,N) = \frac{1}{3}\sum_{I=1}^{6} V_i(L,M,N,I) \tag{9}$$

where V_is are the incident voltages; and the scattered voltages (V_ss) are

$$V_s(L,M,N,I) = \frac{2 * V_i(L,M,N,I) * R + V(L,M,N) * Z}{R+Z} - V_i(L,M,N,I) \tag{10}$$

for I=1 to 6 and with $R = r.h/2$ and $Z = \Delta t/c.h$; where Δt is the time-step. The incident voltages for the next time step become

$$\begin{aligned}V_i(L,M,N,1) &= V_s(L-1,M,N,2)\\ V_i(L,M,N,2) &= V_s(L+1,M,N,1)\\ V_i(L,M,N,3) &= V_s(L,M-1,N,4)\\ V_i(L,M,N,4) &= V_s(L,M+1,N,3)\\ V_i(L,M,N,5) &= V_s(L,M,N-1,6)\\ V_i(L,M,N,6) &= V_s(L,M,N+1,5)\end{aligned} \tag{11}$$

Note that r is a function of v for nonlinear problems and Z is the characteristic impedance of the line represented by l and c. A more detail description of the process can be found in [6].

3 THE TRANSPUTER SYSTEM

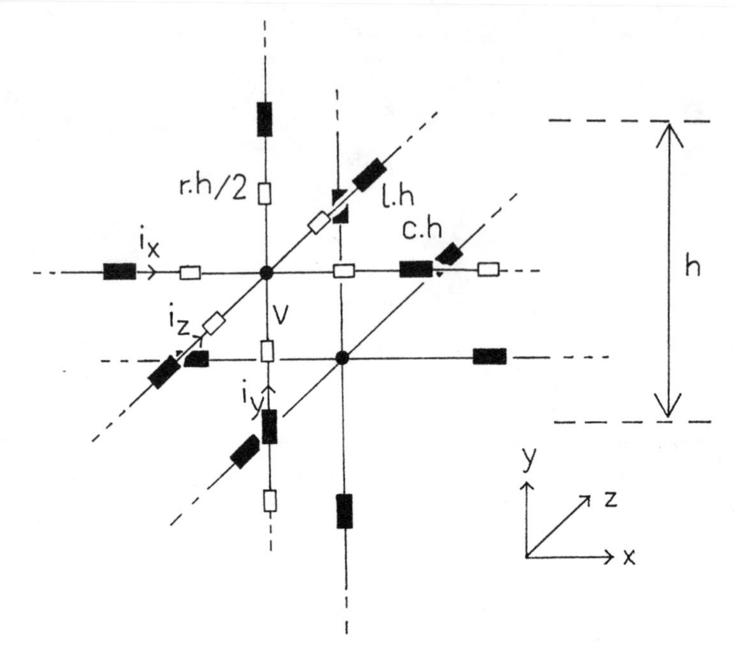

Fig.1 A typical section of a 3-dimensional transmission-line matrix.

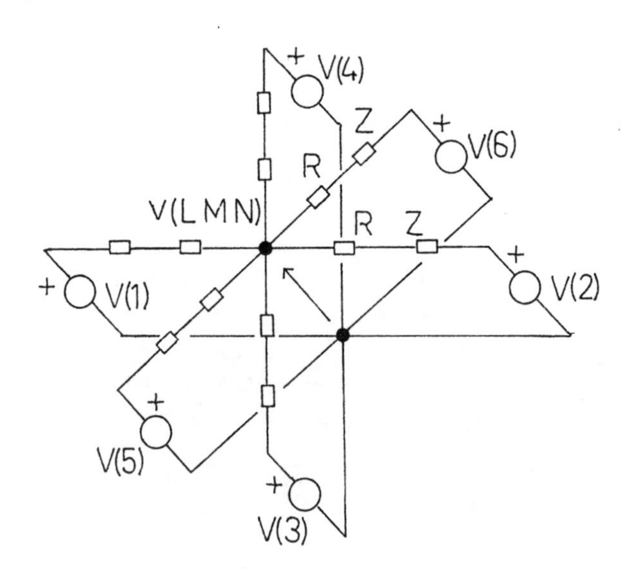

Fig.2 Thevenin's equivalent circuit at a typical node (L, M, N).

```
V(I) = 2*V_i(L,M,N,I)
for I = 1 to 6
```

Figure 3 shows the configuration of the transputer system used in this study. This system consists of four 20 MHz T800 transputers and one 286 based machine. They have been configured to solve some pipeline intensive problems.

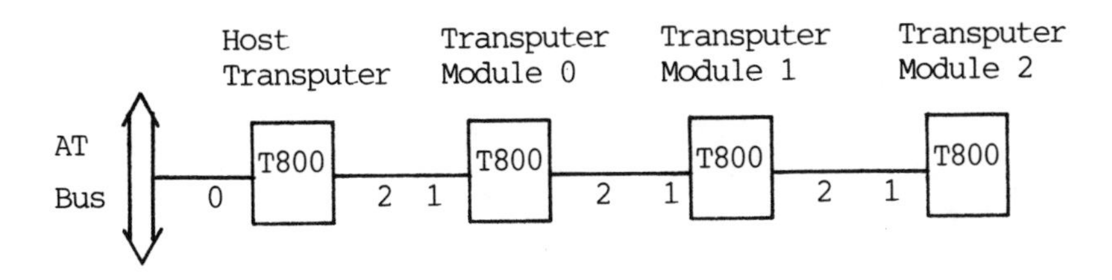

Fig.3 The transputer system.

Eqs(9-11) are the main numerical procedure of the TLM algorithm. When implemented in the transputer system, the problem (the cube of fig.4) is divided into four equal slices and one of which is the shaded volume (due to symmetry, only 1/4 of the volume is needed for computation) of fig.4. For example, if the mesh size = $L/8$, then the value of N will range from 1 to 8 and the first transputer will handle N=1 and 2 and similarly for the other three. Eqs(9-11) are then processed by the four transputers. After one iteration, all necessary data are transferred through the serial links. Eqs(9,10) can be handled independently by the four transputers, however, for solving eq(11) cross-reference data from the adjacent transputer is needed. This will degrade the overall performance of the system slightly.

4 EXAMPLES

The algorithm described in section 2 has been applied to both two- and three- dimensional, linear and nonlinear problems. In this section only three-dimensional nonlinear cases are considered. The problem under discussion is the heating of a cube (fig. 4) with zero initial temperature and the surfaces are kept at unity for time ≥ 0. Exact solutions are available if the medium is linear, however, for nonlinear problems the exact solutions are generated by the numerical algorithm using a fine mesh and time-step. The value of r is given by

$$r(v) = A + B * v^m$$

where A, B and m are constants.

Fig.5 shows the variation of r with v for two different sets of parameters. In the following examples only curve I has been used. Some typical numerical results are shown in figs.6-7 with L (fig.4) =1, A=0.1, B=0.9 and m=2; i.e., $r(0) = 0.1$ and $r(1) = 1$. Fig.6

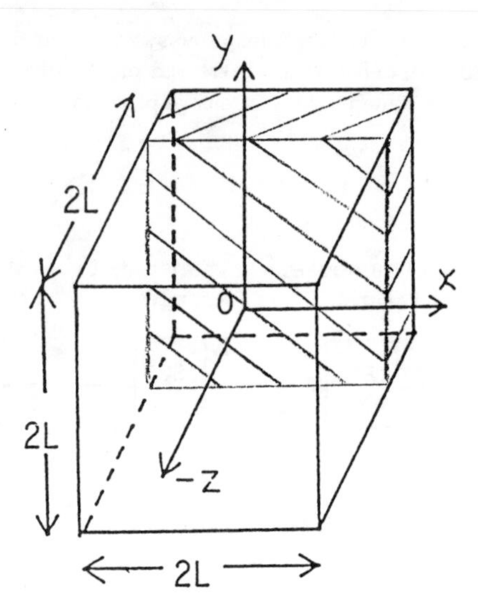

Fig.4
The example under study. The initial temperature is 0 and the surface temperature is suddenly raised to unity at time=0.

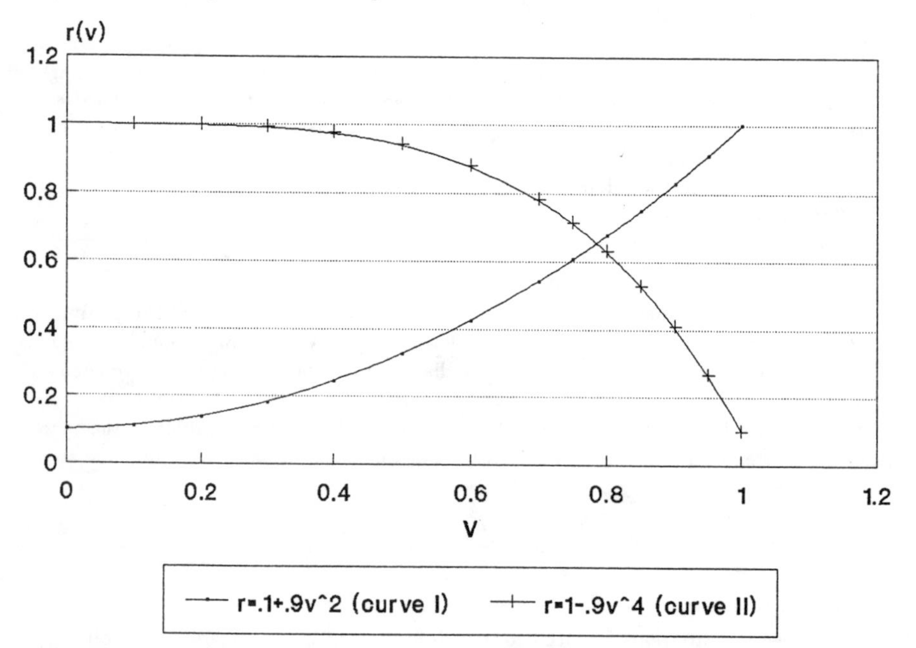

Fig.5 Variation of r with v for two different sets of parameters.

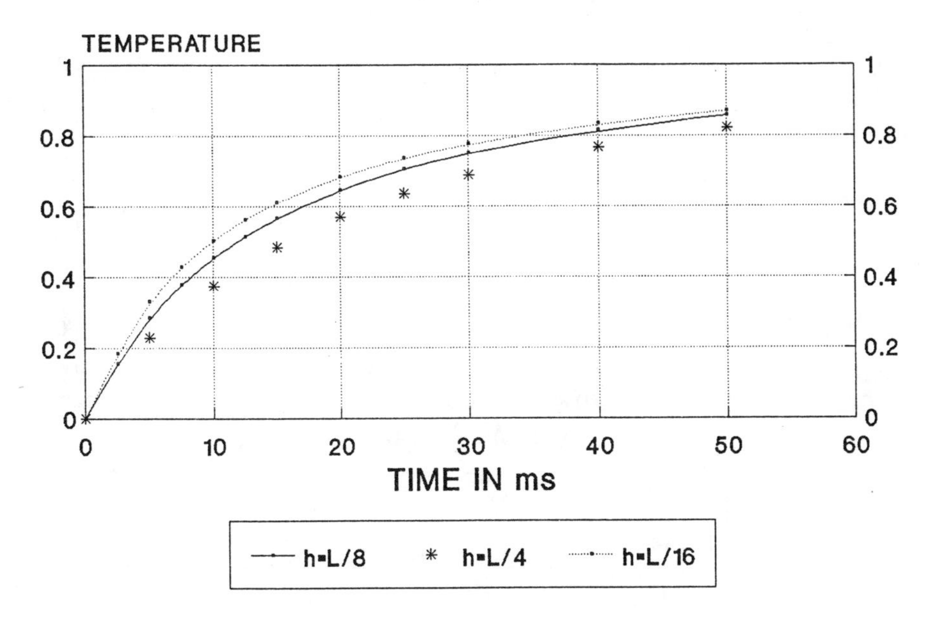

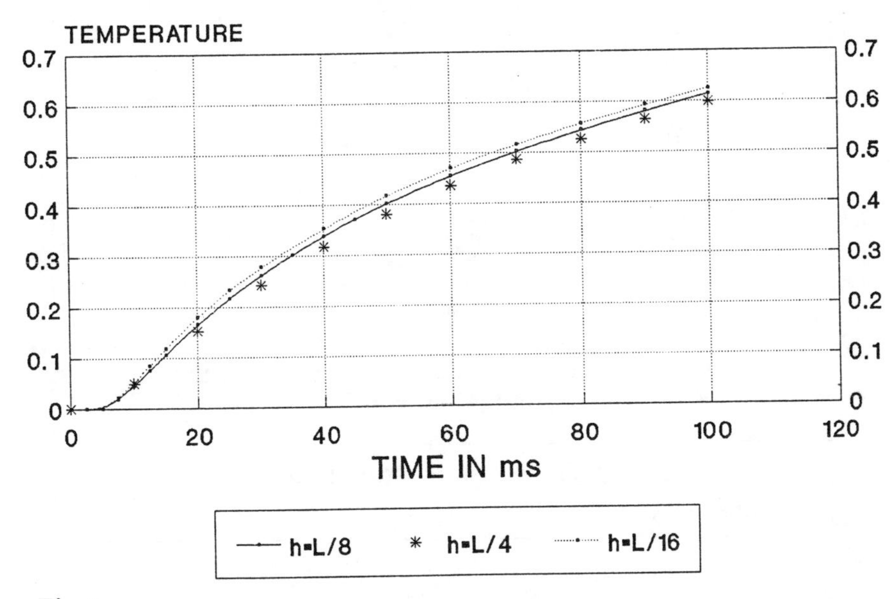

Fig.6

Effect of mesh size on accuracy for the TLM algorithm for two sample positions of the cube.

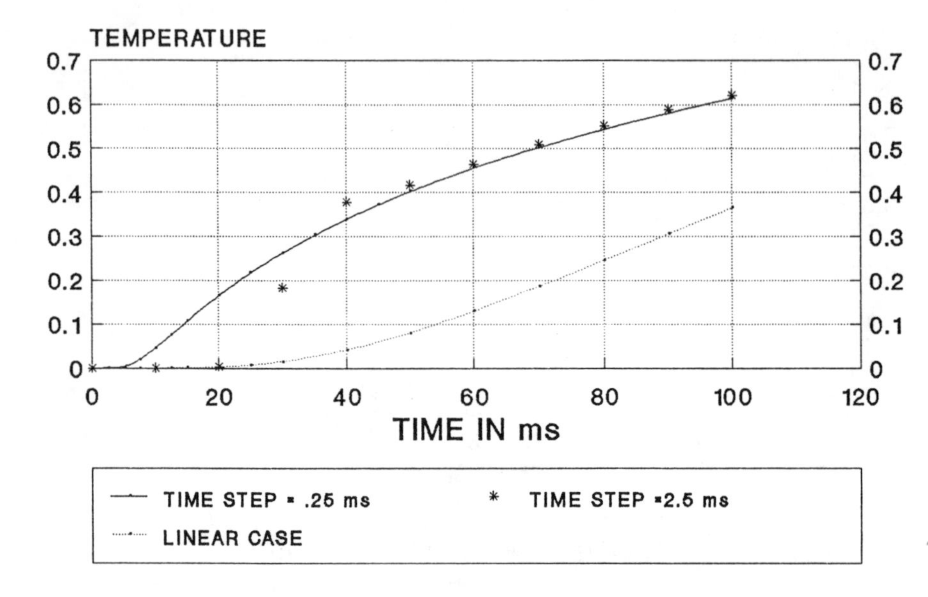

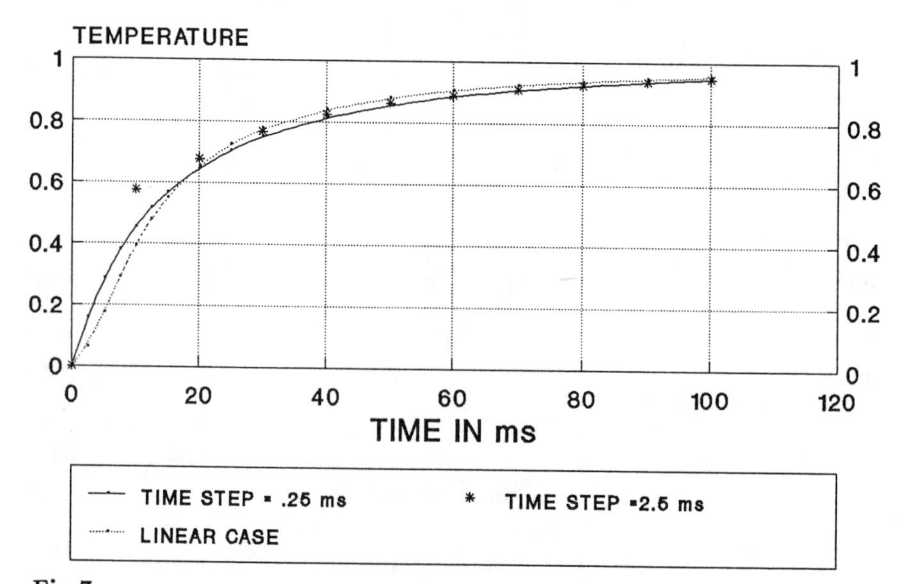

Fig.7

Effect of time-step on accuracy. For comparison purpose, results for a linear problem (r=0.7) are also included.

shows the accuracy of the TLM algorithm using three different mesh sizes and at two sample positions along the major diagonal. The time-steps have been chosen such that the major source of error is due to space discretisation.Fig.7 shows the effect of the time-step on the accuracy; as a comparison, the temperature rise in a linear medium (r=0.7) is also shown. A relatively large time-step can be used, however, as expected, the accuracy is degraded; particularly during the initial part of the transient. In all cases a fixed time-step has been used, however, an adaptive time-stepping algorithm can always be included as in [6].

The problem has been recoded using OCCAM and run on the transputer system described above. The CPU times required to solve a typical nonlinear problem is shown in Table 1. In all cases the number of iteration equals 500 and h=1/4 or 1/8. It can be seen that the T800x4 system performs better (in terms of CPU time) than the VAX8820 and at a substantially reduced computing cost. The performance gain between four T800s and one T800 is better than 3.2. Because of the explicit nature of the TLM algorithm, it is reasonable to expect that the performance gain will increase with the number of transputer used.

$CPUTime(s)$	$VAX8820$	$T800 * 1$	$T800 * 4$
$h = 1/4$	6.87	14.17	4.32
$h = 1/8$	33.87	103.7	28.7

Table 1 Performance comparison between different computing environments. Elapsed time for the VAX system is usually 5-10 times longer.

5 CONCLUSIONS

An approved TLM algorithm for the solution of three-dimensional nonlinear heat conduction problems has been described. The method is explicit and stable such that highly nonlinear problems can be solved efficiently. Further improvements in efficiency can be obtained by implementing the algorithm on a transputer system. Numerical experiments show that a transputer system with four T800s can easily outperform a super-mini-computer such as the VAX8820.

REFERENCES

[1] P B Johns, "A simple explicit and unconditionally stable numerical routine for the solution of the diffusion equation", *IJNME*, Vol-11, pp 1307-1328, 1977.

[2] G Bulter and P B Johns, "The solution of moving boundary heat problems using the TLM method of numerical analysis", in *Numerical Methods in Thermal Problems*, (Eds R W Lewis and K Morgan), Pineridge Press, Swansea, pp 189-195, 1979.

[3] S Pulko, A Mallik and P B Johns, "Application of transmission-line modelling (TLM) to thermal diffusion in bodies of complex geometry", *IJNME*, Vol-23, pp 2302-2312, 1986.

[4] Y K Chan, D Wong and C C Wong, "Solution of the diffusion equation on a multi-floating-point unit", in *Numerical Methods in Thermal Problems*, (Eds R W Lewis and K Morgan), Pineridge Press, Swansea, pp 1208-1216, 1989.

[5] P B Johns and Mark O'Brien, "Use of the transmission-line modelling (TLM) method to solve nonlinear lumped networks", *Radio and Electronic Engineer*, Vol-50, pp 59-70, 1980.

[6] C C Wong and K K Cheng, "Simulation of the transport equation by the network and transmission-line modelling", *This Proceeding.*

A Generalized Finite Difference Solution of the Heat Conduction Equation for Arbitrarily Shaped 3-D Domains

L. De Biase(*), F. Feraudi(*), V. Pennati(**)
(*) Dipartimento di Matematica , Università Statale di Milano, Italy
(**) Centro di Ricerca Idraulica e Strutturale, ENEL, Milano, Italy

ABSTRACT

The aim of this work is the solution of heat diffusion equation in solid homogeneous bodies. At first the boundary is automatically approximated by means of a polyhedron; on this a 3D orthogonal cartesian network is automatically constructed. Finally the solution of the differential problem is afforded by general FDF's coupled with a Crank-Nicolson scheme.

Key words: Boundary Approximation, Heat Conduction, General Finite Differences.

1. INTRODUCTION

In this paper a transient problem of heat conduction (irreversible problem) is studied in homogeneous solid bodies with no heat sources, described by the differential equation:

$$\varrho c \frac{\partial u}{\partial t} - \frac{\partial}{\partial x}\left(\alpha \frac{\partial u}{\partial x}\right) - \frac{\partial}{\partial y}\left(\alpha \frac{\partial u}{\partial y}\right) +$$

$$- \frac{\partial}{\partial z}\left(\alpha \frac{\partial u}{\partial z}\right) = 0 \qquad (1)$$

where ϱ, c, α stand for density, thermal capacity and thermal conductivity of the body, respectively. We are concerned with the problem of heat distribution within the body as time varies, given the initial conditions:

$$u(x,y,z,0) = r(x,y,z) \qquad (2)$$

and boundary conditions:

$$u(x,y,z,t) = s(x,y,z) \text{ on } \delta D \text{ for } t \geq 0 \qquad (3)$$

or boundary conditions involving the normal derivative.

Unfortunately, in spite of existence and uniqueness results, the solution is not obtained in analytical form on every possible domain. To solve this kind of problem, therefore, numerical methods are the field to be investigated. Among the most frequently used numerical methods we remind Galerkin methods, variational methods in a finite element formulation, least squares methods and collocation methods.

Besides these, finite difference methods are very often preferred since they lead to very simple discretization both of the domain of definition and of the differential operator [5]. The awkward problem with these methods is discretizing boundary condition in such a way that stability of the numerical solution be guaranteed. To this aim methods making use of boundary fitted networks have been studied [11] and peculiar finite difference formulas have been devised to cope with this problem [13].

This kind of difficulty is efficiently overcome by means of generalized finite difference formulas since they ensure the desired approximation degree, non constant net spacing, direct discretization of three dimensional domains, with no transformation, easy approximation of any kind of boundary conditions, and the possible definition of local subgrids [7], which makes an adaptive method formulation easier to achieve.

2. GENERALIZED FINITE DIFFERENCE FORMULAS.

Given a function $u: R^3 \rightarrow R$, a grid $G \subset R^3$ and point $P_i \varepsilon G$, derivatives of order s at P_i can be approximated by solving a system of $p \geq s$ truncated Taylor expansion centered at P_i and evaluated at p points P_{ik} surrounding P_i; this yields an approximation degree of p+1-s. Taylor system is formally inverted and the so obtained approximating formulas give derivatives, in one direction or mixed with the same effort, as linear combinations of values of function u at P_i

and P_{ik} (k=1,2,3,....,p). Coefficients in such linear combinations are simple functions of coordinates of P_i and P_{ik}.
The choice of points P_{ik} is very delicate and important in order to obtain simple formulas. Proper stencils can be extracted from the grid for every point P_i and the simple choice criterion implies that grid generation can be done automatically.

By general finite difference formulas, also boundary conditions are easily discretized. In particular flux boundary conditions are dealt with by writing the normal derivative as a combination of its components along the coordinate axes. When the number of points surronding P_i is less then needed to approximate some derivative, this derivative can be obtained by Taylor expansion of the same derivative, centered at the closest internal grid point.

Since approximation of mixed derivatives requires exactly the same effort asked for by other derivatives, in the unlucky situation quoted above the degree of approximation is preserved. Besides this, in order to avoid definition of external fictitious points, completely asymmetric formulas were devised for boundary points.

Since different formulas are available to us and since their choice depends on the position of P_i and of the surrounding points P_{ik} (especially with respect to the boundary), a few different stencils can be constructed around points P_i.

Grid generation will therefore be possible without particular assumptions on the shape of the domain: a rectangle is not anyhow needed.

A wholly automatic generation of 3D orthogonal cartesian networks was presented in [10]. In order that coupling automatic grid generation with general finite difference formulas be efficient, an accurate representation of the boundary is very important.

Approximation of 3D boundaries is a very delicate question especially if we want as reduced as possible input data and reasonable computational effort. Such approximation should represent the boundary precisely enough and if flux boundary condition are given, it should also meet some class of continuity requirement.
A good representation can be obtained by CAGD techniques.

3. APPROXIMATION OF 3D BOUNDARIES.

Many are in the literature the methods proposed for boundary approximation [3]; they can be subdivided in two main classes, local and global ones, a part for a few mixed methods.

Among global methods a classical one is Shepard's: scattered data on the boundary are given; their coordinates are used to define a rational function, together with their distance from the moving point. This method was one of the first proposed; it is rather coarse but very simple.

A mixed method by Foley [8] constructs local approximations of the scattered data, then evaluates such approximations on some two dimensional net of points and, by means of these values, defines a piecewise Hermite function. Errors are subsequently reduced by a corrector deduced from Shepard's method. This method is very flexible but it is computationally expensive, due to its double level of approximation.

A local method by Bézier [6] makes use of Bernstein polynomials to parametrically define every patch. This formulation leads to locating Bézier points which are used as vertices of a control polyhedron. The approximating surface includes some Bézier points and it lies in the convex hull of the polyhedron. In Bézier method, the problem of joining patches has not been solved completely but this method is very interesting because convergence to the real surface is ensured with respect to an increasing number of faces of the control polyhedron.

Another very interesting global hierarchical method is Shepard's modified octree technique [7]. The surface is at first approximated by a cube containing it. Such a cube is then subdivided in octants and only non empty octants are kept. Each octant is then subdivided by the same technique and this proceeds until the approximation is judged satisfactory. The set of non empty cubes thus defined is then smoothed by plane cuts at vertices and at half edges.

Smoothing can also be accomplished by Unicubix [9] which, starting from any polyhedron can produce a smooth surface, union of geometrically continuous surfaces.

Since Bézier result about an increasing number of faces and Unicubix are available, we decided that the boundary of the physical domain could profitably be approximated by means of polyhedra.

In our method a polyhedron is defined as follows: some "mean" planes F_i are generated from input points, which at first we think of as vertices of the various faces. A check on the distances of every point from the corresponding F_i gives us information about correctness of input data. For each F_i we compute the sum of squared distances of the vertices from it. Such sums are ordered higher to lower. The worst would-be faces are examined: if the distance of some point from the corresponding F_i is greater than a fixed threshold, we choose the 3 points closest to F_i, generate a plane G_i crossing them and consequently a triangular face. If, now, some of the other points is close enough to G_i, we modify its coordinates in order to bring it onto G_i. Otherwise other triangular faces connecting the first one to the distant points are generated. Our polyhedron will then be obtained by assembly of all faces.

4. OUR ALGORITHM.

In dealing with heat equation we wanted our procedure to be as automatic as possible and to require minimal input (Table 1 gives input required for problem 2 presented in next section). We therefore adopted general finite difference formulas and coupled them with a Crank-Nicolson scheme. We moreover wanted the space-time discretization to allow variable time steps and non constant net spacing.
The main steps in the algorithm are:

1. Automatic approximation of the boundary: a set of points and a few parameters are input and a polyhedron is constructed on them with triangular and non triangular faces as described in previous section. Equations of mean planes are computed on the basis of input points. Evaluation of the approximating polyhedron is made by computing distances of every point from the corresponding mean plane. If the approximation is not satisfactory, a new polyhedron yielding a more accurate approximation is automatically generated.
2. A 3D orthogonal cartesian network is built by means of Nx, Ny, Nz grid planes whose coordinates are defined by three proper assignement functions evaluated at Nx-1, Ny-1, Nz-1 arbitrary points in [0,1]. With this definition of coordinates, the net spacing needs not be constant.

```
0 1 1 0 0 0 0 0 0 0 0 0 0 0 1 0 0 0 0 1 0 0 0 0 0 0 0 0 0 0 0 0 0 0 0 0 1 1 1
1  14  18  6  4  4  4  4  4  4  4  4  4  4  4  4  6
0.125  0.25  0.125  -0.125  -0.25  -0.125  -0.27  0.25  0.5
0.25  -0.25  -0.5  -0.25  -0.125  0.125  0.25  0.125  -0.125
-0.21650635  0.  0.21650635  0.21650635  0.  -0.21650635  -0.4430127  -0.4430127
0.  0.4330127  0.4330127  0.  0.  -0.21650635  -0.21650635  0.  0.21650635  0.21650635
0.5  0.5  0.5  0.5  0.5  0.5  0.05  0.  0.  0.  0.  0.  -0.5  -0.5  -0.5  -0.5  -0.5  -0.5
  1  2  3  4  5  6
  1  8  9  2
  2  9 10  3
  3 10 11  4
  4 11 12  5
  5 12  7  6
  6  7  8  1
  7 14 15  8
  8 15 16  9
  9 16 17 10
 10 17 18 11
 11 18 13 12
 12 13 14  7
 13 14 15 16 17 18
0
7  7  7
1.00000E+00  1.00000E+00  1.00000E+00  2.00000E-01  8.00000E-02  5.00000E-02
0.01  0.1  0.01  0.01  60  0.01  0.2  0.00001
```

Table 1.   Input for problem 2.

3. Plane sections are constructed by intersecting grid planes orthogonal to the z axis with the polyhedron approximating the boundary. In such sections boundary grid points and M internal grid points are determined. Each section is then represented by a topological matrix recording relative positions of grid points in the section. For grid points not belonging to any section a topological matrix is defined too.
4. By juxtaposing all these 2D topological matrices in their natural order, a 3D topological matrix is obtained and it gives the distribution of internal and boundary points on the domain.
5. The grid is checked by the following tests:
   a) no grid plane should cross vertices of the approximating polyhedron; when this happens for a grid plane, its equation is automatically modified.
   b) No grid plane parallel to a coordinate plane should contain vertices of the polygonal boundary of a plane section parallel to another coordinate plane; when this happens we modify the equation of the "wrong" plane. This check is particularly useful when flux boundary condition are given.
   c) In order to guarantee the desired approximation degree of general FDF's, the ratio between distances of any three consecutive grid planes should lie between 1/3 and 3. If not so the grid is refined or assignment function are modified.
   d) Boundary conditions should not annihilate (because of too close internal points) at more than a fixed number of points. If not so, assignment functions are modified.
   e) At every grid point approximation of derivatives requires a fixed number of surrounding points. If some of them is missing, a new grid plane is included at proper position.

   Every type of modification and consequent grid generation is automatically performed.
6. The differential equation is discretized by a generalized Crank-Nicolson scheme to the second order of precision with respect to time and space components. At every instant $t=t_{n+1}$ a linear system of M equations is solved and values $u_{i,n+1}$ ($i=1,2,..M$; $n=0,1,..$) are computed at the M internal grid points, after appropriate introduction of initial and boundary conditions.
7. Algebraic systems obtained at every time instant are solved by a direct method in which pivots are selected by Markowitz criterion combined with a relative size test [1]. The known vector of the system is updated at every instant. The matrix of coefficients is updated only when the time step

is altered, which means once in a while. The solution is obtained by a backward procedure.

5. NUMERICAL EXPERIENCE.

In order to test our algorithm we solved some heat conduction problems defined on three dimensional non rectangular domains. The numerical solution was compared with the known solution, expressed in analytical form, by means of $\|\cdot\|_2$ and $\|\cdot\|_\infty$.

Problem 1 was a thermal transient problem on a domain defined by juxtaposition of two exagonal truncated pyramids (see Fig. 1). Minimum and maximum coordinates are -0.5 and 0.5 along the x axis, -0.433013 and 0.433013 along the y axis and -0.5 and 0.5 along the z axis, respectively. The assignment function defining coordinates of grid planes was:

$$\Phi = \phi^3 + 1.5\,\phi^2 - 1.5\,\phi \qquad (4)$$

so that a finer grid would surround the origin, internal to the domain.

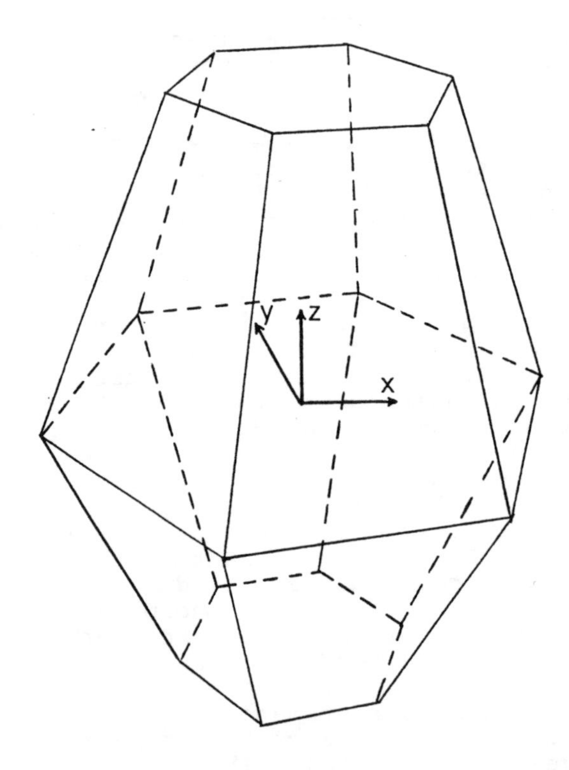

Figure 1. View of the domain of Problem 1.

Initial time was t=0.2 and initial time step was $\Delta t = 0.001$. A transient of 11.778 seconds was studied, in 122 time steps. The time step was automatically tripled seven times: after the 1st, 2nd, 37th, 72nd, 94th, 110th and 121st step.
The real solution was:

$$u=(4t+1)^{-1/2}\left(e^{-x^2/\alpha(4t+1)}+e^{-y^2/\alpha(4t+1)}+e^{-z^2/\alpha(4t+1)}\right) \quad (5)$$

with $\alpha = 0.01$, and it was used to establish initial and boundary conditions. The computational network was composed by nine planes in each directions with 592 grid points (266 internal and 326 on the boundary); the minimum net spacing was 0.082 and the maximum 0.166.
Numerical results are given in Table 2.

step	t	Δt	u max	u min	$\|\cdot\|_\infty$	$\|\cdot\|_2$
1	0.201	0.001	1.1167	7.921E-2	2.51E-4	1.22E-3
3	0.213	0.009	1.1021	8.319E-2	3.11E-3	1.49E-2
38	0.546	0.027	8.406E-1	1.736E-1	2.69E-3	1.27E-1
73	1.491	0.081	5.597E-1	2.657E-1	1.30E-2	5.59E-2
95	3.273	0.243	3.878E-1	2.645E-1	4.38E-3	2.77E-2
122	11.77	2.187	1.788E-1	1.641E-1	9.20E-5	7.53E-4

Table 2. Results for problem 1.

Problem 2 was defined on a very similar domain as for problem 1, but coordinates of one vertex were changed so that the polyhedron had to be regenerated automatically. There were, at the end, 17 faces, 6 of which triangular. The computational network was composed by seven planes in each direction with 299 grid points (117 internal points and 182 on the boundary); the minimum net spacing was 0.135 and the maximum 0.213.
Numerical result are given in Table 3.

step	t	Δt	u max	u min	$\|.\|_\infty$	$\|.\|_2$
1	0.21	0.01	1.10531	3.419E-2	3.95E-2	1.56E-2
2	0.24	0.03	1.07097	4.060E-2	1.42E-2	5.51E-2
26	1.02	0.09	6.654E-1	1.813E-1	4.13E-2	1.08E-1
51	3.99	0.81	3.051E-1	2.211E-1	4.58E-3	2.22E-2
60	12.90	2.43	1.745E-1	1.758E-1	1.39E-4	7.94E-4

Table 3. Results for problem 2.

In both examples we found maximum error at the points where u was maximum.

6. CONCLUDING REMARKS.

The technique presented herein can automatically approximate the boundary of arbitrarily shaped 3D domains; it automatically generates an orthogonal cartesian network and on it uses a Crank-Nicolson scheme coupled with generalized finite difference formulas.

It therefore allows solution of parabolic problems of heat diffusion in homogeneous solid bodies.

From numerical examples presented in previous section, the method shows stability and efficiency. In particular in problem 1 a very good approximation of the real solution was obtained, in spite of a low number of grid points. This was due, mainly, to the possibility of reducing the net spacing in the region where the gradient of the solution was subject to high variation.

A version of the algorithm capable to deal with multiconnected regions is under completion.

A smoother approximation of the domain could be obtained for instance by including Unicubix in our code, after the approximating polyhedron has been built.

We think an extension which should easily be accomplished is discretization of flux boundary conditions, as already done in 2D in [4].

Another extension we are going to try soon will include local and global adaptive refinements of the spatial network as we already did for elliptic problems on domains with known analytical form of the boundary in [12].

7. REFERENCES

1. Wait, R. The Numerical Solution of Algebraic Equations, Wiley, J. and Sons, New York, 1975.

2. Yarry, M.A. and Shepard, M.S. Automatic 3D Mesh Generation by the Modified-octree Technique, Int. J. Numer. Methods Eng. vol 20, pp. 1965-1990, 1984.

3. Bohm, W., Farni, G. and Kahamann, J. A Survey of Curve and Surface Methods in CAGD, Computational Aided Geometric Design 1, pp. 1-60, 1984.

4. Reali, M., Rangogni, R. and Pennati, V. Compact Analitic Expression of Two Dimensional Difference Forms, Int. J. Num. Methods Eng. vol 20, pp. 121-130, 1984.

5. Shih, T.M. Numerical Heat Transfer, H.P.C., Washington, 1984.

6. Farin, G. Triangular Bernstein-Bézier Patches, Computational Aided Geometric Design 3, pp. 83-127, 1986.

7. Reali, M., Dassie, G. and Pennati, V. Direct General Finite Difference Techniques for Elliptic Problems Defined on Bounded or Unbounded Two Dimensional Domains. Chapter 3, Numerical Methods in Transient and Coupled Problems, (Ed. Lewis, R.W., Hinton, E., Bettes, P. and Shrefler, B.A.), pp. 43-58, Wiley, J. and Sons, New York, 1986.

8. Foley, T.A. Scattered Data Interpolation and Approximation with Error Bounds, Computational Aided Geometric Design 3, pp. 163-177, 1986.

9. Shirman, L.A. and Sequin, C.H. Local Surfaces Interpolation with Bézier Patches, Computational Aided Geometric Design 4, pp. 279-295, 1987.

10. De Biase, L., Galli, A. and Pennati, V. Automatic Generation of Orthogonal Cartesian Network for Direct Solution of Differential Problems on General Shape 3D Domains by means FDFG (Ed. Sengupta, S., Hauser, J., Eiseman,P.R. and Thompson, J.F.) pp. 269-276, Proceedings of the 2nd Int. Conf. on Numerical Grid Generation in CFD, Miami Beach, 1988. Pineridge Press, Swansea, 1988.

11. Grandi, G.M. and Ferreri, J.C. On the Solution of Heat Conduction Problems Involving Heat Sources Via Boundary-Fitted Grids, Comm. Appl. Methods, vol 5, pp. 1-6, 1989.

12. Corti, S., De Biase, L. and Pennati, V. A Multistep 3D Generalized Difference Technique, (Ed. Lewis, R.W. and Morgan, K.) pp. 1153-1162, Proceeding of the 6th Int. Conf. on Num. Meth. for Termal Problems, Swansea, 1989. Pineridge Press, Swansea, 1989.

13. Noye, B.J. Five Points FTCS Finite Difference Methods for Heat Conduction,Comm. Appl. Methods, vol 5, pp. 337-345, 1989.

Numerical Solution for Transient Heat Conduction in a Narrow and Long Axisymmetric Domain by Combination of FDM and BEM

H.J. Kang, W.Q. Tao

Department of Power Machinery Engineering, Xi'an Jiaotong University, Xi'an, Shaanxi 710049, China

ABSTRACT

The boundary element method and the finite difference method are combined to solve transient heat conduction in a narrow-and-long domain. To couple the solutions of the two parts, the compatibility and equilibrium conditions are used. Numerical results of two problems and comparison between the solutions of different numerical methods are presented.

INTRODUCTION

In the past two decades, with the rapid developments of computer techniques and numerical methodologies, numerical simulation technique for fluid flow, heat transfer and industrial thermal process has received considerable attention of an increasing large number of investigators due to its low cost, speed, complete information and ability to simulate ideal conditions [1] . Several numerical analysis methods have evolved. The first widely known approximation method is the finite difference method(FDM). The other commonly used methods are finite element metnod(FEM) and boundary element method(BEM). These different numerical methods have their own advantages and drawbacks. In brief, FDM is simple in principle, and easy to use. Its weakness is in poor ability to simulate problems with complex geometries. The finite element method is paricularly suited for problems with complex geometries. However, the data input and the discretization process of FEM are time consuming and easy to evolve errors. In this regard, the boundary element method is very attractive. In this method the discretization is performed only for the boundaries of the computation domain, thereby, reducing the number of dimensionality on the computation domain by one. Thus, considerable savings in data input and computer CPU time can be achieved. The solution accuracy of BEM is, in general, quite high. One weakness of BEM, however, is that for narrow-and-long geometries its solution accuracy may be deteriorated. This is because the grids located at the two ends of the narrow-and-long domain have weak effects on each other. In addition, when one uses BEM to solve axisymmetric problems, the computation of the discretization coefficients for the element near the axis is quite tedious [2]. A common used practice to deal with a narrow-and-long geometry by BEM is to devide the whole domain into several subdomains so that each subdoamin has a adequate geometry.

The present work solves the transient heat conduction in a narrow-and-long axisymmetric domain by the combination of BEM and FDM. The geometry studied has a regular shape near its axis, therefore, the finite difference formulation is used in this part to take advantage of its simplicity. The remainder of the domain is irregular in shape, the boundary element formulation is adopted here. The linkage between these two parts is completed by implementing the compatibility and equiblibrium conditions. To examine the solution accuracy of this approach, the same problem is also solved by FDM. To solve a heat conduction problem in a irregular domain with the convective boundary condition by FDM, a new practice is proposed. Two examples are discussed. The numerical results confirm the validity and applicability of the combination approach.

PROBLEM FORMULATION AND BEM DISCRETIZATION EQUATION

Consider transient heat conduction in a homogeneous, isotropic axisymmetric solid body schematically shown in Fig. 1. In the absence of heat generation inside the domain, the governing equation is

$$k \nabla^2 T = \rho c \frac{\partial T}{\partial \tau} \quad \text{in } \Omega \tag{1}$$

with convective boundary condition:

$$-k \frac{\partial T}{\partial n} = h\,(T - T_f), \text{ or } \quad q_n = h\,(T - T_f) \quad \text{on } \Gamma \tag{2}$$

and initial boundary condition of the type

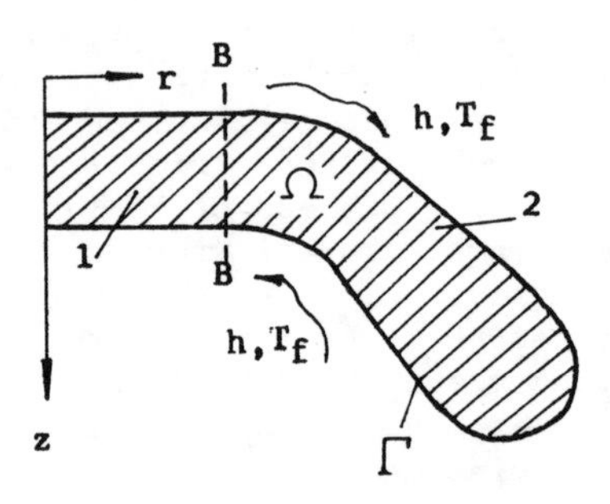

Fig. 1 Sketch of solution domain

$$T = T_o \qquad \text{in } \Omega \tag{3}$$

The boundary element method for the analysis of transient axisymmetric heat conduction problems was developed by Wrobel and Brebbia [2] . In the present paper only a brief description will be given. The details may be found in [2] .

Using the weighted residual method and the technique of integration by parts, we can obtain following boundary integral equation for the above problem

$$C_i T_i + \frac{\alpha h}{k}\int_0^{\tau}\int_S T^{*} T r\,dS\,dt = \frac{\alpha h}{k} T_f \int_0^{\tau}\int_S T^{*} r\,dS\,dt + \int_0^{\tau}\int_S q^{*} T r\,dS\,dt + \left[\int_A T^{*} T r\,dA\right]_{t=0} \tag{4}$$

The fundamental solution T^{*} is

$$T^{*} = \frac{2\pi}{[4\pi\alpha(\tau - t)]^{3/2}} \exp\left[-\frac{r^2 + r_i^2 + (z - z_i)^2}{4\alpha(\tau - t)}\right] I_o\left[\frac{r r_i}{2\alpha(\tau - t)}\right] \tag{5}$$

The flux q^{*} is calculated by

$$q^{*} = -k\frac{\partial T^{*}}{\partial n} = \frac{k}{8\sqrt{\pi}[\alpha(\;\; - t)]^{3/2}} \times \exp\left[-\frac{r^2+r_i^2+(z-z_i)^2}{4\alpha(\tau-t)}\right] \times \left\{\left(r I_o\left[\frac{r r_i}{2\;(\tau - t)}\right] - I_1 r_i\left[\frac{r r_i}{2\alpha(\tau - t)}\right] r_{,n}\right) + (z - z_i) I_o\left[\frac{r r_i}{2\alpha(\tau - t)}\right] z_{,n}\right\} \tag{6}$$

For the numerical solution of equation (4), the boundary is discretized into a series of elements over which the geometry, temperature and heat flux vary according to chosen interpolation functions. In order to perform the integration with time one also needs to assume a certain type of profile on time for T and q. A stepwise profile is used here. This assumption implies that both T and q are constant over each time interval, which should be considered reasonable for small time interval. This assumption makes possible the analytical evaluation of the time integrals in equation (4). The LHS integral becomes:

$$\int T^{*} dt = \frac{1}{2\alpha\sqrt{\pi d}}\int_C^{\infty} I_o(2a\,x) x^{-1/2} e^{-x} dx = \frac{1}{2\alpha\sqrt{\pi d}} \sum_{n=0}^{\infty} \frac{a^{2n}}{n!^2}\Gamma(2n+1/2, C) \tag{7}$$

where

$$d = r^2 + r_i^2 + (z - z_i)^2 \;; \quad a = \frac{rr_i}{d}$$

$$x = \frac{d}{4\alpha(t_2 - t)} \;; \quad C = \frac{d}{4\alpha(t_2 - t_1)} \tag{8}$$

For the RHS time integral in the boundary integral equation, we have

$$\int_{t_1}^{t_2} q^* \, dt = \frac{k}{\alpha d\sqrt{\pi d}} \Bigg\{ \left[rr_{,n} + (z - z_i)z_{,n}\right] \times \sum_{n=0}^{\infty} \frac{a^{2n}}{n!^2} \Gamma(2n + 3/2, C) - r_i r_{,n} \sum_{n=0}^{\infty} \frac{a^{2n+1}}{n!^2(n + 1)} \Gamma(2n + 5/2, C) \Bigg\} \tag{9}$$

By using the following recurrence relation, all the incomplete gamma functions in the above series can be evaluated in terms of $\Gamma(1/2, C)$

$$\Gamma(n + 1, C) = n\,\Gamma(n, C) + C^n e^{-C}$$

$$\Gamma(1/2, C) = \sqrt{\pi}\,(0.3480242 - 0.0958798p + 0.747855p^2)pe^{-C} \tag{10}$$

$$p = 1/(1 + 0.47047\sqrt{C})$$

The final discretized form of the boundary integral equation (4) is

$$[H]\,[T^{(m)}] = [G]\,[q^{(m)}] + [B]\,[T^{(m-1)}] \tag{11}$$

where the matrix $[B]$ accounts for the conditions at the previous time step. The detailed expressions for the coefficients of the matrices $[H]$ and $[G]$ are quite lengthy and will not be enumerated here.

In the evaluation of the matrix coefficients linear element was used. The line integration was calculated via Gaussian integration formulas with six points. The surface integration was based on a triangle. The triangular coordinates and the associated weighting factors were due to Hammer et al. [3] . The integration points was seven.

The time-marching scheme used in this work was as follows: each time step was treated as a new problem, therefore, at the end of each time step temperature values at a sufficient number of internal points were calculated, and these values were served as the pesudo-initial distribution for the next time step. The reasons for choosing this time-marching scheme are twofold. First, according to the background of investigated problems (quenching or tempering of workpieces), the temperature distribution of the body at each time step is required. Second, if a constant time step is adopted, all matrices involved in equation (11) wll also be constant throughout the analysis, so they can be computed once for all.

SOLUTION METHODOLOGY

Attention is now turned to the solution methodology used in the present work for the axisymmetric transient heat conduction in the complex geometry shown in Fig, 1. As it can be seen there, the shape of the center part of the geometry is regular— a circular disk. In this part, as $r_i \rightarrow 0$ the convergence rate in computing the series involved in matrix coefficients decreases. In order to accelerate the convergence rate, small time step is required. However, the small time step may leads to oscilation of series in the region far apart from the symmetric axis. On the other hand, in the center part of the geometry the finite difference method may be implemented with ease. The combination of BEM and FDM is thus adopted: in part 1 of Fig. 1, FDM is used, while in part 2 BEM is adopted. The coupling of the numerical solutions of these two parts is completed by implementing the compatibility and equilibrium conditions. That is, at the common boundary B-B following conditions must be satisfied:

$$T_{B^+} = T_{B^-} \tag{12}$$

$$q_{B^+} = q_{B^-} \tag{13}$$

where T_{B^+} and q_{B^+} stand for the values at boundary B-B calculated in part 2, while T_{B^-} and q_{B^-} are those in part 1 at the same location.

The computation procedure was as follows. First, a temporary temperature distribution for the points located at boundary B-B was assumed for the instant of the end of the first time step. Then all the boundary conditions were known for part 1. The temperature distribution at the end of $\Delta\tau$ in part 1 was then calculated by FDM, using a fully implicit scheme. From this computation, the heat flux at the boundary B-B, q^*, was obtained. The values of q^* were served as the known boundary condition for part 2. For the simplicity of programing, the value of q^* was converted into a corresponding heat transfer coefficient h^* by following equation

$$q^* = h^* (T - T_f) \qquad (14)$$

where T was the temperature at the same location which was to be reevaluated by BEM for part 2. In this way all the boundary conditions were of third kind, thus simplifying programing process. From the computation of part 2, the new values of temperature at boundary B-B , T_B were obtained. These values were regarded as the assumed ones, T_B^*, for the next iteration, and the computation was then repeated. The computation for the first time step was stopped if the absolute difference between T_B^* and T_B was less than 10^{-3}. The computations for the successive time steps were performed in the same manner. To avoid divergence of the iteration, an underrelaxation factor of 0.3 was used in transferring the boundary conditions.

To examine the solutions obtained by the combination method the same problem was also solved by FDM. To deal with the irregular boundary the following practice was used. As shown in Fig. 2, suppose in the Cartesian coordinates we have an irregular domain, which has a curved boundary. In order to use a computer program written for Cartesian coordinates the domain is extended to be a rectangle. If the true boundary is isothermal, then the nominal boundary may take the same value of temperature and the extended region(shaded area in Fig. 2) should be assigned a very large thermal conductivity, so that the prescribed boundary temperature may be propagated to the interior. This idea was first proposed by Patanker [4] . In [5] this practice was extended to the case with second kind boundary condition. It was further extended to the convective boundary(third kind) in [6] . It may be outlined as follows. Suppose that an irregular boundary is cooled(or heated) by a fluid with temperature T_f and heat transfer coefficient h(Fig. 3a). The true boundary is simulated by a succession of steps. For each control volume bounded with the true boundary, say control volume P in Fig. 3, following additional source terms should be added to its corresponding source terms:

$$S_{c,ad} = \frac{A}{\Delta V} \frac{T_f}{1/h + \delta/k} \quad ; \quad S_{p,ad} = \frac{1}{1/h + \delta/k} \frac{A}{\Delta V} \qquad (15)$$

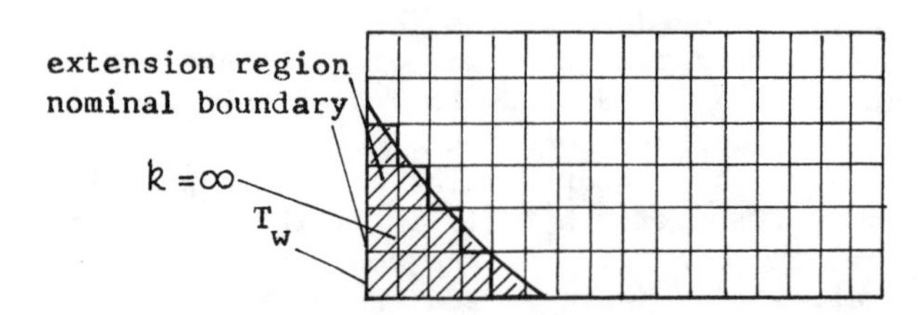

Fig. 2 Schematic diagram of domain extension pratice(first kind boundary condition)

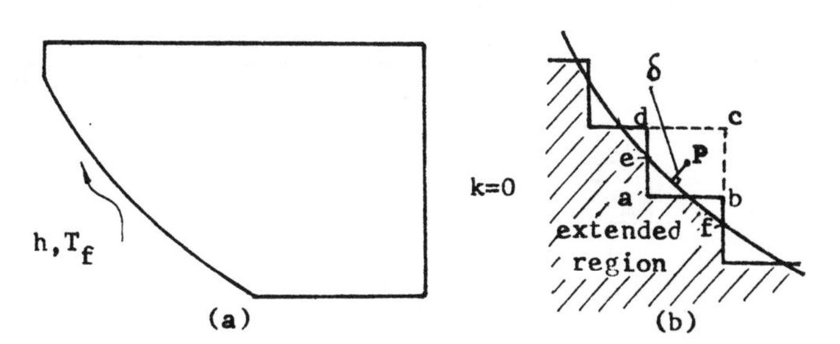

Fig. 3 Schematic diagram of domain extension practice(convective boundary condition)

where δ is the distance between grid point P and the true boundary, ΔV is the volume of control volume P, and A is the area of the true boundary next to control volume P. With unit thickness in z direction, A is equal to the length of arc ef.

Thus, the final discretization equation for control volume P becomes

$$a_P T_P = a_{nb}T_{nb} + (S_c + S_{c,ad})\Delta V \tag{16}$$

where $a_P = a_{nb} - (S_{p,ad} + S_p)\Delta V$, and S_c, S_p are the original linearized source terms [1].

Upon adding the two additional source terms, the extended region should be set inactive, that is, the thermal conductivity of this region should be set equal to zero. In this way the heat energy added into the wall-adjacent control volmes can not be conducted through the extended region, and both the total and local energy conservation conditions can be satisfied.

The domain extension method described above is not limited the Cartesian coordinates, but can be used in any orthogonal coordinate system. The implementation procedure is the same and need not be discussed.

RESULTS AND DISCUSSION

In order to show the numerical accuracy of the methodology presented above, two examples were analyzed. Due to the symmetry with respect to the z-axis, only half of the cross-section needed to be discretized in the two problems.

Transient heat conduction of a disc

The problem analyzed was that a solid disc at constant initial temperature T , subjected to the convective boundary condition along all its surfaces. The adopted discretization is

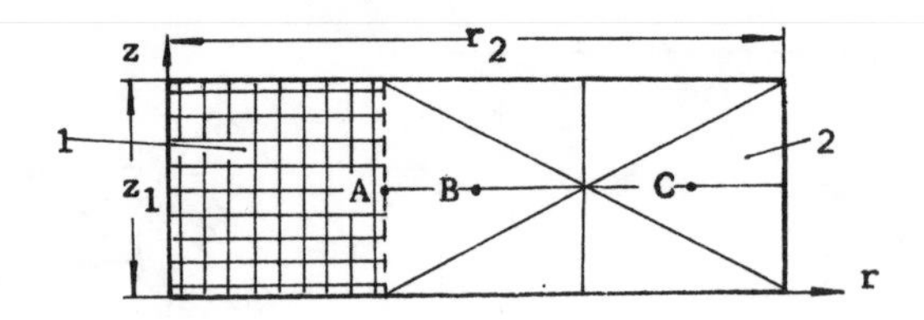

Fig. 4 Discretization of problem 1

shown in Fig. 4. The parameters adopted were $T_0=500°C$, $T_f=30°C$, $h_1=h_2=h_3=500\ W/(m^2 \cdot °C)$. The thermal properties were $k=40.5\ W/(m \cdot °C)$, $\alpha=7.22\times10^{-6}\ m^2/s$.

The analysis was conducted with a time step $\Delta\tau=10$ S. The results from different numerical methods are compared in Fig. 5. As it can be seen there, the agreement is good.

Transient heat conduction of a cap-type body

The second problem studied a cap-type body initially at uniform temperature T_0 and subjected to the convective boundary condition at t=0. The thermal physical properties and T_f were the same as for problem 1. The initial temperature T_0 was assumed to be 550°C, and $h_1=h_2=400\ W/(m^2 \cdot °C)$.

The analysis was performed with a time step $\Delta\tau=10$ S. For the finite difference method, a 52x42 grid was used. The curved boundary was replaced by a succession of steps. This is shown in Fig. 6(a). The adopted discretization for the combination method is shown in Fig. 6(b).

The temperature results for point A ,B and C are compared in Fig. 7. The maximum relative difference between the solution of FDM and that of BEM + FDM was less than 6%, which occured at the initial stage. This difference was in the same order as that reported in [1] , where the solutions of BEM and FEM or analytical method were compared.

CONCLUDING REMARKS

For an axisymmetric narrow-and-long region the transient heat conduction can be solved by the combination of the finite difference formulation and the boundary element formulation, with FDM used for the center part and BEM for the remainder of the solution domain. The compatibility and equilibrium conditions are used to couple the two solutions. This combination can take advantages of each numerical method. Despite the fact that an iterative solution procedure is needed, the total computation time is found not large.

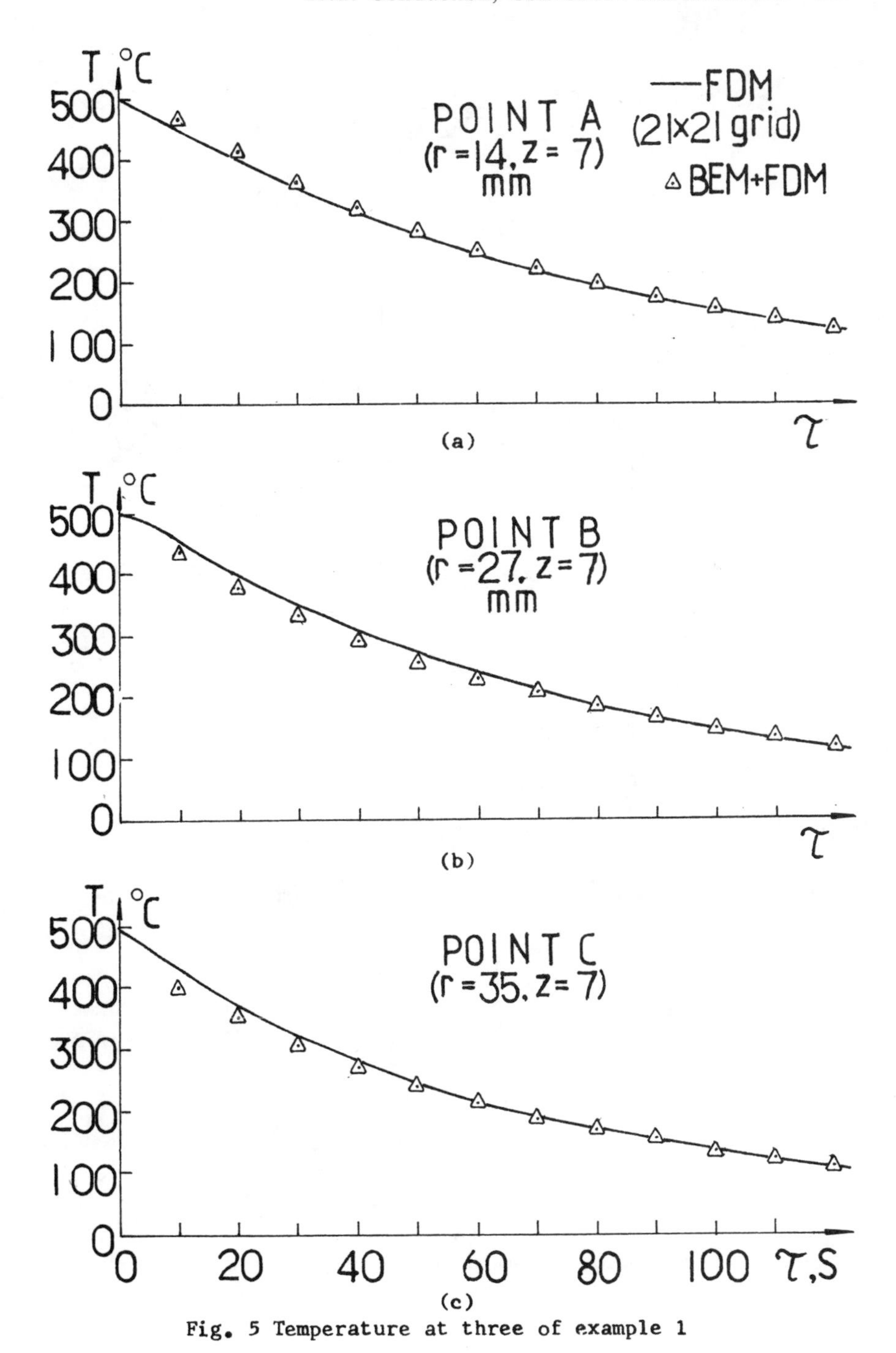

Fig. 5 Temperature at three of example 1

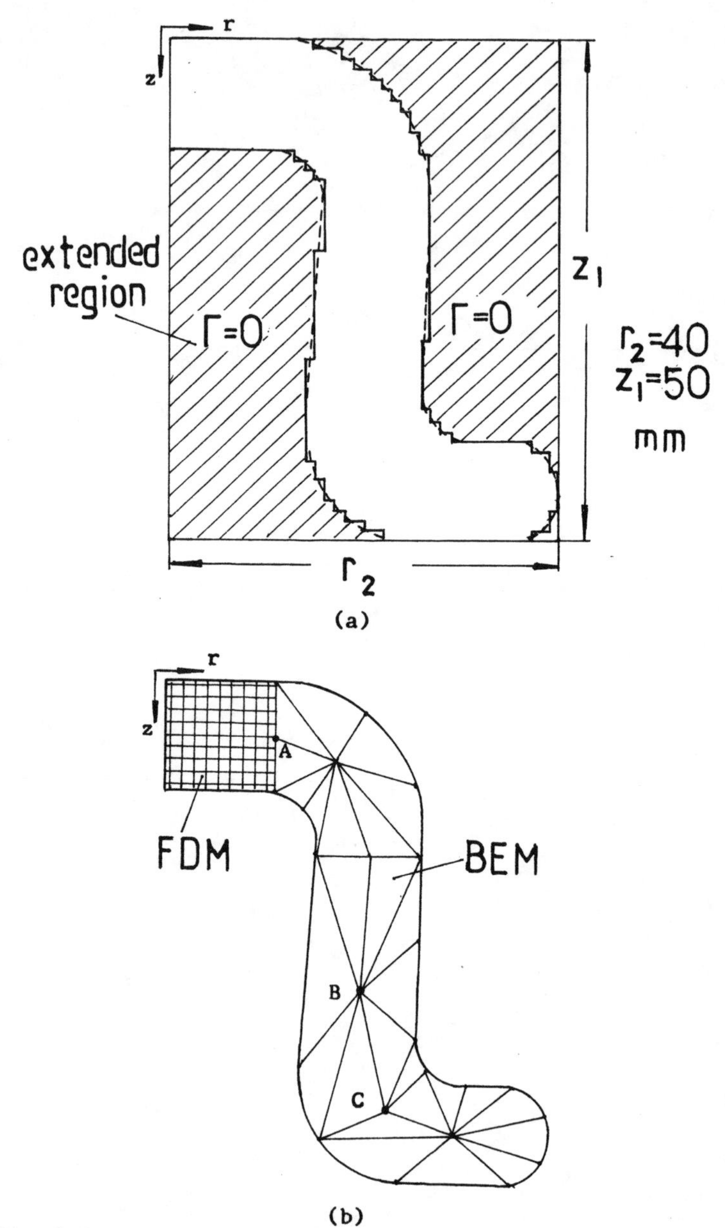

Fig. 6 Computational domain and descretization domain of example 2

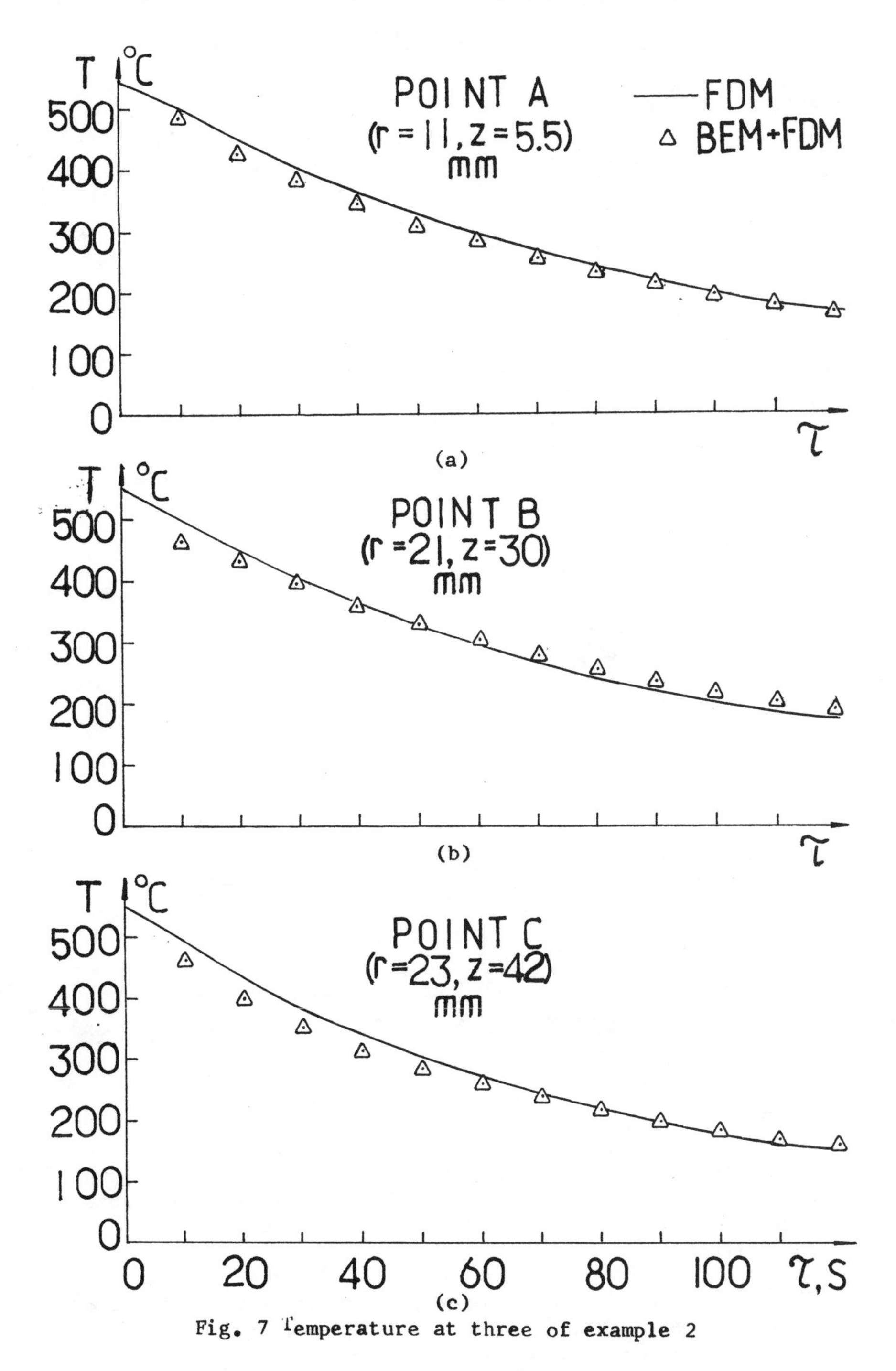

Fig. 7 Temperature at three of example 2

NOMENCLATURE

A	= area
a_p, a_{nb}	= coefficients of finite difference equation
c	= specific heat
h	= heat transfer coefficient
k	= thermal conductivity
n	= outward normal to boundary
q	= heat flux
r, z	= cylindrical coordinates of reference point
r_i, z_i	= cylindrical coordinates of source point
$r_{,n}, z_{,n}$	= direction consines of the outward normal
S_c, S_p	= constant and coefficient in linearized source term
$S_{c,ad}, S_{p,ad}$	= additional constant and coefficient in linearized source term
T	= temperature
T_f	= temperature of surrounding fluid
T_o	= initial temperature

Greek symbols

α	= thermal diffusivity
Γ	= boundary surface
$\Gamma(a,x)$	= incomplete gamma function
δ	= distance
$\Delta\tau$	= time step
ΔV	= volume of the control volume
Ω	= volume of the studied body
ρ	= density
τ	= time

ACKNOWLEDGMENT

This work was supported by the National Natural Science Fundation of China.

REFERENCES

1. Patankar, S. V., Numerical Heat Transfer and Fluid Flow, Hemisphere, Washington, D. C., 1980
2. Wrobel, L. C. and Brebbia, C. A., A Formulation of the Boundary Element Method for Axisymmetric Transient Conduction, Int. J. Heat Mass Transfer, vol. 24, pp. 843-850, 1981
3. Brebbia, C. A., Telles, J. C. F. and Wrobel, L. C., Boundary Element Techniques, Springer-Verlag, Berlin, 1984
4. Patankar, S. V., A Numerical Method for Conduction in Composite Materials, Flow in Irregular Geometries and Conjugate Heat Transfer, Proceedings of Sixth Int. Heat Transfer Conf., vol. 3, pp. 297-302, 1978
5. Prata, A. T. and Sparrow, E. M., Heat Transfer and Fluid Flow Characteristics for an Annulus of Periodically Varying Cross Section, Numer. Heat Transfer, vol. 7, pp. 285-304, 1984
6. Tao, W. Q., Numerical Heat Transfer(in Chinese), Xi'an Jiaotong University Press, Xi'an, 1988

Numerical Modelling of the Heat Transfer in a Liquid Gas Tank

H. Rolfes, J.A. Visser

Department of Mechanical Engineering, University of Pretoria, Pretoria 002, Republic of South Africa

ABSTRACT

By modelling the unsteady heat transfer in liquid gas tanks, the temperature distribution in the tank as well as the heat flux reaching the liquid gas can be predicted. Knowledge of the temperature distribution and heat flux can be used to predict evaporation losses from the tank. By minimizing the evaporation losses, the thermal design of a gas tank can be optimized. This paper presents a finite difference simulation of the unsteady three–dimensional heat transfer in gas tanks. The numerical procedure accounts for radiation from the sun as well as radiative and convective heat transfer with the environment. A non–homogeneous grid is used because the tank consists of several different materials of varying dimensions and properties. Geometrical effects such as variations in the thickness of the isolation material and the diameter and height of the tanks are also studied in an attempt to optimize the design configuration. Predicted evaporation losses correlate favourably with available measured data.

INTRODUCTION

Several gases like nitrogen, oxygen and argon are frequently used for different industrial applications. Due to the large quantities of gas required, the gas is liquefied to reduce the volume needed for storage. As a liquid, the temperature of the gas is much lower than the ambient temperature. Heat is therefore constantly transferred to the liquid gas which causes evaporation losses from the storage tanks. The present competitive market is forcing distributors of liquid gases to minimize evaporation losses in order to reduce the consumer price. This can be achieved by optimizing the thermal design of the storage tanks.

The thermal optimization of a liquid gas tank includes several important aspects of which the most important is the reduction of the heat flux to the liquid gas. The designer must further ensure that no ice will accumulate on the outer surface of the tank during cold winter nights. The temperature on the inside of the concrete slab must also be high enough to prevent brittling of the concrete. To achieve these goals, complete information of the temperature distribution through the tank and the resultant heat transfer is required. The unsteady three–dimensional heat transfer problem cannot be solved analytically, resulting in the development of a numerical model.

A literature survey did not reveal any publications on the numerical prediction of heat flux to, or temperature distributions in liquid gas tanks. Several similar unsteady problems had however been solved in the past, using numerical simulations. A number of papers present one–dimensional and two–dimensional models [1,2,3] for predicting unsteady heat transfer and temperature distributions. Fu et al. [4] presented a three–dimensional model with which the time–varying temperature distribution in composite material bridges can be predicted. Van der Walt et al. [5] used a similar numerical technique to predict the temperature distribution in steel bars during reheating. The model is limited to a homogeneous material and a fixed grid and could therefore not be used to model a gas tank consisting of several different materials. Visser [6] developed a two–dimensional model for predicting temperature distribution in bodies existing of different materials and using a variable grid. This model formed the basis for the development of the present numerical model.

This paper outlines a numerical method for predicting the three–dimensional unsteady heat transfer and the resultant temperature distribution and heat flux in liquid gas tanks.

CONFIGURATION OF A GAS TANK

A schematic representation of the construction of a liquid gas tank is shown in Figure 1. The tank is constructed on a reinforced concrete base about one metre above ground level. Several layers of foam glass are built on top

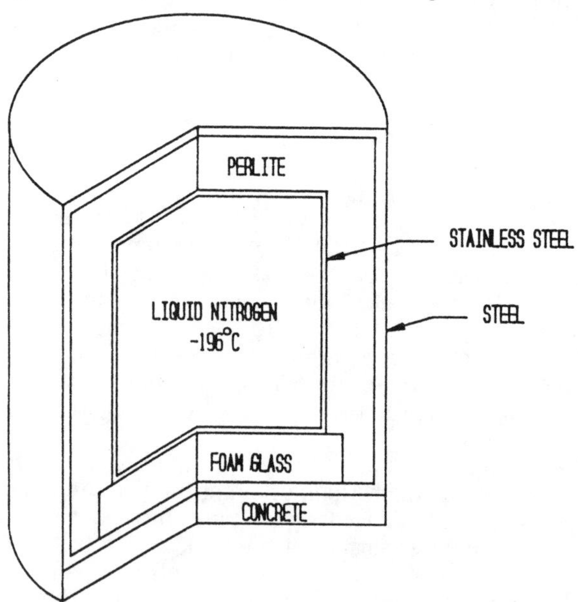

Figure 1. Schematic representation of a liquid gas tank

of this concrete slab to support the liquid gas tank. Possessing good isolation characteristics, the foam glass also serves to isolate the inner tank from the environment. A stainless steel tank, in which the liquid is stored, is positioned on top of the foam glass. A thick layer of perlite isolates the inner tank along the top and sides. The entire construction is enclosed in a steel drum. Typical dimensions of such a drum are: diameter 16m and height 20m.

OUTLINE OF THE MODELLING PROCEDURE

Generation of the grid

The size of the grid representing the tank depends mainly upon the construction, composition and dimensions of the gas tank. Where the dimensions of the different materials in the tank differ, the size of the grid also changes. In the thin steel the grid points are spaced at small intervals while in the thick perlite (isolation material) the points are spaced much further apart. Interface boundaries between different materials always occur midway between grid points. Each material used in the construction of the tank is awarded a numerical value to identify the detail material composition of the tank.

Governing and finite difference equations

The partial differential equation describing heat transfer in the structure of a liquid gas tank, is the three–dimensional, unsteady, conductive heat transfer equation. In cylindrical co–ordinates the equation is:

$$\frac{1}{r}\frac{\partial}{\partial r}\left[rk\frac{\partial T}{\partial r}\right] + \frac{1}{r}\frac{\partial}{\partial \phi}\left[k\frac{\partial T}{\partial \phi}\right] + \frac{\partial}{\partial z}\left[k\frac{\partial}{\partial z}\right] + \dot{q} = \rho C\frac{\partial T}{\partial \tau} . \qquad (1)$$

The finite difference equations used in the numerical procedure are derived by integrating the partial differential equation (1) over control volumes surrounding the grid points [7]. The general finite difference equation for equation (1) can be written as:

$$a_p T_p = \sum a_{np} T_{np} + b . \qquad (2)$$

The coefficient a_p refers to the central grid point and a_{np} to the neighbouring grid points. The numerical value of the constants depends on the material properties, grid dimensions and the time step used in the solution procedure. At the boundaries of the tank the term b and the coefficient a_p in equation (2) contain information on the different heat transfer boundary conditions.

Boundary conditions

The numerical model accounts for heat transfer in the form of radiation from the sun, as well as radiative and convective heat transfer with the environment. These boundary heat transfer processes are accounted for in the modelling procedure by imposing the following boundary conditions. Interfaces between different materials in the structure are also considered.

Convection boundaries: The general finite difference equation (2) is also valid at the convective boundaries of the tank. Natural convection is assumed, so the effect of wind is not included. In this case the coefficient a_p contains information on the convective heat transfer coefficient, while the term b contains information on the surrounding fluid temperature and the heat transfer coefficient. The convective heat transfer coefficient is defined from Newton's law of cooling. The equation used to calculate the heat transfer at each boundary grid point is

$$Q = \bar{h} A (T_s - T_\infty) \ . \tag{3}$$

The natural convective heat transfer coefficient between the steel tank and the surroundings can be calculated for a variety of applications from the following empirical equation [8]:

$$Nu = C_c (Gr_f Pr_f)^m \ . \tag{4}$$

For vertical cylinders the equation is valid if the thickness of the thermal boundary layer is relatively small compared with the diameter of the cylinder. For GrPr numbers from 10^{13} to 10^{19} the constants C_c and m were obtained as 0.10 and 0.33 respectively [8]. The following equation is then used to calculate the mean heat transfer coefficient $\bar{h}$ on the surface:

$$\bar{h} = \frac{4}{3} \frac{Nu \cdot k}{d} \ , \tag{5}$$

where d is the characteristic dimension, in this case the diameter of the tank, and k is the thermal conductivity value of the steel wall.

Sun radiation boundaries: Due to the outdoor location of liquid gas tanks it is exposed to sun radiation. The intensity of the radiation is strongly dependent upon the location of the tank, the atmospheric conditions, time of year and effective surface area. Published values of sun radiation intensities for the location of the tank [9] were used to determine the sun radiation heat transfer to the surface of the tank. The heat transfer is calculated from:

$$Q = G A \epsilon \ . \tag{6}$$

In the above equation ϵ represents the emissivity of the outside surface of the tank which was taken as 0.84 for light colour paints. G represents the sun radiation intensity and A the surface area. Sun radiation intensities for horizontal surfaces and surfaces facing in the four main wind directions were used in the simulation.

Radiation boundaries with the environment: Using the concept of a radiation heat transfer coefficient, the effect of radiative heat transfer with the environment can be treated in a similar way as convection heat transfer. The heat transfer coefficient can be defined as:

$$h_r = \frac{(T_S^2 + T_\infty^2)(T_S + T_\infty)}{[(1-\epsilon_S)/\epsilon_S] + (1/F_{S\infty})} , \quad (7)$$

where T_S is the wall temperature of the tank, T_∞ the ambient temperature, ϵ_S the surface emissivity of the tank and $F_{S\infty}$ the geometric view–factor which is unity in this case [10].

The equation for radiative heat transfer with the environment thus reduces to

$$Q = h_r A(T_S - T_\infty) \ . \quad (8)$$

Interface conduction: Interface conduction occurs in the structure of the tank between two different materials in contact. The grid was generated in such a way that interfaces between different materials occur midway between grid points. Equation (2) can then be used to calculate the heat transfer at the interface between these two materials. The interface conductivity value between the two materials in contact is obtained from the harmonic mean of the conductivity values of the two materials. The harmonic mean formulation for the interface conductivity k_{int} between materials (1) and (2) is:

$$k_{int} = 2k_1 k_2/(k_1 + k_2) \ . \quad (9)$$

If there is a significant difference between the conductivity values of the two materials in contact, the harmonic mean formulation gives a better description of the interface conductivity than the arithmetic mean [7]. This is of special importance at contact surfaces between perlite and steel.

In this study the assumption was made that the contact at the interface between the different materials is perfect. It is acceptable as it results in a higher heat flux and therefore a lower efficiency of the isolation material around the inner gas tank.

SOLUTION OF THE GRID

Solution procedure

Due to the cyclic environmental conditions, the temperature distribution and heat flux in a liquid gas tank varies over a twenty–four hour period. For the purpose of this study a steady state temperature distribution through the tank was initially assumed. From here the solution was allowed to progress in time steps of ten seconds. The Gauss–Seidel iteration scheme was used to solve the finite difference equations (2) at each time step. The solution was fully converged at each time level before moving to the next. This process was continued until the predicted temperature distribution was constant over a twenty–four hour period.

Stability, convergence and computing effort

The solution procedure used is unconditionally stable because the finite difference equations are fully implicit. The solution is assumed to have converged if the cumulative temperature difference over the grid differ with less than a prescribed value from the distribution of the previous iteration.

A typical grid–independent solution for the liquid gas tank, using a 54x37x10 grid (19980 grid points), was reached when modelling the twenty–four hour period between 72 and 96 hours. On a Persetel PS8/90–2 main frame computer results were obtained within 392 cpu minutes.

RESULTS AND DISCUSSION

In Figure 2 the heat flux through the outside wall of the tank is compared with the heat flux that reaches the liquid gas over a twenty–four hour period on a typical summer's day. After six in the morning heat is transferred to the tank as a result of sun radiation and increasing ambient temperature. As the surface temperature of the tank increases above the

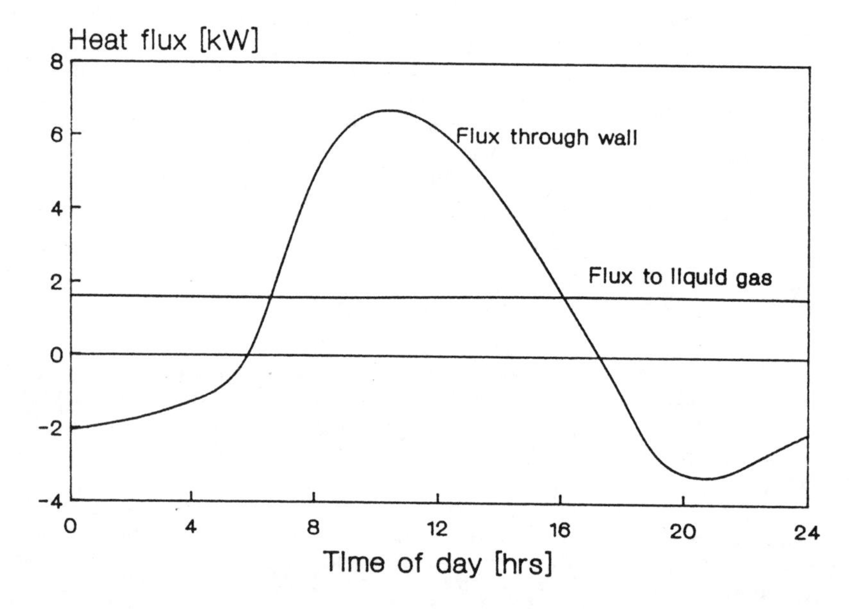

Figure 2. *24 Hour cycle of heat flux through outer and inner tanks*

ambient temperature, heat is transferred from the tank to the environment in the form of convection and radiation. This opposes radiative heat transfer from the sun to the tank and reduces the tempo of total heat transfer. During the afternoon the intensity of the sun radiation decreases which leads to a reduction in the heat flux through the surface. After 4 pm the convective and radiative losses to the environment exceed gains from sun radiation. The net heat transfer to the tank thus becomes negative. This negative heat flux to the tank continues until six am the next morning. The heat transferred to the liquid over a twenty–four hour period is therefore the sum of the positive and negative heat flux through the wall.

The heat flux through the inner tank to the liquid gas is however constant. The isolation material used in this simulation therefore effectively dampens the fluctuating heat flux through the wall. Furthermore, the fluctuation of the wall temperature has a relatively small influence on the resultant heat transfer to the liquid gas, as this fluctuation is small when compared with the total temperature difference between the liquid and the ambient temperature.

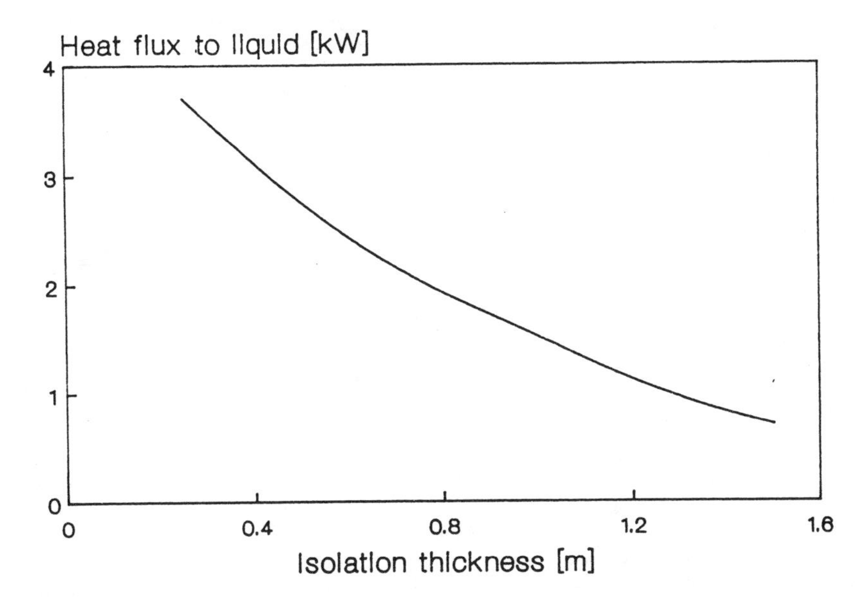

Figure 3. *Heat flux to liquid as a function of isolation material thickness*

Figure 3 shows the relation between the heat flux that reaches the liquid gas and the thickness of the isolation material (perlite). It applies specifically to a liquid gas tank with an inner tank diameter of 13,35 m and a height of 18 m. For this 2477 m^3 tank, filled with liquid nitrogen at $-196\,^0C$, the isolation thickness varied from 0.4 to 1.4 m. The figure indicates how the heat flux reaching the liquid gas decreases as the isolation thickness increases. From this figure accurate quantitative information can be obtained that is essential for the optimization of the design of gas tanks. This result, together with the economic implications, can be used to determine the optimum isolation thickness for the liquid gas tank.

With the isolation material thickness known, the next step in the design process of the tank is to optimize the shape of the tank. The shape is optimized when the heat flux to the liquid gas is minimized. The theoretical optimum shape is obtained when the surface–to–volume ratio is a minimum. Figure 4 exhibits the area–to–volume ratio as a function of the tank diameter. From this the optimal theoretical diameter for a 2477 m^3 tank is found to be 13,5 m.

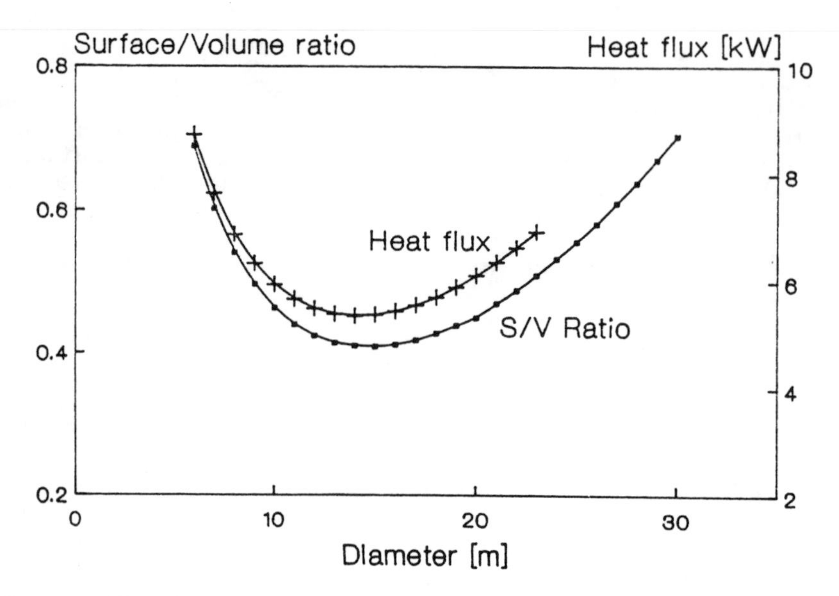

Figure 4. *Heat flux and surface–to–volume ratio for different configurations of a 2477 m^3 tank*

Figure 4 also displays the relation between total heat flux and tank diameter. The total heat flux is made up of the sum of the heat flux through all the different surfaces of the tank. For an isolation thickness of 1,2 m the various heat flux values are given in Table 1.

	Top	Side	Bottom
Heat flux [W/m^2]	5.86	5.79	9.29

Table 1. *Heat flux through the different surfaces*

The heat flux differs as a result of differences in boundary conditions at the surfaces. Knowledge of these fluxes is used to thermally optimize the shape of the tank (2477 m^3). From Figure 4 the diameter of the thermally optimized tank is found to be 14,5 m. This differs from the theoretical optimum diameter of 13,5 m, which did not take into account the variation in heat flux through the different surfaces. The real optimum tank diameter is therefore 14,5 m.

Apart from calculating the heat flux, the model also predicts the temperature distribution throughout the tank. Figure 5 shows the temperature distribution through the centre of the tank at 24h00 on a typical summer's day. The temperature distribution in the tank is calculated at each time step and is therefore available for any time of the day.

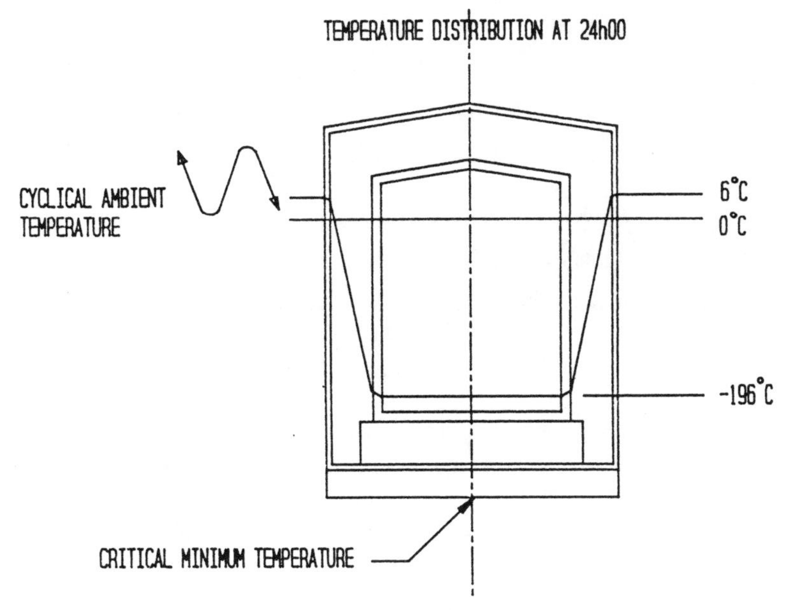

Figure 5. *Temperature distribution at 24h00*

SUMMARY AND CONCLUSIONS

The numerical modelling of the gas tank provides complete information on the temperature distribution and heat flux in the tank. From the temperature distribution in the tank it can be determined whether or not brittling or crumbling of the concrete slab can occur. Edge temperatures and especially the temperature at the bottom of the concrete slab can be predicted to determine whether the accumulation of ice at that point is possible. The knowledge of the heat flux to the gas tank is used to predict the thermal performance of the gas tank. From this information the percentage evaporation losses are calculated. If any of the above aspects are insufficient, the design of the gas tank can be altered until the set limits are complied with. Heat fluxes through the different surfaces of the gas tank are determined and used to optimize the tank geometry for a certain volume and fixed isolation material thickness.

With the proposed procedure it is now possible to model any size and shape of gas tank for any environmental conditions. A useful tool when designing and optimizing gas tanks is therefore supplied.

NOMENCLATURE

A	Surface area for heat transfer (m^3)
a	Coefficient in discretized equation (W/K)
b	Source term in the discretization equation (W)
C	Specific heat capacity of the material (J/kgK)
C_c	Empirical constant
d	Tank diameter (m)

G	Sun radiation intensity (W/m^2)
Gr_f	Grashof number for natural convection
h	Convection heat transfer coefficient (W/m^2K)
h_r	Radiation heat transfer coefficient (W/m^2K)
k	Conductivity of material (W/mK)
k_{int}	Interface conductivity (W/mK)
k_1,k_2	Conductivity for two materials (W/mK)
m	Empirical constant
Nu_f	Nusselt number evaluated at fluid temperature
Pr_f	Prantl number evaluated at fluid temperature
Q	Heat transferred (W)
r,ϕ,z	Space co–ordinates in cylindrical system (m)
T	Temperature (K)
ϵ	Emissivity
ρ	Material density (kg/m^3)
τ	Time (s)

Subscripts

np	General neighbouring grid point
p	Central grid point under consideration
s	Evaluated at the surface
∞	Evaluated at environmental conditions

REFERENCES

1. Sunden, B. A numerical investigation of the transient temperature distribution in a three–layered solid with time–varying boundary conditions, Numerical Methods in Thermal Problems, Vol.6, Part 1, pp. 111–119, 1989.
2. Sunden, B. Numerical prediction of transient heat conduction in a multi–layered solid with time–varying surface conditions, Numerical Methods in Thermal Problems, Vol.5, Part 1, pp. 207–218, 1987.
3. Claes, L. Determination of temperature gradients in a circular concrete shaft, Numerical Methods in Thermal Problems, Vol.5, Part 2, pp. 1518–1529, 1987.
4. Fu, H.C., Ng S.F. and Cheung M.S. Temperature distributions on concrete–steel composite bridges, Numerical Methods in Thermal Problems, Vol.6, Part 1, pp. 202–215, 1989.
5. Van der Walt, J.C., Visser, J.A. and Mathews, E.H. Numerical simulation of the heat transfer in and around a steel bar during reheating and hot rolling, Numerical Methods in Thermal Problems, Vol.6, Part 1, pp. 1456–1465, 1989.
6. Visser, J.A. A two–dimensional numerical model for predicting temperature distribution in a casting channel, Technical Report, Dept. of Mech. Eng., University of Pretoria, 1989.
7. Patankar, S.V. Numerical Heat Transfer and Fluid Flow, Hemisphere Publication Corporation, New York, 1980.
8. Holman, J.P. Heat Transfer, Sixth edition, McGraw–Hill,

Singapore, 1981.

9. Van Deventer, E.N. Climatic and other design data for evaluating heating and cooling requirements of buildings, CSIR research report 300, Pretoria, South Africa, 1971.

10. Visser, J.A. The numerical prediction of temperature distribution in steel billets during hot rolling, M.Eng Thesis, Dept of Mech. Eng., University of Pretoria, 1986.

Solving Transient Heat Conduction by BEM with Global Energy Balance Incorporated

A.J. Nowak(*), C.A. Brebbia(**)
(*) Institute of Thermal Technology, Silesian Technical University, 44-101 Gliwice, Konarskiego 22, Poland
(**) Computational Mechanics Institute, Wessex Institute of Technology, Ashurst Lodge, Ashurst, Southampton SO4 2AA, England

ABSTRACT

Paper deals with transient heat conduction problems solved by BEM. Standard formulation with time independent fundamental solution is completed by global energy balance. Both, integral equation and heat balance are transformed into boundary only form. Different solution strategies are discussed.

INTRODUCTION

Boundary Elements Method consists in transforming boundary value problem into integral equation which is then solved numerically, e.g. [1],[2]. For transient heat conduction the Fourier-Kirchhoff's partial differential equation [3] is the governing equation. This equation physically combines local energy balance with the Fourier's low of heat conduction.

Applying the reciprocity theorem to the governing equation and to the definition of fundamental solution integral equation is obtained. Since this relationship is fully equivalent to primary boundary value problem, the global energy balance should be fulfilled automatically. Unfortunately, integral equation cannot be solved exactly. Boundary elements concept with locally defined shape functions results in approximate solution. As a consequence global energy balance does not hold exactly. Generally, the more accurate solution of integral equation is obtained, more exactly global energy balance is fulfilled.

The global equilibrium condition is frequently used to verify an accuracy of the approximate solution and it is applicable to measurement of convergence in many iteration procedures. However, in numerical heat transfer it is very desirable to obtain solution which satisfies the global energy balance. This feature has, for example, Finite Differences Method [3].

In this paper heat balance equation is transformed into boundary only form. Two possible procedures of incorporating this additional equation into analysis are discussed.

PROBLEM FORMULATION

Equations governing transient heat conduction
Assuming constant thermal properties of the region Ω with a boundary Γ transient heat conduction is governed by the following differential equation

$$\nabla^2 u(x,t) = \frac{1}{a}\,\frac{\partial u(x,t)}{\partial t} = \frac{1}{a}\,\dot{u}(x,t) \qquad \text{in } \Omega \qquad (1)$$

where $u(x,t)$ is a temperature at point x for time t, $\dot{u}$ represents temporal derivative of temperature and $a = k/\rho c$ stands for thermal diffusivity. Heat conductivity is represented by k, ρ stands for density and c is specific heat.

On the boundary Γ an arbitrary kind of boundary condition is described. As the most general one boundary condition of Robin's type is considered in this paper

$$q = -k\,\frac{\partial u}{\partial n} = h\,(u - u_f) \qquad \text{on } \Gamma \qquad (2)$$

where q is a heat flux, h represents heat transfer coefficient, u_f is a known fluid temperature and $\partial()/\partial n$ stands for outward normal derivative.

To obtain unique solution of the problem (1),(2) initial condition has to be specified

$$\phi(x) = u(x,t=0) \qquad \text{in } \Omega \qquad (3)$$

where $\phi(x)$ is known function of space.

Transformation to integral equation
In this paper time independent fundamental solution

is used to transform boundary value problem (1)-(3) into the integral equation. The fundamental solution u_o^* satisfies the Poisson's equation

$$k \nabla^2 u_o^* = \Delta_i \qquad (4)$$

$$u_o^* = \frac{1}{2\pi k} \ln r + B_o \qquad (5)$$

where Δ_i is the Dirac's delta function acting at point i; r is the geometrical distance between points and B_o is an arbitrary constant.

Applying the second Green's theorem to equations (1) and (4) one obtains the following well known integral equation, i.e.

$$c_i\, u_i + \int_\Gamma q_o^*\, u\, d\Gamma = \int_\Gamma q\, u_o^*\, d\Gamma + \frac{k}{a} \int_\Omega \dot{u}\, u_o^*\, d\Omega \qquad (6)$$

where c_i is a position dependent constant and heat flux analog q_o^* is defined as

$$q_o^* = -k \frac{\partial u_o^*}{\partial n} \qquad (7)$$

Energy balance for transient problem

Let us consider the global heat flux Q_g leaving the region Ω through the boundary Γ

$$Q_g = \int_\Gamma q\, d\Gamma \qquad (8)$$

Applying divergence theorem and taking into account Eqs (1) and (2) one can write

$$Q_g = \int_\Gamma q\, d\Gamma = -k \int_\Gamma \frac{\partial u}{\partial n}\, d\Gamma =$$

$$= -k \int_{\Omega} \nabla^2 u \, d\Omega = - \frac{k}{a} \int_{\Omega} \dot{u} \, d\Omega \qquad (9)$$

Relationship (9) implies that global energy balance can be expressed as

$$\int_{\Gamma} q \, d\Gamma + \frac{k}{a} \int_{\Omega} \dot{u} \, d\Omega = 0 \qquad (10)$$

It is important to point out that any value B_o can be selected in Eq.(5), as in Eq.(6) this constant is multiplied by Eq.(10). Thus

$$B_o \left\{ \int_{\Gamma} q \, d\Gamma + \frac{k}{a} \int_{\Omega} \dot{u} \, d\Omega \right\} = 0 \qquad (11)$$

and B_o does not matter.

In such formulation, constant B_o plays the role of factor defining the contribution of the global equilibrium condition to the solution of the problem. The larger this value is, the better global heat balance is fulfilled. It should be however stressed, that such formulation can lead to non-unique solutions [4],[5].

Now integral equation (6) as well as global energy balance (10) are expressed in terms of boundary integrals only. In order to accomplish this transformation the Multiple Reciprocity technique is applied [6],[7].

The Multiple Reciprocity Technique

The Multiple Reciprocity method is a general method of transforming domain integrals arising in BEM into equivalent boundary integrals. When Eq.(6) is transformed procedure utilizes the set of higher order fundamental solutions satisfying the recurrence formula

$$\nabla^2 u^*_{j+1} = u^*_j \qquad \text{for } j=0,1,2,\ldots \qquad (12)$$

Taking into account relationship (12) and integrating domain integrals by parts as many times as required one can obtain the exact boundary only formulation of the problem (1)-(3), [7]

$$c_i \; u_i + \sum_{j=0}^{\infty} \frac{1}{a^j} \int_{\Gamma} q_j^* \frac{\partial^j u}{\partial t^j} \, d\Gamma = \sum_{j=0}^{\infty} \frac{1}{a^j} \int_{\Gamma} u_j^* \frac{\partial^j q}{\partial t^j} \, d\Gamma \quad (13)$$

where derivative of zero order is the function itself.

Equation (13) is the integral equation with respect to space and differential one with respect to time. Introducing boundary elements influence matrices **H** and **G** one obtains

$$\mathbf{H}_0 \; \mathbf{U} + \frac{1}{a} \mathbf{H}_1 \; \dot{\mathbf{U}} + \frac{1}{a^2} \; \mathbf{H}_2 \; \ddot{\mathbf{U}} + \ldots =$$

$$= \mathbf{G}_0 \; \mathbf{Q} + \frac{1}{a} \mathbf{G}_1 \; \dot{\mathbf{Q}} + \frac{1}{a^2} \mathbf{G}_2 \; \ddot{\mathbf{Q}} + \ldots \quad (14)$$

Domain integral appearing in energy balance can be treated in a similar manner. This requires to introduce a set of functions

$$\nabla^2 f_{j+1} = f_j \qquad \text{for } j=0,1,2,.. \quad (15)$$

with initial condition $f_0 = 1$.

Following the Multiple Reciprocity procedure one obtains for the heat balance equation

$$\sum_{j=1}^{\infty} \frac{1}{a^j} \int_{\Gamma} w_j \frac{\partial^j u}{\partial t^j} \, d\Gamma = \sum_{j=0}^{\infty} \frac{1}{a^j} \int_{\Gamma} f_j \frac{\partial^j q}{\partial t^j} \, d\Gamma \quad (16)$$

where

$$w_j = - k \frac{\partial f_j}{\partial n} \quad (17)$$

Functions f_j and w_j create one row matrices **F** and **W**. Though Eq. (16) is written in the form similar to Eq. (14)

$$\frac{1}{a} \mathbf{W}_1 \; \dot{\mathbf{U}} + \frac{1}{a^2} \; \mathbf{W}_2 \; \ddot{\mathbf{U}} + \ldots =$$

$$= F_o Q + \frac{1}{a} F_1 \dot{Q} + \frac{1}{a^2} F_2 \ddot{Q} + \ldots \quad (18)$$

Higher order fundamental solutions and set of f functions

Both Eqs(12) and (15) can be easily solved analytically when Laplace's operator is written in terms of cylindrical (for 2-D problems) or spherical (for 3-D problems) coordinate system. For example, for 2-D problems the general form of function u_j^* is given by the following expression

$$u_j^* = \frac{1}{2\pi k} r^{2j} (A_j \ln r - B_j) \quad \text{for } j = 1,2,\ldots \quad (19)$$

where r represents distance between points and coefficients A_j and B_j are obtained recurrently from the relationship

$$A_{j+1} = \frac{A_j}{4 (j + 1)^2} \quad (20a)$$

for $j = 0,1,2,\ldots$

$$B_{j+1} = \frac{1}{4 (j + 1)^2} \left(\frac{A_j}{j + 1} + B_j \right) \quad (20b)$$

Usually recommended values $A_o = 1$ and $B_o = 0$ [6] result from the form of classical fundamental solution u_o^*

$$u_o^* = \frac{1}{2\pi k} \ln r \quad (21)$$

Heat flux analog q_j^* is calculated from relationship

$$q_j^* = - \frac{k}{2\pi} \left[(2 j \ln r + 1) A_j - 2 j B_j \right] r^{2j-1} \frac{\partial r}{\partial n} \quad (22)$$

Notice that formula (20) introduces the factorials into the denominators of the coefficients A_j and B_j and this guarantees convergence of series (13). It should be also stressed that the functions u_j^* and q_j^* have no singularities (for $j = 1,2,\ldots$) thus, the integration does not require any special technique.

Integration of Eq.(15) gives

$$f_j = C_j\, R^{2j} \qquad j=0,1,\ldots \tag{23}$$

$$w_j = -\, k\, 2\, j\, C_j\, R^{2j-1} \qquad j=1,\ldots \tag{24a}$$

$$w_o = 0 \tag{24b}$$

where R is a distance of the point from the origin of coordinate system and coefficient C_j satisfies Eq.(20a)

SOLUTION PROCEDURES

As was stated previously, only approximate solution of Eq.(6) can be obtained. Simplifications are introduced in two stages; first when Eq.(14) is obtained, and then when it is solved cf. [5]. Hence Eq.(10) is also not satisfied exactly and, as a consequence, B_o constant affects results.

The simplest possible approach of taking energy balance into account is to replace one equation of system (14) by Eq.(18). However, there is no general rule which equation should be replaced. Therefore this attempt to the problem is not recommended.

Instead, two alternative procedures can be proposed [4]. The first one is to include into solution vector the quantity R_g being the residuum of global heat balance when approximate solution is obtained

$$R_g = \left\{ \int_\Gamma q\, d\Gamma + \frac{k}{a} \int_\Omega \dot{u}\, d\Omega \right\} \tag{25}$$

Eq.(6) can be rewritten in the following form

$$c_i\, u_i + \int_\Gamma q_o^{*}\, u\, d\Gamma - \int_\Gamma q\, u_o^{*}\, d\Gamma -$$

$$- \frac{k}{a} \int_\Omega \dot{u}\, u_o^{*}\, d\Omega + B_o\, R_g = 0 \tag{26}$$

Simultaneously the set of equations to be solved has to be completed by the definition (25).

The second proposal is based on the observation that global energy balance does not introduce any new unknowns. Thus problem defined by Eqs(6) and (10) is overdetermined and its solution can be obtained by least squares method. This approach, although being more time consuming, seems to present some advantages over the first one. It is expected that solution of transient problem will be more accurate and also more stable.

Problem requires numerical verification and appropriate research is currently carried out.

REFERENCES

1. C.A. Brebbia, J.C.F. Telles and L.C. Wrobel, Boundary Element Techniques: Theory and Applications in Engineering, Springer - Verlag, Berlin, 1984.

2. C.A. Brebbia and J. Dominguez, Boundary Elements - An Introductory Course, Comp. Mech. Publications, Mc Graw-Hill Book Co., 1988.

3. M.N. Özişik: Boundary Value Problems of Heat Conduction, International Textbook Company, Scranton, 1968.

4. G. Kuhn: Boundary element technique in elastostatics and linear fracture mechanics. Lecture notes in course: Finite Element Method and Boundary Element Method from the mathematical and engineering point of view. International Centre of Mechanical Sciences, Udine (Italy) - September 1986.

5. M.A. Jaswon and G.T. Symm: Integral equation method in potential theory and elastostatics. Academic Press, London, 1977

6. A.J. Nowak, C.A. Brebbia: The Multiple Reciprocity Method - A new approach for transforming BEM domain integrals to the boundary. Eng. Analysis with Boundary Elements, vol. 6, No. 3, 1989

7. A.J. Nowak : The Multiple Reciprocity Method of solving heat conduction problems. Proc. 11th BEM Conference, Cambridge, Massachusetts, USA, (ed. C.A. Brebbia & J.J. Connor) Springer-Verlag, vol. 2, 1989, pp 81-95

An Implicit Spline Method of Splitting for the Solution of Multi-Dimensional Transient Heat Conduction Problems

S.P. Wang, Y. Miao, Y.M. Miao

Department of Power Machinery Eng., Xi'an Jiaotong University, Xi'an 710049, China

ABSTRACT

An implicit spline method of splitting is proposed for the solution of two- and three-dimensional transient heat conduction problems. The features of this method are that in each computation step the problem is treated as a one-dimensional case in the implicit form, and only one tridiagonal system is evaluated. The accuracy and stability of the method are discussed. A two- and a three-dimensional transient heat conduction problems are solved, the results are in good agreement with the corresponding exact solutions and other numerical data.

1. INTRODUCTION

Many different methods of numerical analysis, such as the finite difference, finite element, boundary integral equation and Laplace transform have been presented in references(1-8) for solving transient heat conduction problems. Bhattacharya[1] and Lick[2] applied the improved finite-difference method(FDM) to time-dependent heat conduction problems with step-by-step computation in the time domain. The finite-element method(FEM) based on a variational principle was used by Gurtin[3] to analyze the unsteady problem of heat transfer. Emery and Carson[4] and Visser[5] applied variational formulations in their finite-element solutions of nonstationary temperature distribution problems. Bruch and Zyvoloski[6] solved transient linear and nonlinear two-dimensional heat conduction problems using the finite-element weighted residual process. Rources and Alarcon[7] presented a formulation for a two-dimensional isotropic continuous solid using the boundary integral equation method(BIEM) with a finite-difference approach in the time domain. Chen et al[8] successfully applied a hybrid method based on the Laplace transform and the FDM to transient heat conduction problems.

The main disadvantages of previous methods are the requirement of complicated procedure, longer computer time and larger storage.

A cubic spline method has been developed in the numerical integration of partial differential equations since the pioneering work of Rubin, Graves and Khosla[9-10]. This method has the advantages of simple procedure, shorter computer time, smaller storage and high-order accuracy as well as direct representation of gradient boundary conditions.

In this paper, an implicit spline method of splitting is proposed for the solution of two- and three-dimensional transient heat conduction problems. In section 2, a spline scheme of splitting is presented. In section 3, the stability of the method is assessed. In order to illustrate the numerical accuracy and efficiency of the present method, different examples are analyzed in section 4. The results are compared with the corresponding exact solutions and other numerical data, and the agreement is good.

2. ANALYSIS

The governing differential equation for transient heat conduction in a homogeneous and isotropic solid body is given by

$$k_x \frac{\partial^2 \theta}{\partial x^2} + k_y \frac{\partial^2 \theta}{\partial y^2} + k_z \frac{\partial^2 \theta}{\partial z^2} + Q = \rho c \frac{\partial \theta}{\partial t} \qquad (1)$$

with boundary conditions of the following types:

$$\theta = \theta_s \qquad \text{on } S_1 \qquad (2)$$

$$k_x \frac{\partial \theta}{\partial x} l_x + k_y \frac{\partial \theta}{\partial y} l_y + k_z \frac{\partial \theta}{\partial z} l_z + q = 0 \quad \text{on } S_2 \qquad (3)$$

$$k_x \frac{\partial \theta}{\partial x} l_x + k_y \frac{\partial \theta}{\partial y} l_y + k_z \frac{\partial \theta}{\partial z} l_z + h(\theta - \theta_\infty) = 0 \quad \text{on } S_3 \qquad (4)$$

and the initial condition

$$\theta(x,y,z,0) = \theta_o(x,y,z) \qquad (5)$$

where the union of the surfaces S_1, S_2, and S_3 forms the complete boundary of the solid of volume V; S_1 is the part of the boundary on which the temperature θ is prescribed, S_2 is the part on which the heat flux q is prescribed, and S_3 is the part on which the convective heat transfer $h(\theta - \theta_\infty)$ is prescribed.

The implicit spline schemes of splitting for the numerical solution of equation(1) are

$$\theta_{ijk}^{n+\frac{1}{3}} = \theta_{ijk}^{n} + \Delta t \cdot \frac{1}{\rho c}(k_x M_{ijk}^{n+\frac{1}{3}} + \frac{1}{3}Q) \tag{6a}$$

$$\theta_{ijk}^{n+\frac{2}{3}} = \theta_{ijk}^{n+\frac{1}{3}} + \Delta t \cdot \frac{1}{\rho c}(k_y L_{ijk}^{n+\frac{2}{3}} + \frac{1}{3}Q) \tag{6b}$$

$$\theta_{ijk}^{n+1} = \theta_{ijk}^{n+\frac{2}{3}} + \Delta t \cdot \frac{1}{\rho c}(k_z P_{ijk}^{n+1} + \frac{1}{3}Q) \tag{6c}$$

where θ_{ijk}, M_{ijk}, L_{ijk}, P_{ijk} are the spline approximations of $(\theta)_{ijk}$, $(\theta_{xx})_{ijk}$, $(\theta_{yy})_{ijk}$, $(\theta_{zz})_{ijk}$ respectively, and there are spline relations as follows:

$$\Delta x_{ijk} \cdot M_{i-1,jk}/6 + (\Delta x_{ijk} + \Delta x_{i+1,jk}) \cdot M_{ijk}/3 + \Delta x_{i+1,jk} \cdot M_{i+1,jk}/6 = (\theta_{i+1,jk} - \theta_{ijk})/(\Delta x_{i+1,jk}) - (\theta_{ijk} - \theta_{i-1,jk})/(\Delta x_{ijk}) \tag{7a}$$

$$\Delta y_{ijk} \cdot L_{i,j-1,k}/6 + (\Delta y_{ijk} + \Delta y_{i,j+1,k}) \cdot L_{ijk}/3 + \Delta y_{i,j+1,k} \cdot L_{i,j+1,k}/6 = (\theta_{i,j+1,k} - \theta_{ijk})/(\Delta y_{i,j+1,k}) - (\theta_{ijk} - \theta_{i,j-1,k})/(\Delta y_{ijk}) \tag{7b}$$

$$\Delta z_{ijk} \cdot P_{ij,k-1}/6 + (\Delta z_{ijk} + \Delta z_{ij,k+1}) \cdot P_{ijk}/3 + \Delta z_{ij,k+1} \cdot P_{ij,k+1}/6 = (\theta_{ij,k+1} - \theta_{ijk})/(\Delta z_{ij,k+1}) - (\theta_{ijk} - \theta_{ij,k-1})/(\Delta z_{ijk}) \tag{7c}$$

where

$$\Delta x_{ijk}=x_{ijk}-x_{i-1,jk}>0, \quad \Delta y_{ijk}=y_{ijk}-y_{i,j-1,k}>0$$

$$\Delta z_{ijk}=z_{ijk}-z_{ij,k-1}>0.$$

If boundary conditions are given, we can obtain spline solutions of Equation (1) from Equations (6) and (7).

For convenience, we now reduce Equations (6) and (7) to a scalar set of equation that only contains: θ_{ijk}.

As a first step, Equation (6a) is written in the following form:

$$\theta_{ijk}^{n+\frac{1}{3}} = F_{ijk} + S_{ijk}M_{ijk}^{n+\frac{1}{3}} \tag{8}$$

From Equations (7a) and (8), we find

$$A_i\theta_{i-1,jk}^{n+\frac{1}{3}}+B_i\theta_{ijk}^{n+\frac{1}{3}} + C_i\theta_{i+1,jk}^{n+\frac{1}{3}} = D_i \quad , \quad i=2,N-1 \tag{9}$$

where

$$A_i = \frac{\Delta x_{ijk}}{6S_{i-1,jk}} - \frac{1}{\Delta x_{ijk}}$$

$$B_i = \frac{\Delta x_{ijk} + \Delta x_{i+1,jk}}{3\, S_{ijk}} + \frac{1}{\Delta x_{ijk}} + \frac{1}{\Delta x_{i+1,jk}}$$

$$C_i = \frac{\Delta x_{i+1,jk}}{6\, S_{i+1,jk}} - \frac{1}{\Delta x_{i+1,jk}}$$

$$D_i = \frac{\Delta x_{ijk}F_{i-1,jk}}{6\, S_{i-1,jk}} + (\Delta x_{ijk}+\Delta x_{i+1,jk})F_{ijk}/(3S_{ijk})$$

$$+ \Delta x_{i+1,jk}F_{i+1,jk}/(6S_{i+1,jk})$$

Under proper conditions this system can be solved with the

Thomas algorithm.

Using the same techniques, we can derive equations for the determination of $\theta_{ijk}^{n+\frac{2}{3}}$ and θ_{ijk}^{n+1} that are similar to Equation (9).

3. TRUNCATION ERROR AND STABILITY

In the work of Rubin and Khosla [10], the spatial accuracy of the spline approximation for interior points has been estimated.

For second derivatives or diffusion terms, cubic spline approximation has second-order accuracy, which can be maintained even with rather large nonuniformities in mesh width.

For the linear heat conduction Equation (1) with Q=0, the interior point stability of scheme (6) can be assessed with the Von Neumann Fourier decomposition for $\Delta x_{ijk} = \Delta y_{ijk} = \Delta z_{ijk} = h =$ constant.

Let
$$\theta_{ijk}^{n+\frac{1}{3}} = \bar{\theta}_{k_1k_2k_3}^{n+\frac{1}{3}} \cdot e^{I(k_1ih + k_2jh + k_3kh)},$$

$$\theta_{ijk}^{n+\frac{2}{3}} = \bar{\theta}_{k_1k_2k_3}^{n+\frac{2}{3}} \cdot e^{I(k_1ih + k_2jh + k_3kh)},$$

$$\theta_{ijk}^{n+1} = \bar{\theta}_{k_1k_2k_3}^{n+1} \cdot e^{I(k_1ih + k_2jh + k_3kh)},$$

$$\theta_{i\pm1,jk}^{n+\frac{1}{3}} = \bar{\theta}_{k_1k_2k_3}^{n+\frac{1}{3}} \cdot e^{I[k_1(i\pm1)h+k_2jh+k_3kh]},$$

$$\theta_{i,j\pm1,k}^{n+\frac{2}{3}} = \bar{\theta}_{k_1k_2k_3}^{n+\frac{2}{3}} \cdot e^{I[k_1ih+k_2(j\pm1)h+k_3kh]},$$

$$\theta_{ij,k\pm1}^{n+1} = \bar{\theta}_{k_1k_2k_3}^{n+1} \cdot e^{I[k_1ih+k_2jh+k_3(k\pm1)h]},$$

$$\rho_1 = \bar{\theta}_{k_1k_2k_3}^{n+\frac{1}{3}} / \bar{\theta}_{k_1k_2k_3}^{n},$$

$$\rho_2 = \bar{\theta}_{k_1k_2k_3}^{n+\frac{2}{3}} / \bar{\theta}_{k_1k_2k_3}^{n+\frac{1}{3}} ,$$

$$\rho_3 = \bar{\theta}_{k_1k_2k_3}^{n+1} / \bar{\theta}_{k_1k_2k_3}^{n+\frac{2}{3}} ,$$

$$\rho = \bar{\theta}_{k_1k_2k_3}^{n+1} / \bar{\theta}_{k_1k_2k_3}^{n} , \quad I^2 = -1.$$

From Equations (6a) and (7a), we find

$$\rho_1 = \frac{1}{1+\dfrac{6\alpha(1-\cos k_1h)}{2+\cos k_1h}}$$

where $\alpha = \Delta t\cdot k_x/(\rho ch^2)>0.$

Using the same techniques, we can obtain

$$\rho_2 = \frac{1}{1+\dfrac{6\beta(1-\cos k_2h)}{2+\cos k_2h}}$$

$$\rho_3 = \frac{1}{1+\dfrac{6\gamma(1-\cos k_3h)}{2+\cos k_3h}}$$

where $\beta = \Delta t\cdot k_y/(\rho ch^2)>0, \qquad \gamma = \Delta t\cdot k_y/(\rho ch^2)>0.$

From $\rho = \rho_1 \cdot \rho_2 \cdot \rho_3$ we have

$$|\rho| \leq |\rho_1| \cdot |\rho_2| \cdot |\rho_3|$$

$$= \frac{1}{1+\dfrac{6\alpha(1-\cos k_1h)}{2+\cos k_1h}} \cdot \frac{1}{1+\dfrac{6\beta(1-\cos k_2h)}{2+\cos k_2h}} \cdot \frac{1}{1+\dfrac{6\gamma(1-\cos k_3h)}{2+\cos k_3h}} \tag{10}$$

The Von Neumann condition necessary for the suppression of all error growth requires that the spectral radius $|\rho| \leq 1$. It is obvious from formula (10) that $|\rho| \leq |\rho_1| \cdot |\rho_2| \cdot |\rho_3| \leq 1$, so the scheme (6) is unconditionally stable with arbitrary values of Δt, h, k_x, k_y and k_z.

4. NUMERICAL APPLICATION

Example 1 (Two Dimensions)

In this example the heat dissipation of a square plate is analyzed (Fig.1). The governing equation, boundary conditions, and initial condition for this example are given by

$$\frac{\partial \theta}{\partial t} = \frac{\partial^2\theta}{\partial x^2} + \frac{\partial^2\theta}{\partial y^2} + Q \ , \ Q = \sin(\frac{\pi x}{L})\cdot\sin(\frac{\pi y}{L})\cdot e^{-\nu t},$$

$$0 \leq x,\ y \leq 1,\ t \geq 0 \tag{11}$$

$$\theta(0,y,t) = \theta(1,y,t) = 0$$
$$\theta(x,0,t) = \theta(x,1,t) = 0 \tag{12}$$

and $$\theta(x,y,0) = \sin(\frac{\pi x}{L})\sin(\frac{\pi y}{L}) \tag{13}$$

where $L=L_x=L_y=1$, $\nu=1$.

The spline splitting scheme for solving Equation (11) is

$$\theta_{ij}^{n+\frac{1}{2}} = \theta_{ij}^{n} + \Delta t(M_{ij}^{n+\frac{1}{2}} + \frac{1}{2}Q^{n+\frac{1}{2}}) \tag{14a}$$

$$\theta_{ij}^{n+1} = \theta_{ij}^{n+\frac{1}{2}} + \Delta t(L_{ij}^{n+1} + \frac{1}{2}Q^{n+1}) \tag{14b}$$

where M_{ij}, L_{ij} are the spline approximations of $(\partial^2\theta/\partial x^2)_{ij}$, $(\partial^2\theta/\partial y^2)_{ij}$, respectively.
Equation (14a) or (14b) may be written as Equation (8), then can be solved with Thomas algorithm.

The exact solution for this example is given by

$$\theta(x,y,t) = \sin(\frac{\pi x}{L})\cdot \sin(\frac{\pi y}{L})\cdot e^{-\nu t} \tag{15}$$

The nodal temperatures at t=0.1s are listed in Table 1 using 121 nodes. Table 1 shows a comparison of the present solutions with exact solutions. The maximum deviation of the solutions is

0.85% when the element size $\Delta x = \Delta y = 0.1$ is used. Decreasing the element size would make the present solutions convergent toward the exact solutions.

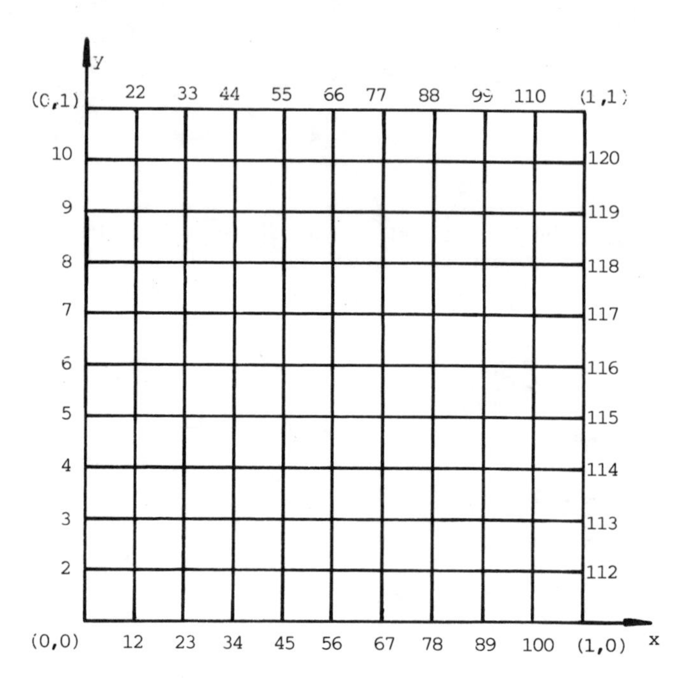

Fig. 1 Location of nodes for the first example

Table 1 Comparison of present solutions with exact solutions at t=0.1

Node position	Present solution Δt=0.005	Exact solution
17	0.2773	0.2796
31	0.3101	0.3126
50	0.8535	0.8606
53	0.5017	0.5058
55	0.0000	0.0000
94	0.52748	0.53185
102	0.1630	0.1644
108	0.1630	0.1644

Example 2 (Three Dimensions)

The heat dissipation of a solid plate is analyzed. Its initial temperature is 10°C, and the temperatures on the boundary surfaces of this solid are kept constant. For simplicity, it is assumed that $k_x = k_y = k_z = 1$ W/m·°C, $L_x=L_y=L_z=1.0$m, $\Delta x=\Delta y=\Delta z=0.125$m, and $\rho c=1$ J/m³·°C. Thus, the governing equation, boundary conditions, and initial condition for this example are given by

$$(\partial^2\theta/\partial x^2)+(\partial^2\theta/\partial y^2)+(\partial^2\theta/\partial z^2) = \partial\theta/\partial t \tag{16}$$

$$\theta(0,y,z,t)=\theta(1,y,z,t)=0 \tag{17a}$$

$$\theta(x,0,z,t)=\theta(x,1,z,t)=0 \tag{17b}$$

$$\theta(x,y,0,t)=\theta(x,y,1,t)=0 \tag{17c}$$

and

$$\theta(x,y,z,0)=10 \tag{18}$$

where L_x, L_y and L_z are the lengths of the solution domain in the x,y, and z directions, respectively.

The exact solution is given by[8]

$$\theta(x,y,z,t) = \sum_{i=1}^{\infty}\sum_{j=1}^{\infty}\sum_{k=1}^{\infty} C_n \sin(i\pi x)\sin(j\pi y)\sin(k\pi z) \cdot \exp[-(i^2+j^2+k^2)\pi^2 t] \tag{19}$$

where

$$C_n = \frac{80}{ijk\pi^3}[(-1)^i-1][(-1)^j-1][(-1)^k-1] \tag{20}$$

A comparison of the present solutions with exact solutions and the solutions of Chen et al.[8] is shown in Table 2. The present solutions are not only in agreement with exact solutions but also more accurate than those of Chen et al.[8]. The maximum deviation from the exact solutions is about 1.88% when the element size $\Delta x=\Delta y=\Delta z=0.125$m and $\Delta t=0.006$ s is used. Decreasing the element size and time increment would make the present solution convergent toward the exact solution.

Table 2 Solutions corresponding to t=0.06 s

Node Position	Present solutions		exact solution	Ref.[8]
	$\Delta t=0.006$	$\Delta t=0.003$		
(0.375,0.25,0.5)	2.312	2.257	2.279	2.241
(0.5,0.5,0.5)	3.492	3.427	3.462	3.389
(0.75,0.5,0.25)	1.784	1.736	1.751	1.726
(0.375,0.375,0.375)	2.779	2.717	2.744	2.695

5. CONCLUSION

An implicit spline method of splitting for the solution of multidimensional transient heat conduction problems has been proposed in this paper. Two different examples have been analyzed. The solutions of these examples are in good agreement with the corresponding exact solutions. In addition, the solutions are more accurate than those of Chen et al. [8].

The accuracy and stability are discussed. For linear transient heat conduction problems, the method is unconditionally stable.

The present method can also be applied to problems with time-dependent boundary conditions. The procedure is general, unconditionally stable and accurate. In the present study the procedure has been applied to heat conduction problems, and it can also be generally applied to most parabolic partial differential equations.

NOMENCLATURE

c	specific heat, J/Kg·°C	t	time
h	heat transfer coefficient, constant step increment	ν	constant
		ρ	density, Kg/m^3
k_x	thermal conductivity in x direction, W/m·°C	θ	temperature
k_y	thermal conductivity in y direction, W/m·°C	θ_o	initial temperature
k_z	thermal conductivity in z direction, W/m·°C	θ_∞	surrounding temperature
l	direction cosine normal to surface	Q	heat generator
q	normal heat flux	S_1, S_2, S_3	boundary surfaces

Subscripts

i	nodal point value	z	in z direction
j	nodal point value		
k	nodal point value		
s	on boundary surface		
x	in x direction		
y	in y direction		

REFERENCES

1. Bhattacharya, M.C. An Explicit Conditionally Stable Finite Difference Equation for Heat Conduction Problems, Int. J. Numer. Methods Eng., Vol.21, pp.239-265, 1985.
2. Lick, W. Improved Difference Approximations to the Heat Equation, Int. J. Numer. Methods Eng., Vol.21, pp.1957-1969, 1985.
3. Gurtin, M.E. Variational Principles for Linear Initial-Value Problems, Q. Appl. Math., Vol.22, pp.252-256, 1964.
4. Emery, A.F. and Carson, W.W. An Evaluation of the use of the Finite Element Method in the Computation of Temperatur ASME J. Heat Transfer, Vol.39 pp.136-145, 1971.
5. Visser, W. A Finite Element Method for the Determination of Non-Stationary Temperature Distribution and Thermal Deformations, Proc. Conference on Matrix Methods in Structural Mechanics, pp.925-943, Air Force Institute of Technology, Wright Patterson Air Force Base, Dayton, Ohio, 1965.
6. Bruch, J.C. and Zyvoloski, G. Transient Two-Dimensional Heat Conduction Problems Solved by the Finite Element Method, Int. J. Numer. Methods Eng., Vol.8, pp.481-494, 1974.
7. Rources, V. and Alarcon, E. Transient Heat Conduction Problems Using B.I.E.M., Comput. Struct., Vol.16, pp.717-730, 1983.
8. Chen et al., Application of Hybrid Laplace Transform/Finite-Difference Method to Transient Heat Conduction Problems, Numerical Heat Transfer, Vol.14, pp.343-356, 1988.
9. Rubin, S.G. and Graves, R.A. Cubic Spline Approximation for Problems in Fluid Mechanics, NASA TR R-436, 1975.
10. Rubin, S.G. and Khosla, P.K. Higher Order Numerical Solutions Using Cubic Splines, AIAA J., Vol.14, pp.851-858, 1976.
11. Wang, S.P. A Spline Method for the Solution of Mathematical and Physical Equations, TR85-119, Xi'an Jiaotong University, 1985.
12. Wang, S.P. An Application of Spline Method in Fluid Mechanics, TR85-173, Xi'an Jiaotong University, 1985.
13. Wang, S.P. et al., Spline-Difference Solution of Turbulent Boundary Layers on Blades of A Centrifugal Impeller, Proc. of 4th Chinese National Computational Fluid Dynamics Conference, Shanghai, Oct. 1988.
14. Wang, S.P. et al., Spline Splitting Method for the Solution of Unsteady Convection-Diffusion Problems, Power Machinery and Thermophysical Eng., Xi'an Jiaotong University Press, 1989.
15. Wang, S.P. et al., Spline Method for the Solution of Unsteady Convection-Diffusion Equations, Proc. of 4th Chinese National Fluid Mechanics Conference, Beijing University Press, 1989.
16. Wang, S.P. et al., Theory of Spline/Multigrid Method and Its Application, Proc. of 4th Chinese National Fluid Mechanics Conference, Beijing University Press, 1989.

SECTION 2: RADIATIVE HEAT TRANSFER

Numerical Modelling of Radiative and Convective Heat Transfer for Flows of a Non-Transparent Gas in a Tube with a Grey Wall

M.W. Collins, J. Stasiek(*)

Thermo-Fluids Engineering Research Centre, City University, London, EC1V OHB, England

ABSTRACT

An analytical investigation is presented of the influence of radiative heat transfer on the complex heat exchange relating to flow of a non-transparent (optically active) gas inside a tube of diffuse grey properties. Utilising the separable-kernel method and surface transformation, a set of non-linear differential equations is developed which are solved by the Runge-Kutta method with Hamming modification. Numerical results for short tubes with a length - diameter ratio of 5 are given.

INTRODUCTION

Heat transfer by forced convection to a gas flowing in a tube has received almost exhaustive study in the literature but little consideration has been given to the added effects caused when significant thermal radiation is also present. Such radiation transfer becomes especially important at the high temperature levels encountered in advanced types of power-plants such as the nuclear rocket. In some instances, the radiation will impose an additional heat load on a part which is to be kept cool, and hence this exchange must be estimated when the cooling requirements are computed. In other cases, the radiation will help reduce the temperature of a region operating at high temperature [1]

This paper is concerned with the energy exchange that occurs inside a tube when heat is being transferred by radiant interchange between elements on the tube wall and gas volumes,

---

*On leave of absence from Technical University of Gdansk, Poland.

and by forced convection to a gas which absorbs and emits radiation. The tube is heated by an externally applied uniform or nonuniform heat flux, and there is a constant or variable convective heat transfer coefficient at the inner surface. The proposed method, which includes the influence of gas emission and absorption, is based on the zone division approach first formulated by Hottel and developed by Siegel and Perlmutter [1] [2]. In this, the non-isothermal gas and surface are divided into infinitely small isothermal elements. Also, it involves the transformation-zone technique, where the emission of the gas body is replaced with an equivalent surface emission [3].

ANALYSIS

The system to be analysed is shown schematically in Fig. 1. An optically active gas at a specified inlet temperature $T_{g,i}$ flows into the tube and is heated to an average exit temperature $T_{g,e}$. A nonuniform heat flux q(X) is supplied to the tube wall by external means, and the outside surface of the tube is assumed to be insulated. Each end of the tube is exposed to an outside environment or reservoir at specified temperatures, $T_{r,i}$ and $T_{r,e}$ at inlet and exit of the tube respectively. The inside of the tube wall is a diffuse grey surface with an emissivity ε. The Planck mean volume absorption coefficient α is constant and the extinction coefficient $\chi \ll 1$. It is assumed that there is no axial conduction in the tube wall or in the gas and that the convection heat transfer coefficient h(X) is nonuniform throughout the tube.

The analytical relation between temperatures and heat fluxes can be obtained from an energy balance for the elemental surface dA_i and gas volume dV_i. According to the net radiation method of Poljak [2], and other workers [1] [3] the energy balance for an elementary surface dA_x a distance X from the tube inlet equals.

$$q_i^*(X) + q(X) = q_o(X) + h(X)[T_w(X) - T_g(X)] \qquad (1)$$

The terms on the left are respectively the total incoming radiation and the flux due to wall heating.

On the right hand side the respective terms are the radiative heat flux leaving the surface element and the heat flowing by convection from the wall to gas.

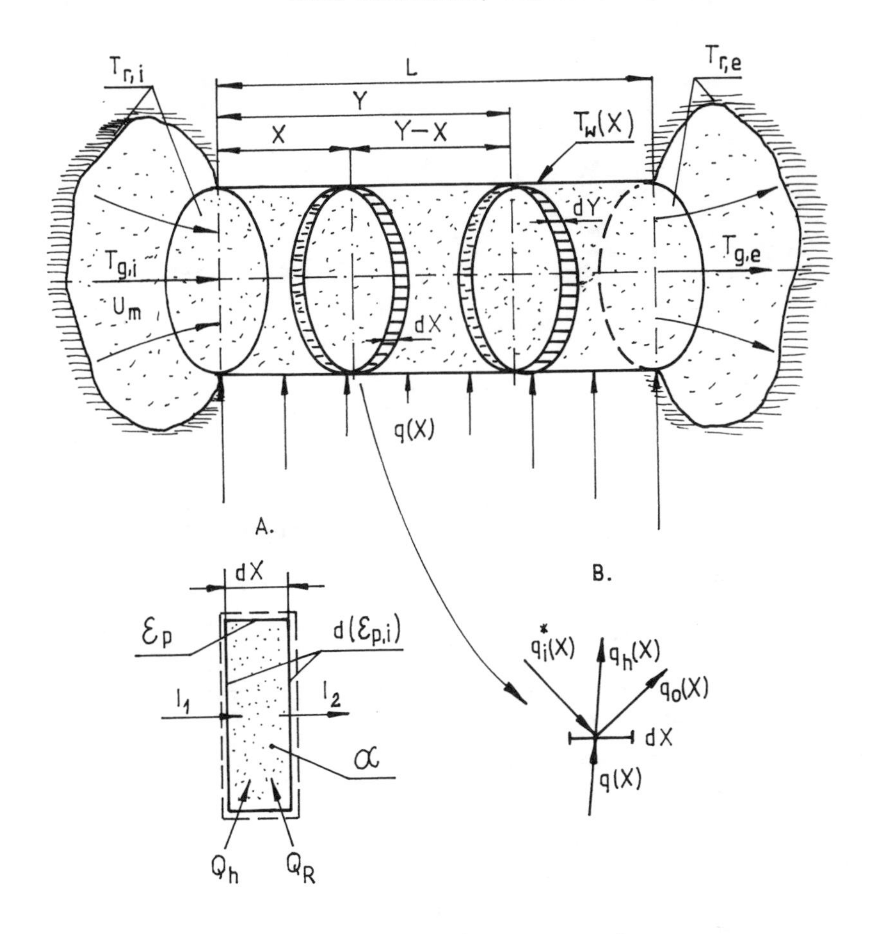

Fig. 1 Circular tube geometry and gaseous volume.

The radiation terms are now considered in detail. $q_o(X)$ is composed both of direct emission, $\varepsilon\ \sigma\ T_w^{\ 4}$, and of reflected radiation which is (1-ε) times the total incoming radiation:

$$q_o(X) = \varepsilon\ \sigma\ T_w^{\ 4}(X) + (1-\varepsilon)q_i^*(X) \tag{2}$$

The total incoming radiatve heat flux $q_i^*(X)$ is composed of three types of terms, the radiation coming from the reservoirs at the ends of the tube, the radiation arriving as a result of the outgoing radiation from the other elements on the internal tube surface, and the radiation arriving from elementary gas bodies in the form of cylindrical slices (Fig. 1) [1] [3].

These quantities can be written respectively as paris of terms as follows, where the first and second term in each pair is for inlet and exit facing radiation respectively:

$$q_i^*(X) = \sigma\, T_{r,i}^4 \tau(X)F(X) + \sigma\, T_{r,e}^4 \tau(L-X)\, F(L-X) +$$

$$+ \int_0^X q_o(Y)\, K(X-Y)\, \tau(X-Y)\, dY + \int_X^L q_o(Y)\, K(Y-X)\, \tau\, (Y-X)\, dY +$$

$$+ \int_0^X e_{g,b}\, d(\varepsilon_{p,i})\, F(X-Y)\, \tau(X-Y) + \int_X^L e_{g,b}\, d(\varepsilon_{p,i})\, F(Y-X)\, \tau(Y-X) \quad (3)$$

The functions F and K are the geometric configuration factors for the system under consideration and τ is the gas transmissivity of thickness X or (L-X) [1] [3].

In order to solve the set of Eqns (1), (2), and (3), an additional heat balance is written for the flowing gas. Since the gas is nontransparent to radiation, heat transferred to the wall is by convection Q_h and radiation Q_R. For a cylindrical volume element of length dX and diameter D, the heat transferred is, [1] [2] [3]:

$$Q_h = \Pi\, D\, h(X)\, [T_w(X) - T_g(X)] \quad (4)$$

and

$$Q_R = \frac{\sigma\, \varepsilon_p}{a_p\left\{\frac{1}{\varepsilon} + \frac{1}{a_p} - 1\right\}} [AT_w^4(X) - T_g^4(X)] \quad (5)$$

where

$$A = \frac{a_p}{\varepsilon_p}\,; \quad \varepsilon_p = \frac{\alpha\, D}{4} = \frac{k}{4}\,; \quad k = \alpha\, D \quad (6)$$

This quanitity (Q_h + Q_R) is equal to the net heat removed from the volume element by the flowing gas, which is:

$$J_2 - J_1 = U_m \frac{\Pi D^2}{4} \rho C_p \frac{dT_g(X)}{dX} dX \tag{7}$$

The mean fluid velocity U_m is assumed constant so that kinetic energy changes of the gas are neglected. These three quantities are equated and the result is rearranged into the form [2] [3]:

$$\frac{dT_g(X)}{dX} = \frac{4}{U_m \rho C_p} \Bigg\{ h(X) \Big[T_w(X) - T_g(X)\Big] + \frac{\sigma \varepsilon_p}{a_p \left\{\frac{1}{\varepsilon} + \frac{1}{a_p} - 1\right\}} [AT_w^4(X) - T_g^4(X)] \Bigg\} \tag{8}$$

According to the method presented in [1], [2] and [3], integro-differential equaitons (1), (2), (3) and (8) can be transformed into differential ones which are integrable numerically. These equations can be reduced by an approximation of configuration factors F, K and transmissivity τ involving exponential functions (Separable Kernel Method) [2]. Using such an analysis the following dimensionless set of differential equations is obtained which requires simultaneous solution:

$$\frac{d^2t_w}{dx^2} \left[\frac{H + H_1x + H_2x^2}{\varepsilon} + 4\, t_w^3\right] + 12\, t_w^2 \left(\frac{dt_w}{dx}\right)^2$$

$$- \frac{1}{\varepsilon} \Big[(H + H_1x + H_2x^2)\, (S + S_1x + S_2x^2 + 4\, RAt_w^3)$$

$$- 2\, (H_1 + H_2x)\Big] \frac{dt_w}{dx} = - 4E\, (1 + Mx + Cx^2) + \frac{2C}{\varepsilon}$$

$$+ \frac{(t_w - t_g)}{\varepsilon} \left\{ (H + H_1 x + H_2 x^2) \left[4E\varepsilon + S_1 + 2\, S_2 x \right.\right.$$

$$\left.\left. + (S + S_1 x + S_2 x^2)(S + S_1 x + S_2 x^2 + 4Rt_g^3) \right] \right\} + \frac{1}{\varepsilon} t_w^4 \left[\varepsilon k(2+k \right.$$

$$\left. + 2(H_1 + H_2 x)RA - (H + H_1 x + H_2 x^2)(S + S_1 x + S_2 x^2 + 4\, Rt_g^3)RA \right]$$

$$- \frac{1}{\varepsilon} t_g^4 \left[2\varepsilon(2 + k)\varepsilon_{p,s} - (H + H_1 x + H_2 x^2)(S + S_1 x + S_2 x^2 + \right.$$

$$\left. + 4\, Rt_g^3)R + 2\,(H_1 + H_2 x)R \right] \tag{9}$$

and

$$\frac{dt_g}{dx} = (S + S_1 x + S_2 x^2)(t_w - t_g) + R(At_w^4 - t_g^4) \tag{10}$$

where

$$S = \frac{4h}{U_m \rho C_p} \; ; \; S_1 = \frac{4h_1 D}{U_m \rho C_p} \; ; \; S_2 = \frac{4h_2 D^2}{U_m \rho C_p} \; ; \; t = T \left(\frac{\sigma}{q} \right)^{1/4}$$

$$H = \frac{h}{q} \left(\frac{q}{\sigma} \right)^{1/4} \; ; \; H_2 = \frac{h_1 D}{q} \left(\frac{q}{\sigma} \right)^{1/4} \; ; \; H_2 = \frac{h_2 D^2}{q} \left(\frac{q}{\sigma} \right)^{1/4}$$

$$R = \frac{4\,\sigma\,\varepsilon_p}{U_m \rho C_p a_p \left\{ \frac{1}{\varepsilon} + \frac{1}{a_p} - \varepsilon \right\}} \left(\frac{q}{\sigma} \right)^{1/4} ; \; x = \frac{X}{D} \; ; \; 1 = \frac{L}{D}$$

$$\varepsilon_{p,s} = 0,75k \; ; \; E = \frac{k(2+k)(1-\varepsilon)+(2+k)^2\varepsilon}{4\varepsilon} \; ; \; M = \frac{mD}{q} \; ; \; C = \frac{cD^2}{q} \tag{11}$$

Eqns. (9) and (10) were derived on the assumption of nouniform abstractive heat flux and convective heat transfer coefficient:

$$q(X) = q + mX + c\, X^2 \tag{12}$$

and

$$h(X) = h + h_1 X + h_2 X^2 \tag{13}$$

that is parabolic distributions.

Eqn. (10) is first order and requires only one boundary condition. This condition is that at the inlet of the tube the gas temperature has a specified value $t_{g,i}$:

$$t_g = t_{g,i} \text{ at } \underline{x = 0} \tag{14}$$

Eqn (9) is second order and requires two boundary conditions. These are found by use of the approximation for configuration factors and transmissivity. At $\underline{x = 0}$ this gives:

$$\left.\frac{dt_w}{dx}\right|_{x=0} = \frac{1}{\left[H + 4\varepsilon t_w^3(0)\right]} \left\{ H(S + 2 + k - \frac{H_1}{H})(\,t_w(0) - t_{g,i}) \right.$$

$$- (2+k) + M + \varepsilon\,(2+k)\; t_w^4\,(0) - \varepsilon\,(2+k) t_{r,i}^4 +$$

$$\left. + HR\left[A t_w^4(0) - t_{g,i}^4\right] \right\} \tag{15}$$

and for $x = 1$,

$$\frac{(1-\varepsilon)}{\varepsilon(2+k)} + \left[1 + k + e^{-(2+k)1}\right] + \frac{1}{2}\; t_{r,i}^4 \;\; \bar{e}^{\,(2+k)1} \;\; - \;\; t_w^4(1)$$

$$= \frac{1}{\varepsilon}\left[H + H_1 1 + H_2 1^2\right]\left[t_w(1) - \;\; t_g(1)\right] - \frac{1}{2}\, t_{r,e}^4 - \varepsilon_{p,s}\, e^{-(2+k)1}$$

$$\int_0^1 t_g^4 \, e^{(2+k)x} \, dx - e^{-(2+k)l} \int_0^1 \left[\frac{1-\varepsilon}{\varepsilon} (H + H_1 x + H_2 x^2)(t_w - t_g)\right.$$

$$\left. + t_w^4\right] e^{(2+k)x} \, dx + M \left\{\left[\frac{1}{(2+k)} - \frac{1}{(2+k)^2} + \frac{1}{(2+k)^2} e^{-(2+k)l}\right] \cdot\right.$$

$$\left.\frac{(1-\varepsilon)}{\varepsilon} - \frac{l}{\varepsilon}\right\} + C \left\{\left[\frac{l^2}{(2+k)} - \frac{2l}{(2+k)^2} + \frac{2}{(2+k)^3} - \frac{2}{(2+k)^3} e^{-(2+k)l}\right] \cdot\right.$$

$$\left.\frac{(1-\varepsilon)}{\varepsilon} - \frac{l^2}{\varepsilon}\right\} \qquad (16)$$

The procedure which yields the boundary conditions (15) and (16) is presented in [1] and [3]. However, the solution procedure involved assuming $t_w(0)$ and calculating the first derivative with respect to x at the wall. The boundary condition for $x = l$ (equation (16)) is used for verification of the calculation, which comes to an end when the condition is fulfilled.

In research practice, despite the formal satisfaction of boundary conditions, a numerical solution also requires the overall thermal balance of the system to be correct [1] [2]. According to [1] and [3] this heat balance takes the dimensionless form:

$$1 + \frac{Ml^2}{2} + \frac{Cl^3}{3} + \frac{H}{S} t_{g,i} + \frac{1}{2(2+k)} t_{r,i}^4 \left[1 - e^{-(2+k)l}\right] + \frac{(1-\varepsilon)}{\varepsilon(2+k)} \cdot$$

$$\left[1 - e^{-(2+k)l}\right] + \frac{1}{2(2+k)} t_{r,e}^4 \left[1 - e^{-(2+k)l}\right] = \frac{H}{S} t_{g,e}$$

$$+ \frac{1}{2} \int_0^1 \left[\frac{(1-\varepsilon)}{\varepsilon} (H + H_1 x + H_2 x^2)(t_w - t_g) + t_w^4\right]\left[e^{-(2+k)x}\right.$$

$$\left. + e^{-(2+k)(1-x)}\right] dx + \frac{1}{2} \varepsilon_{p,s} \int_0^1 t_g^4 \left[e^{-(2+k)x}\right.$$

$$\left. + e^{-(2+k)(1-x)}\right] dx + \Phi \qquad [17]$$

where

$$\Phi = \frac{(1-\varepsilon)}{2\varepsilon} \left\{ - \frac{Ml}{(2+k)} (1-e^{-(2+k)l}) + C \left(- \frac{l^2}{(2+k)} + \frac{2l}{(2+k)^2} - \frac{4}{(2+k)^3} \right) + C \left[e^{-(2+K)l} \left(\frac{l^2}{(2+k)} + \frac{2l}{(2+k)^2} + \frac{4}{(2+k)^3} \right) \right] \right\} \quad [18]$$

From the preceding set of nonlinear differential equations, together with the two boundary conditions and the overall heat balance (17), it is necessary first to determine the parameters H_i; H_1; H_2; S; S_1; S_2; R; k; ℓ; $t_{r,i}$; $t_{g,i}$ and then assuming $t_w(0)$ calculate dt_w/dx for x=0. This calculation enables further solution of Eqns. (9) and (10), a procedure described in detail in [1], [3].

NUMERICAL RESULTS

The set nonlinear differential Eqns. (9) and (10) was solved by the the Runge-Kutta method with Hamming modification, also using the IBM standard library. The calculations were performed on an IBM PC computer. In [1], [2], and [3], the overall problem, treated by various calculations and analyses has already been prosented for a flow system with transparent and optically active gas. In this paper, therefore, we concentrate on certain specific results. These illustrate the influence of (i) the radiative properties of the gas and wall, (ii) the nonuniformity of a convection heat transfer coefficient on the wall, t_w, and gas, t_g, temperature distribution. Tables 1a and b show the set of dimensionless numbers and physical quantities which were used in the calculation. These are based on data from [1] supplemented with additional quantities describing an optically active gas i.e. dimensionless numbers R,k and A [3].

These also cover a practical range of values for the influence of (ii) and (iii) above. The run numbers 1-4 and 2a, 3a are defined with reference to a comprehensive set of heat flux distributions accommodated within.

Table 1a. SUMMARY OF DATA USED IN PREDICTIVE STUDY

$\ell = 5$ $H = 0.8$ $S = 0.01$ $t_{r,i} = t_{g,i} = 1.5$ $A = 0.85$ $k = 0.1$
$\varepsilon_{p,s} = 0.075$

	$\varepsilon = 1.0$ $E = 1.1025$ $R = 3.12 \cdot 10^{-4}$			
	1	2	3	4
$t_w(0) = t_{w,i}$	1.7065	1.7239	1.6889	1.7390
$t_{w,e} = t_w(1)$	1.7237	1.7431	1.7038	1.8484
$t_{r,e} = t_{g,e}$	1.5266	1.5305	1.5225	1.5372
Δb [%]	0.141	0.26	0.036	0.314

(Table 1a continued)

	$\varepsilon = 0.01$ $E = 6.3$ $R = 1 \cdot 10^{-4}$			
	1	2	3	4
$t_w(0) = t_{w,i}$	2.4085	2.4102	2.4070	2.4121
$t_{w,e} = t_w(1)$	2.4478	2.4556	2.4401	3.0750
$t_{r,e} = t_{g,e}$	1.5581	1.5487	1.5487	1.5854
Δb[%]	0.57	0.635	0.517	0.82

Table 1b. SUMMARY OF DATA USED IN PREDICTIVE STUDY (CONTINUED)

$\ell = 5$ $H = 0.8$ $S = 0.01$ $t_{r,i} = t_{g,i} = 1.5$ $A = 0.85$

	$\varepsilon = 1.0$ $E = 1.0$ $\varepsilon_{p,s} = 0.0$ $k = 0.0$ $R = 0.0$			
	2	3	2a	3a
$t_g(0) = t_{w,i}$	1.7660	1.7890	1.7202	1.7095
$t_{w,e} = t_w(1)$	1.7917	1.8276	1.7236	1.7075
$t_{r,e} = t_{g,e}$	1.5132	1.5051	1.5274	1.5305
Δb [%]	0.01	0.113	0.05	0.075

(Table 1b continued)

	ε = 1.0 E = 1.1025 k = 0.1 $\varepsilon_{p,s}$ = 0.075 R = $3.12 \cdot 10^{-4}$			
	2	3	2a	3a
$t_w(0) = t_{w,i}$	1.7192	1.7295	1.6961	1.6901
$t_{w,e} = t_w(1)$	1.7462	1.7662	1.7059	1.6954
$t_{r,e} = t_{g,e}$	1.5207	1.5155	1.5310	1.5336
$\Delta b[\%]$	0.21	0,284	0.09	0.058

equation (12) and given by:

$$\begin{aligned} &1. \quad q(x)/q = 1 \\ &2. \quad q(x)/q = 1 + 0.2x - 0.04x^2 \\ &3. \quad q(x)/q = 1 - 0.2x + 0.04x^2 \\ &4. \quad q(x)/q = 1 + 0.2x \end{aligned} \tag{19}$$

Similarly, a set of alternative data for h, accommodated by equation (13) result in the following run definitions for the non-dimensional convective parameters S and H:

$$\begin{aligned} &1. \quad S(x) = 0.01 && H(x) = H = 0.8 \\ &2. \quad S(x) = 0.0035x+0.0005x^2 && H(x)\ 0.8-0.28x+0.04x^2 \\ &3. \quad S(x) = 0.01-0.0058x+0.0008x^2 && H(x)\ 0.8-0.464x+0.064x^2 \\ &2a. \quad S(x) = 0.01+0.0035x-0.0005x^2 && H(x)\ 0.8+0.28x-0.04x^2 \\ &3a. \quad S(x) = 0.01+0.0058x-0.0008x^2 && H(x)\ 0.8+0.464x-0.064x^2 \end{aligned} \tag{20}$$

Also in Tables 1a and b is an expression of the accuracy of the method $\Delta b\%$. It has already been mentioned that the outlet boundary condition given by equation (16) is a termination criterion. For this a percision parameter was defined as being the ratio of the difference between the left and right hand sides of (16) divided by the r.h.s. For this study, the parameter was set at 0.01%. A similar parameter (Δb) was defined for the overall energy balance, equation (17), the values for the various calculation runs being given in Tables 1a and b. These values, all well under 1%, are partly a reflection of the parameter for (16), but mainly an expression of overall accuracy of the method.

Fig. 2 illustrates the Stanton number S(x) distributions along the inner surface of the wall. Solutions were obtained for a short tube having a length-diameter ratio of 5.

F

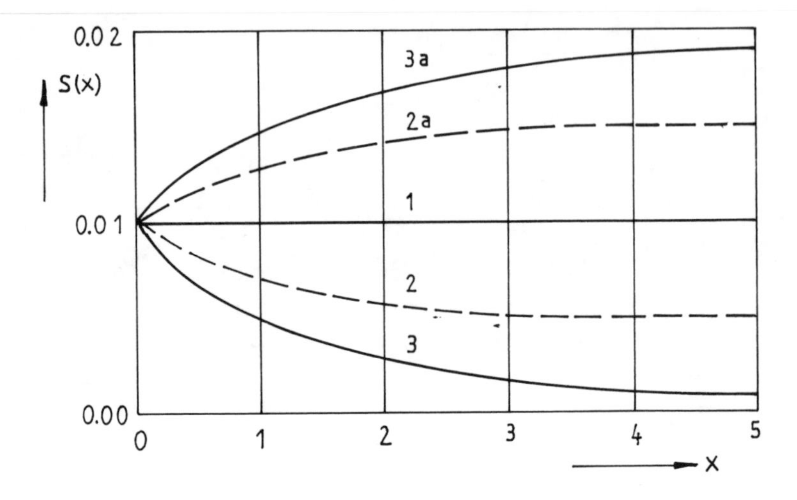

Fig. 2 Stanton number S(x) distributions along the inner surface wall.

The results for different values of nonuniform heat flux q(x)/q are shown in Figs. 3 and 4 for emissivities of 1 and 0.01 and dimensionless aborptivity k = 0.1 (data from Table 1a). Additionally, the figure includes temperature distribution ofr pure convection and constant heat flux. [1] The results show the correct and significant influence of dimensionless absroptivity k and wall emissivity ε on the temperature distribution t_w and t_g. It is demonstrated that there is substantial influence of heat flux distribution for a low value of wall emissivity.

In Figs. 5 and 6, the results for different values of Stanton number S and of H are plotted for ε = 1.0 and k = 0.0 and 0.1. For the same external heat flux, an increase in the Stanton number (examples 3-3a) tends to decrease the axial temperature distribution along the tube and increase the axial gas temperature gradient. The present results show that the effect of nonuniform convective heat transfer coefficient cannot be neglected. As demonstrated in [1], [3] radiative heat transfer significantly changes local values of heat transfer coefficient from a wall to a transparent gas and this is far more evident in the case of an optically active gas. Even for an emissivity as low as 0.01 the effect of radiation cannot be ignored in comparison with the convective transfer.

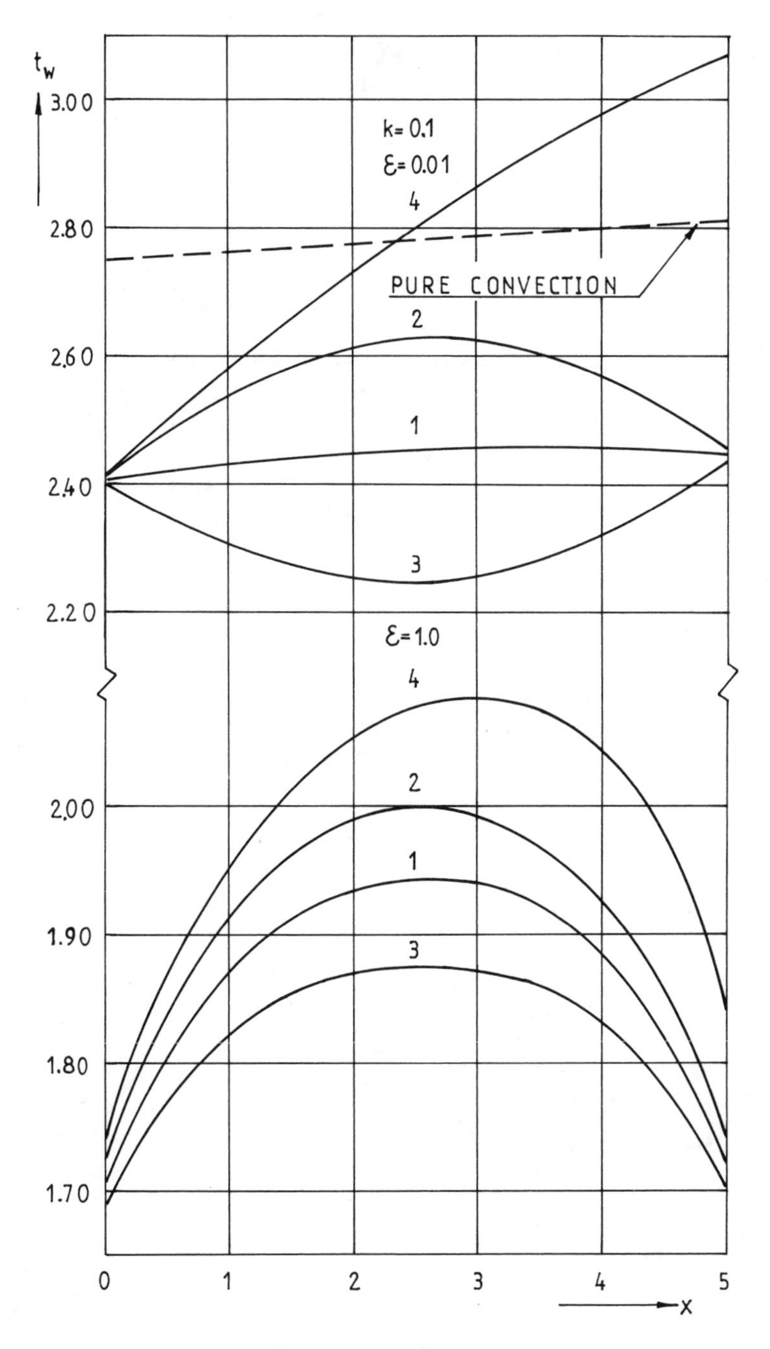

Fig. 3 The effect of dimensionless nonuniform heat flux on temperature distribution in the tube wall.

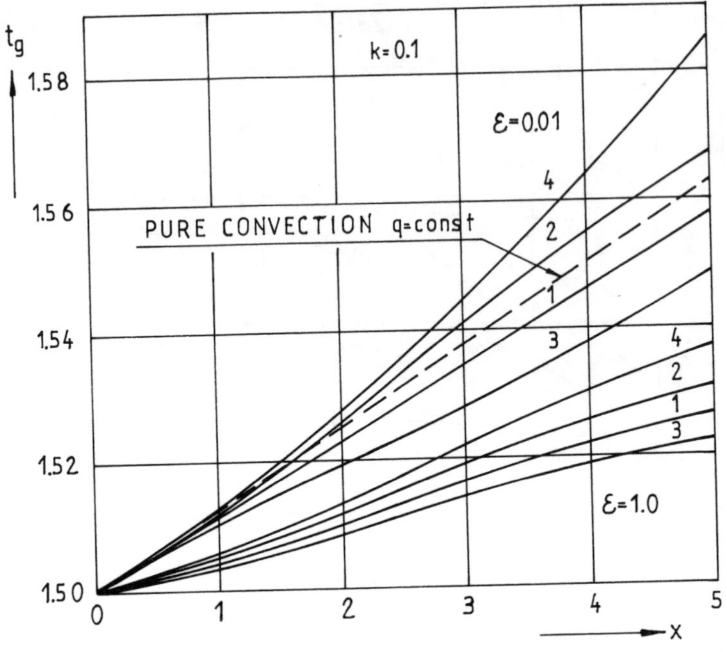

Fig. 4 The effect of dimensionless nonuniform heat flux on temperature distribution in the gas.

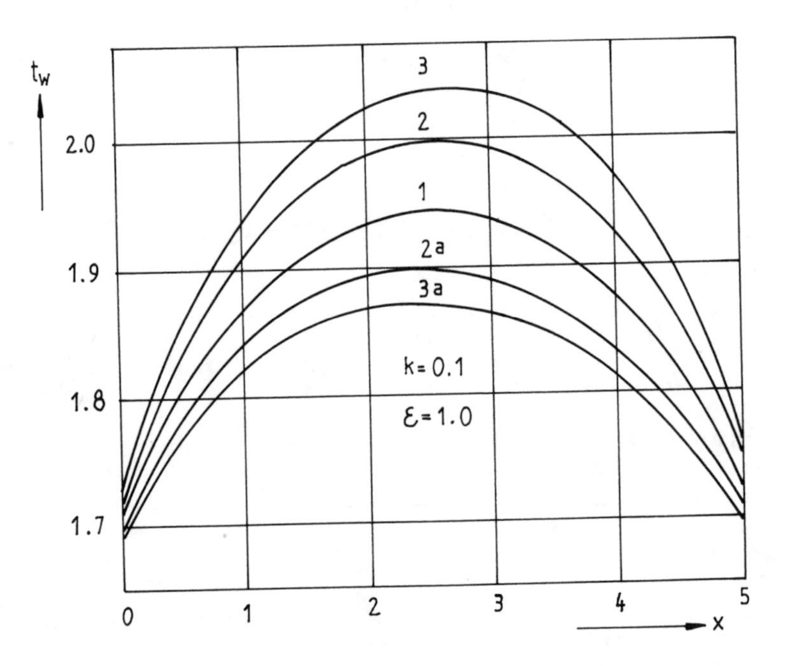

Fig.5a Effect of nonuniform convection heast transfer coefficient on the temperature distributions in the tube wall.

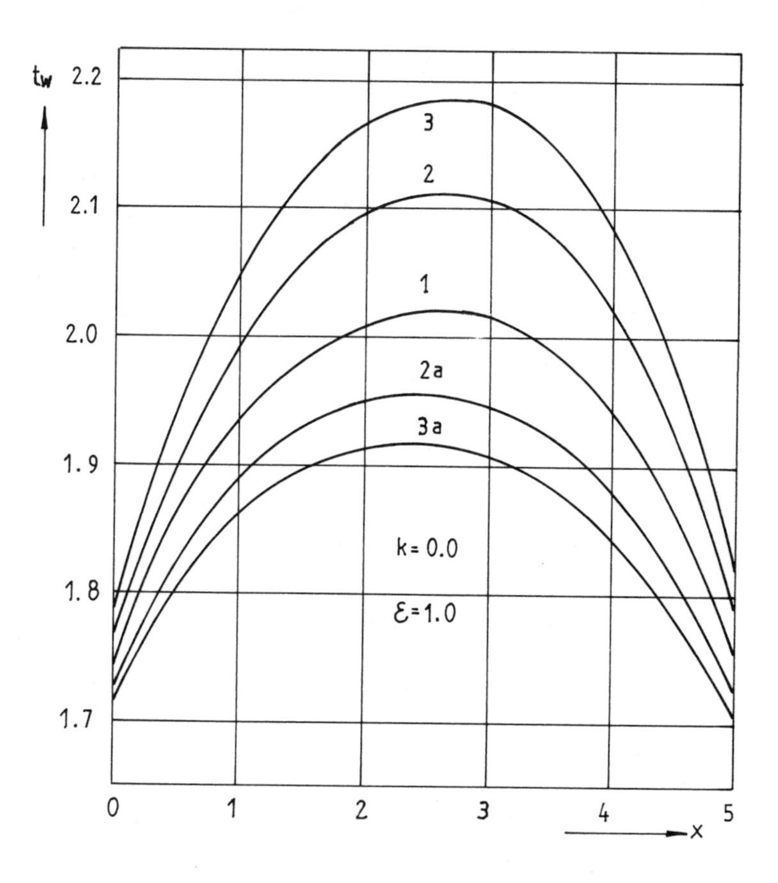

Fig. 5b Effect of nonuniform convection heat transfer coefficient on the temperature distributions in the tube wall.

CONCLUSIONS

A satisfactory treatment has been developed of the complex heat exchange problem of combined radiation and convection in a short tube containing an optically active gas. The wall heat flux and convective heat transfer coefficient may be non-uniform, and parabolic distributions can be accommodated. The results show that even when either mode is small, its effect should not be ignored. The overall energy balance is accurate in this study to within 1%.

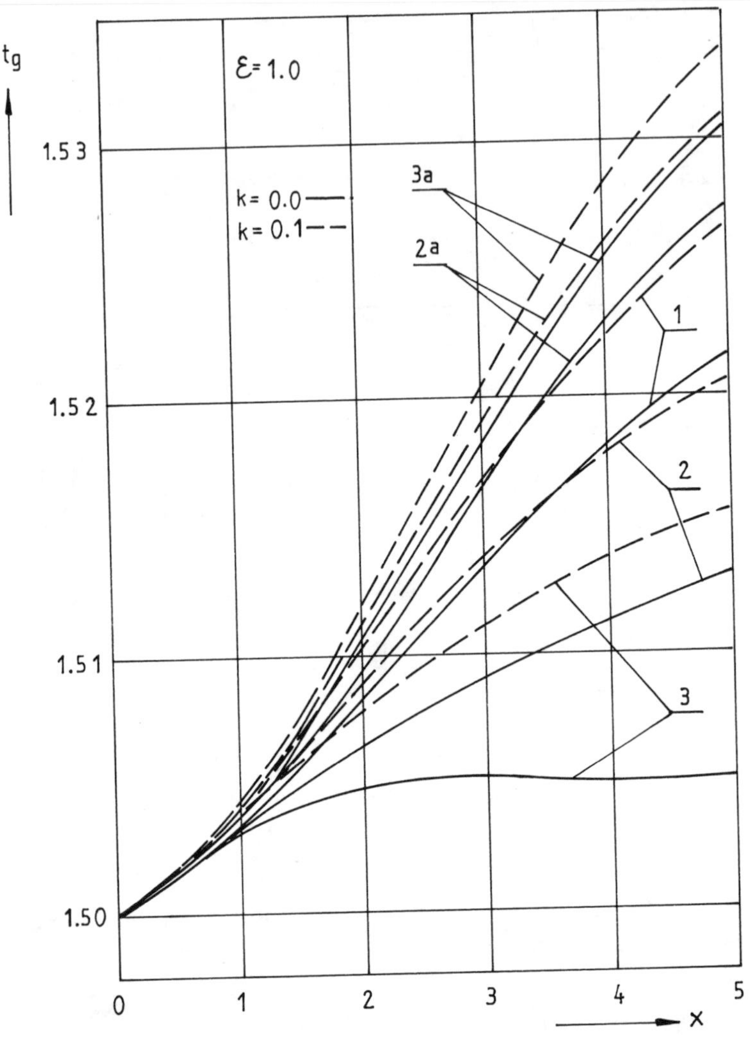

Fig.6 Effect of nonuniform convection heat transfer coefficient on the temperature distributions in the gas.

REFERENCES

1. Siegel, R. and Perlmutter, M. Convective and Radiant Heat Transfer for Flow of Transparent Gas in a Tube with a Grey Wall, Int. J. Heat Mass Transfer, Vol.5, p.p. 639-660, 1962.

2. Siegel, R. and Howell, J.R. Thermal Radiation Heat Transfer, McGraw-Hill, New York, 1981.

3. Stasiek, J. Transformational-zone Method of Calculation of Complex Heat Exchange During Flow of Optically Active Medium Inside Tube of Diffuse Grey Surface. Wärme und Stoffübertragung Vol.22, p.p. 129-139, 1988.

Natural Convection-Radiation Interactions in a Cube Filled with a Non-Gray Gas

T. Fusegi(*), K. Ishii(*), B. Farouk(**), K. Kuwahara(***)

(*) The Institute of Computational Fluid Dynamics, 1-22-3 Haramachi, Meguro, Tokyo 152, Japan

(**) Department of Mechanical Engineering and Mechanics, Drexel University, Philadelphia, Pennsylvania 19104, USA

(***) The Institute of Space and Astronautical Science, 3-1-1 Yoshinodai, Sagamihara, Kanagawa 229, Japan

ABSTRACT

A three-dimensional numerical study was performed on interactions of natural convection and radiation in a cubical enclosure, which is filled with carbon dioxide gas. The enclosure was heated differentially by a set of opposing side walls. The remaining four surfaces were thermally insulated. Gas radiation was analyzed by the P_1-differential approximation method. The non-gray behavior of carbon dioxide gas was handled by the weighted sum of gray gas model. Computations were carried out over a range of the Rayleigh number, Ra, between 10^5 and 10^9. The Prandtl number and the overheat ratio were held fixed at 0.68 and 1.0, respectively. Unsteady transitional flows were computed by a direct simulation method, without using any turbulence model. From the predictions, a mean heat transfer correlation has been proposed as $Nu = 0.323Ra^{0.342}$ in the surface/gas radiation mode, where Nu is the time and spatially averaged Nusselt number at the isothermal walls. Comparing with two-dimensional correlation obtained previously, the three-dimensional results for a cube exhibit 15 to 20 % reduction in the heat transfer rate. With radiation, strong end effects were evident for the predictions of flow and temperature fields.

INTRODUCTION

Interactions of natural convection and radiation occur within an enclosure in a variety of thermal engineering applications, such as cooling of electronic equipment, solar collectors and fires in compartments. In spite of the technological importance, relatively little research effort has been devoted to this class of problems due to its complexity. The analysis must consider the three modes of heat transfer (conduction, convection and radiation), which are strongly coupled in the process. In addition, the integro-differential nature of gas radiation requires considerable computational efforts. Hence, gas radiation models are employed for engineering analysis of natural convection-radiation interactions to reduce the computational effort to a tractable amount [1].

The present paper examines interactions of natural convection and radiation in a non-gray gas filled cubical enclosure, which is heated differentially by opposing side walls. The schematic of the enclosure is depicted in *Figure* 1. The isothermal surfaces are the planes located at $x^* = 0$ and L_o. The remaining walls are thermally insulated. All the inner surfaces are assumed to be black for radiation. Carbon dioxide gas is considered as the medium. The geometry and boundary conditions of the present problem are mathematically well-posed and model various thermal engineering systems properly.

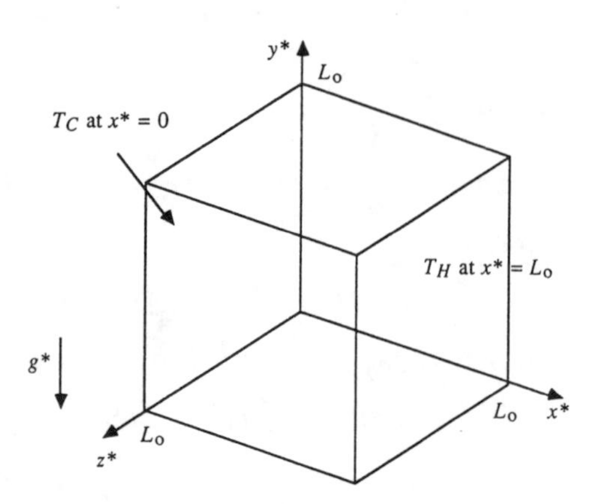

Figure 1 Differentially heated cubical enclosure

Previous works on this class of problems were limited to two-dimensional configurations [2-5]. These numerical studies predict significant changes in characteristics of flow fields and heat transfer rates across enclosures due to radiation when the overheat ratio, δ, becomes large. In three-dimensional enclosures, the presence of end walls will have large effects on the temperature and flow fields through radiative exchange among surfaces. Therefore, three-dimensional analysis should be considered to study situations of practical importance.

In order to analyze gas radiation, a non-gray spherical harmonics (P_1) approximation method is employed, which incorporates the weighted sum of gray gas (WSGG) model. This approach reduces the transfer equation for the radiative intensity of a non-gray gas to a set of differential equations for the irradiance of *component gray gases*. Hence, the resultant equations are compatible in their form with the governing equations for flows. This method has been utilized for a turbulent convection-radiation interaction analysis of flows in a furnace by Song and Viskanta [6]. In their paper, the absorptivity of combustion gases (mixtures of carbon dioxide, water vapor and air) is approximated by that of two component gray gases. In the present study, three component gray gases are considered for carbon dioxide gas [7]. The same WSGG model was used earlier in a two-dimensional analysis [5].

Computations were performed over a wide range of the Rayleigh number, that extends from 10^5 to 10^9. Predictions of the flow and temperature fields were obtained by directly integrating the time-dependent governing equations. Calculations were carried out by using a higher order upwind scheme for the convection terms of the Navier-Stokes equations and sufficiently small time steps. No turbulence model was employed for any of the cases. In two-dimensional natural convection flow analysis, the direct simulation method has been used successfully to predict onsets of instabilities [8, 9].

MATHEMATICAL MODEL

The governing equations consist of the time-dependent Navier-Stokes equations, the energy equation and the transfer equation for the radiative intensity. The last equation is transformed to three differential equations for the irradiance of the component gray gases by using the P_1 approximation method and the WSGG model. The non-dimensionalized governing equations can be expressed in tensor form as:

$$\frac{\partial \rho}{\partial t} + \frac{\partial}{\partial x_j}(\rho u_j) = 0 \tag{1}$$

$$\frac{\partial}{\partial t}(\rho u_i) + \frac{\partial}{\partial x_j}(\rho u_j u_i) = -\delta_{i2}\frac{1}{\mathrm{Fr}}(\rho - 1) - \frac{\partial p}{\partial x_i} + \frac{1}{\mathrm{Re}}\frac{\partial}{\partial x_j}\left(\frac{\partial u_i}{\partial x_j} + \frac{\partial u_j}{\partial x_i} - \delta_{ij}\frac{2}{3}\frac{\partial u_k}{\partial x_k}\right) \tag{2}$$

$$\frac{\partial}{\partial t}(\rho T) + \frac{\partial}{\partial x_j}(\rho u_j T) = \frac{1}{\mathrm{Re\,Pr}}\frac{\partial^2 T}{\partial x_j \partial x_j} + \frac{1}{3\,\mathrm{Bo}}\sum_{m=1}^{3}\frac{1}{\tau_m}\frac{\partial^2 J_m}{\partial x_j \partial x_j} \tag{3}$$

$$\frac{\partial^2 J_m}{\partial x_j \partial x_j} + 3\tau_m^2 (4W_m T^4 - J_m) = 0, \quad m = 1, 2, 3 \tag{4}$$

where δ_{ij} is the Kronecker delta ($\delta_{ij} = 1$ if $i = j$, and $\delta_{ij} = 0$ otherwise). The fluid properties are assumed to be invariant except for the density. In order to study flows at a large overheat ratio, density variation is accounted for in all the terms of the governing equations. The ideal gas law is used for the density.

The physical variables are non-dimensionalized in the following manner:

$$(x, y, z) = (x^*, y^*, z^*) / L_o, \ (u, v, w) = (u^*, v^*, w^*) / u_o,$$
$$t = t^* u_o / L_o, \ p = (p^* - p_o) / \rho_o u_o^2, \ \rho = \rho^* / \rho_o, \ \mu = \mu^* / \mu_o,$$
$$k = k^* / k_o, \ g = g^* / g_o, \ T = T^* / T_o, \ J = J^* / \sigma T_o^4$$

The boundary conditions are stated as follows:

$$u = v = w = 0 \tag{5}$$

$T = (2 - \delta)/2$ at $x = 0$, $T = (2 + \delta)/2$ at $x = 1$, and

$$\frac{\partial T}{\partial n} = \frac{\mathrm{Re\,Pr}}{\mathrm{Bo}}\left(\pm q_s - \sum_{m=1}^{3}\frac{1}{3\tau_m}\frac{\partial J_m}{\partial n}\right)$$

at $y = 0$ and $z = 0$ (with the positive sign),
and at $y = 1$ and $z = 1$ (with the negative sign) (6)

$$\frac{\partial J_m}{\partial n} = \pm 3\tau_m \frac{\varepsilon_w}{2(2 - \varepsilon_w)}(J_m - 4W_m T^4), \quad m = 1, 2, 3$$

at $x = 0$, $y = 0$ and $z = 0$ (with the positive sign),
and at $x = 1$, $y = 1$ and $z = 1$ (with the negative sign) (7)

where n denotes the coordinate normal to the surface and q_s is the surface radiative flux, which is computed as $q_s = W_0 \Sigma_i F_{1\text{-}i} (T_w^4 - T_i^4)$, where $F_{1\text{-}i}$ is the shape factor for radiation between surface elements 1 and i. The summation is taken for all the surfaces that the element 1 can see. In the expression, W_0 is defined as $W_0 = 1 - \Sigma^3_{j=1} W_j$.

SOLUTION METHOD

The governing equations (1) - (4) are discretized by a control-volume based finite difference procedure. Together with the boundary conditions (5) - (7), approximate solutions are obtained by an iterative method based on the pressure correction algorithm, SIMPLE [10]. The present method employs the *S*trongly *I*m*p*licit Scheme [11] to improve convergence characteristics of solutions. SIP is applied to the planes of constant z in order to simultaneously determine dependent variables in the x and y directions on each plane.

The convection terms in the momentum equation (2) are approximated by the QUICK scheme [12], while those in the energy equation (3) are discretized by a hybrid scheme [10]. The QUICK scheme is a third-order accurate upwind scheme, which possesses the stability of the first-order upwind differencing and is free from numerical diffusion experienced with the first-order scheme. The entire enclosure is considered as the computational domain. The number of grid points for computations is 41 x 41 x 41 for lower Rayleigh number cases ($Ra < 10^7$), and it is increased to 51 x 51 x 51 for computations with higher Rayleigh numbers. Variable grid spacing is used to resolve steep gradients of the velocity and the temperature near the walls. Care is taken to distribute several grid points inside the boundary layer formed along the isothermal walls. In this manner, the minimum value of grid spacing is reduced to approximately 10^{-4} near the isothermal walls.

A time step of as small as 10^{-3} second is used for the computations. At each time level, convergence is assumed when the following convergence criterion is satisfied:

$$\frac{|\phi_n - \phi_{n-1}|}{|\phi_n|_{maximum}} \leq 10^{-4} \quad \text{for all } \phi \tag{8}$$

where ϕ denotes any dependent variable, and n is the value of ϕ at the n-th iteration level.

In order to calculate the surface radiative flux, the shape factor for an elemental surface with respect to other elemental surfaces needs to be evaluated. This is accomplished by computing the shape factors between a control volume surface of the finite difference mesh system located on the boundary and any other control volume surfaces that it can see. Simpson's formula for numerical integration is employed for multiple integration of the formula for the shape factor between two differential elements taken on the control volume surfaces, from which the desired shape factors between two control volume surfaces are computed.

RESULTS AND DISCUSSION

Computations are performed to investigate interactions of natural convection and radiation in a cubical enclosure by employing the mathematical model described in a previous section. Carbon dioxide gas at the reference pressure of 1 atm is considered as the medium. Solutions are acquired over the Rayleigh number range of $1.67 \times 10^5 \sim 10^9$. In order to isolate effects of the two radiation modes (surface and gas radiation) on the flow and temperature fields, several computations were performed by neglecting either of the radiation modes under the same conditions for the remaining parameters. In the computations, the temperature difference of the isothermal walls is set at 555 K. The same value is chosen as the reference temperature, at which value the fluid properties are evaluated. Consequently, the overheat ratio, δ, is unity and the cooled and heated side wall temperatures are set at $T_C = 278$ K and $T_H = 833$ K, respectively. The reference Prandtl number is equal to 0.68 for all the cases. Under these conditions, the height of the enclosure varies from 2.57×10^{-2} m to 4.66×10^{-1} m and the Boltzmann number (interaction parameter) ranges from approximately 50 to 200. The optical thickness for carbon dioxide gas is approximately $0.3L_o$, where L_o is the enclosure length.

The computations were carried out on the Hitachi S820 supercomputer at the Institute of Computational Fluid Dynamics. The system has the maximum computational speed of 3 GFLOPS. The computer code was vectorized for the supercomputer. The typical CPU time for obtaining converged solutions was approximately 30 minutes *per* run with grid points of 51 x 51 x 51 in the natural convection cases. However, the CPU time is increased to approximately 3 hours in the radiation cases. About 200 MB of memory space was used for a computation. The graphic display of the results was obtained by a three-dimensional interactive graphic software [13].

Characteristics of the Temperature and Flow Fields ($Ra = 4.5x10^6$)
Steady state conditions could be predicted up to the Rayleigh number of 10^8. In this sub-section, results for Ra = 4.5×10^6 are examined as a representative case. In the obtained results, the fields in the half domain in $z = 0.5 \sim 1$ are the mirror images of those in the remaining flow field in $z = 0 \sim 0.5$. *Figure* 2 presents cross-sectional views of isotherms and velocity vectors on the plane of symmetry located along $z = 0.5$.

If radiation is neglected (the natural convection mode), the temperature and flow fields in this plane are almost symmetric with respect to the point ($x = 0.5$, $y = 0.5$). Slight deviations from the exact centro-symmetry are due to the density variation effect. Hydraulic and thermal boundary layers develop along the isothermal walls. The temperature field is stratified in the remaining part of the enclosure. The flow field exhibits a distinct boundary layer-stagnant core structure.

When the surface and gas radiation modes are included, significant changes occur in the fields. The field in this plane is no longer symmetric due to an increase in the temperature of the thermally insulated planes by the surface radiation exchange. As seen in the downward shift of the location of the isotherm which corresponds to the average temperature ($T = 1.0$), the overall fluid temperature is higher than the previous case. The boundary layer-stagnant core structure of the flow field is still retained in the surface/gas radiation mode. However, deviation from the centro-symmetry of the flow field is pronounced.

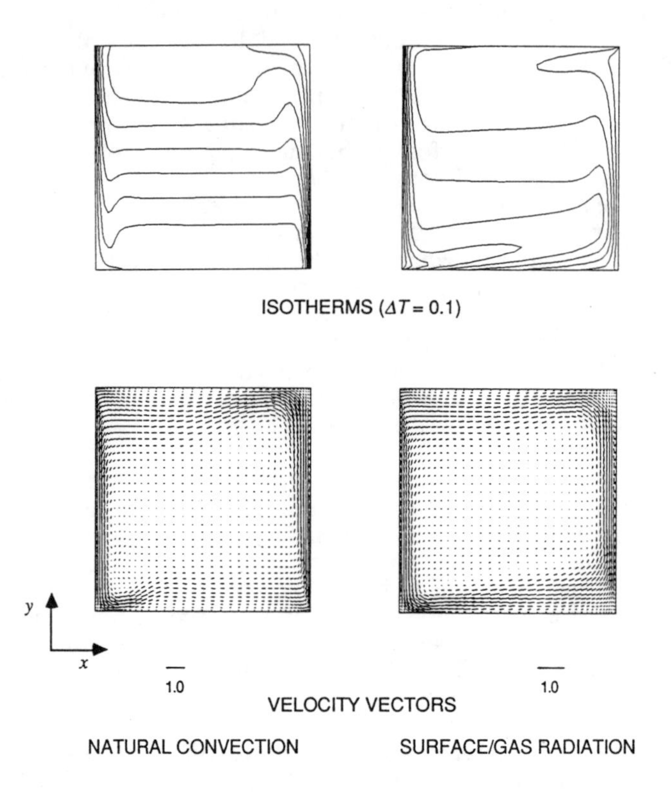

Figure 2 Isotherms and velocity vectors in the plane located at $z = 0.5$ (Ra = 4.5×10^6)

The three-dimensionality of the field characteristics is examined in *Figure* 3, which depicts isotherms and velocity vectors on the plane along the enclosure mid-height ($y = 0.5$). In the natural convection mode, the temperature field varies only slightly in the z direction. In the velocity field, weak secondary flows appear at the corners. The magnitude of the velocity vector in this plane is overall one order of magnitude lower than that in the plane along $z = 0.5$. These results are in good qualitative agreement with the numerical study of Lankhorst and Hoogendoorn [14]. They considered flows in an air-filled cubical enclosure under the Boussinesq approximation.

With radiation, variation in the temperature field near the back walls at $z =$ 0 and 1 is pronounced for three-dimensional cases. This is due mainly to an increase in the wall temperature through surface radiative exchange. As seen in the velocity vectors, the intensity of the secondary flows at the corners increases considerably. Near the heated wall, the secondary vortex centers move toward the symmetry plane located at $z = 0.5$, while those near the cooled wall remain at the corners. The destruction of the symmetric field structure by radiation is evident in this cross-sectional plane.

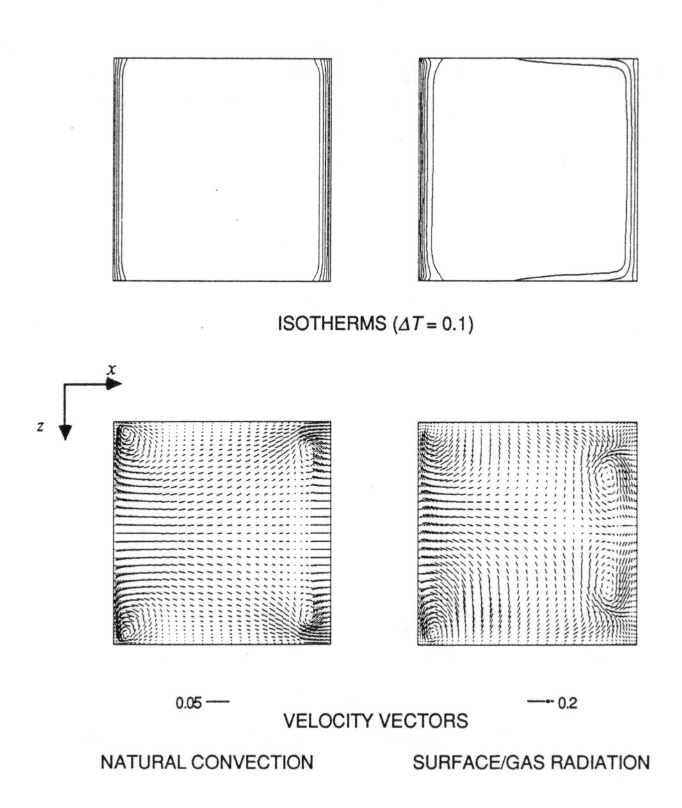

Figure 3 Isotherms and velocity vectors in the plane of the enclosure mid-height at $y = 0.5$ (Ra = 4.5×10^6)

Characteristics of the Temperature and Flow Fields (Ra = 10^9)

At Ra = 10^9, the minimum (dimensional) values of time step and grid spacing used for the computation was 0.2×10^{-3} sec and 0.4×10^{-4} m, respectively. No steady state was reached in the natural convection mode and fluctuations in the fields were observed, indicating that the flow undergoes transition to turbulence. The maximum fluctuation of the average Nusselt number at the isothermal wall is approximately 0.1 percent of its time averaged value. On the other hand, a steady solution could be acquired when radiation was included in the computation. For the two-dimensional natural convection of air (Pr = 0.71), the results of Paolucci and Chenoweth [9] obtained by the direct simulation method predict that transition from laminar to turbulent flow takes place at around Ra = 3×10^8. In another numerical investigation for natural convection flow by the direct simulation method [8], good agreement with an experiment is found in the temperature field fluctuations at Ra = 6×10^8 for water (Pr = 7). The present results indicate that radiation delays the flow transition.

Figure 4 present instantaneous isotherms and velocity vectors at the plane of symmetry located at $z = 0.5$. Comparing with the lower Rayleigh number case discussed previously, the thickness of the boundary layers is considerably reduced. In the natural convection mode, a distinct boundary layer-stagnant core structure is formed in the enclosure. In the flow field along each isothermal surface, the wall boundary layer develops and strikes the horizontal wall. The direction of the flow is bent inward the flow domain at the departing corner. In the numerical study [9], this 'hydraulic jump' has been confirmed as the cause of unsteady flow generation. Weak secondary flows appear in the region formed between the horizontal walls and the main flows.

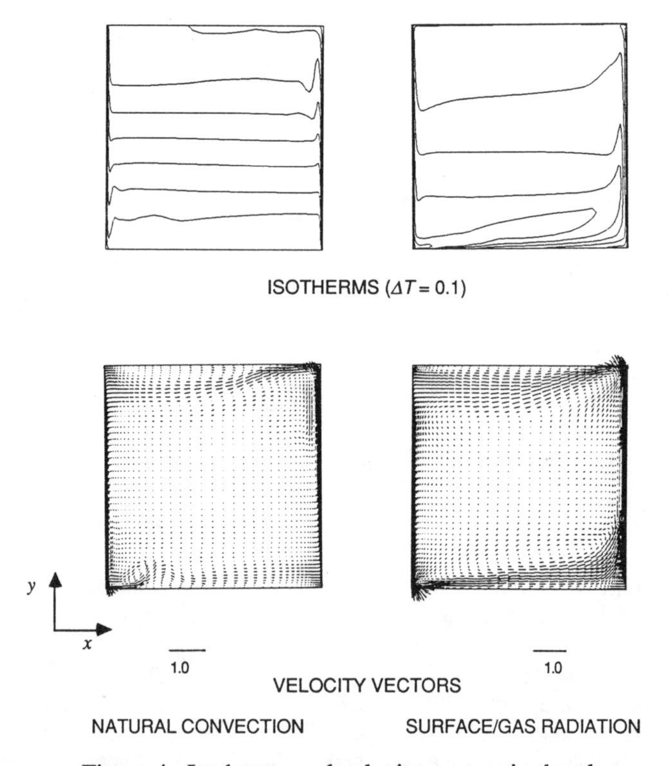

Figure 4 Isotherms and velocity vectors in the plane located at $z = 0.5$ (Ra = 10^9)

With radiation, destruction of the symmetric fields is evident as the lower Rayleigh number case. As in the natural convection flow, the hydraulic jumps occur at the departing corners. No flow separation is seen along the horizontal surfaces in the present case.

In order to study three-dimensional characteristics of the flow and temperature fields, isotherms and velocity vectors on the enclosure mid-height are presented in *Figure* 5. The fields formed by natural convection exhibits a similar pattern to those at the lower Rayleigh number, except for significant reduction in the thermal boundary thickness and shift of the second vortex centers toward the corners. Large variation in the temperature field takes place in the half domain of the enclosure, which includes the heated wall. Intense flows are generated along the end walls near the heated wall due to an increase in the surface temperature due to radiation.

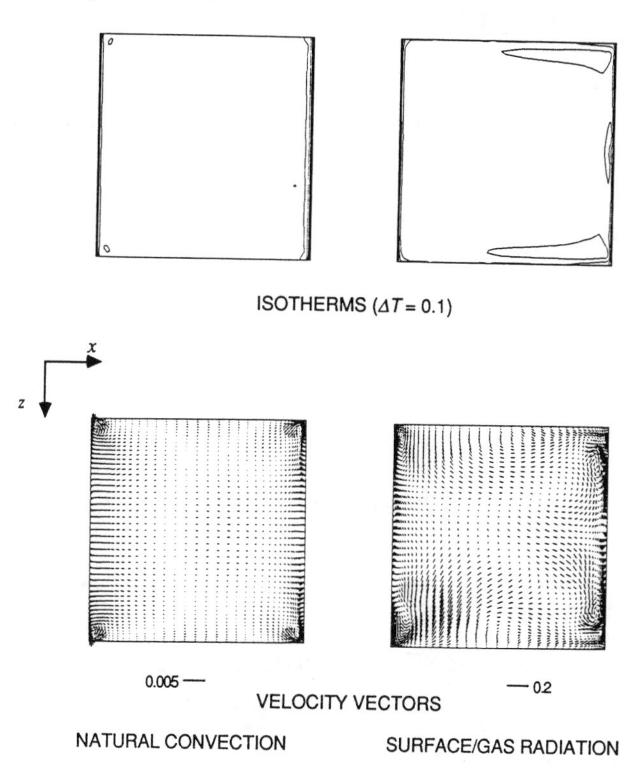

Figure 5 Isotherms and velocity vectors in the plane of the enclosure mid-height at $y = 0.5$ (Ra = 10^9)

Heat Transfer Correlation

The heat transfer rate at the isothermal walls is computed from temperature fields. *Table* 1 presents changes in the average Nusselt number due to radiation. The Nusselt number (Nu_{total}) at the heated wall located at $x = 1$ is defined as

$$\mathrm{Nu}_{total} = \frac{1}{S}\left[\frac{1}{\delta}\int_S \frac{\partial T}{\partial x}\Big|_{x=1} dS + \frac{\mathrm{Re\,Pr}}{\delta\,\mathrm{Bo}}\left(\int_S q_s\big|_{x=1} dS + \int_S \sum_{m=1}^{3} \frac{1}{3\tau_m}\frac{\partial J_m}{\partial x}\Big|_{x=1} dS\right)\right] \qquad (9)$$

where S is the area of the isothermal plane. Each term in the right-hand side of the equation represents, from the first to the third terms, the Nusselt number for conduction, surface radiation and gas radiation. Steady state values are used for calculation of the Nusselt number except for the natural convection mode at Ra

= 10^9, at which value no steady state has been reached. In that case, instantaneous values of the Nusselt number are averaged over a large time interval compared to the characteristic time.

Table 1 The average Nusselt number at the isothermal walls

Ra	L_o [x10^{-1} m]	Bo	Natural Convection	Radiation modes Surface	Radiation modes Surface/Gas
1.67x10^5	0.254	52.6	5.45	20.1	19.3
4.50x10^6	0.762	91.0	14.6	64.6	62.9
10^8	2.17	153	32.4	176	171
10^9	4.66	224	65.0	397	384

With the surface radiation mode, the heat transfer rate increases approximately four times the values of the natural convection cases. The Nusselt number decreases slightly due to absorption of energy by the gas when gas radiation is also accounted for.

From the above data, heat transfer correlations are determined over the investigated Rayleigh number range as

$$Nu = 0.187\ Ra^{0.282} \quad \text{(natural convection)} \tag{10}$$

$$Nu = 0.323\ Ra^{0.342} \quad \text{(surface/gas radiation)} \tag{11}$$

The above correlations are valid for $10^5 < Ra \leq 10^9$ with $\delta = 1$ and $T_o = 555$ K.

For two-dimensional enclosures, which correspond to enclosures with infinite depth in the z direction, heat transfer correlations are available in the literature. For natural convection of a Bussinesq fluid of Pr = 0.71 (air) over $10^6 < Ra \leq 10^{12}$ with $\delta = 0.068$, Markatos and Pericleous [15] have obtained the following expression:

$$Nu = 0.082\ Ra^{0.324} \tag{12}$$

The difference between *Equations* 10 and 12 is insignificant when calculated Nusselt numbers are compared at the same Rayleigh number. As seen in *Figures* 3 and 5, only slight variation occurs in the z direction of the temperature fields due to the thermally insulated end walls. Consequently, the two-dimensional approximation is satisfactory for determination of the heat transfer rate in the natural convection case.

In the surface/gas radiation mode, for the fluid and the boundary condition identical to those considered in the present study, Fusegi and Farouk [5] have proposed a heat transfer correlation for a two-dimensional enclosure as

$$\text{Nu} = 0.343\ \text{Ra}^{0.307}\ \text{Bo}^{0.151} \tag{13}$$

over $10^7 \leq \text{Ra} \leq 10^{10}$ and $150 \leq \text{Bo} \leq 300$ at $\delta = 1$ and $T_o = 555$ K. It should be noted that the Rayleigh and Boltzmann numbers are dependent on each other; if one of them is altered, the other changes accordingly, as presented in *Table* 1. Hence, only the Rayleigh number is considered in *Equation* 11. The heat transfer rate for the three-dimensional enclosure is approximately 15 to 20 percent less than that for the two-dimensional counterpart. This difference is mainly attributed to the thermally insulated boundary condition of the end walls. The average end-wall temperature reaches a high value, which results in a decrease in the total radiative flux of the isothermal walls. Strong three-dimensional effects are present when radiation is accounted for.

CONCLUSIONS

A three-dimensional numerical study has been carried out to examine interactions of natural convection and radiation in non-gray gas flows in a cubical enclosure. The applicability of the present mathematical model for gas radiation analysis, the P_1 approximation method and the WSGG model, for natural convection-radiation interaction problems has been demonstrated. This approach provides efficient solution for this class of problems due to its compatibility with the governing transport equations for the flow. Variety of non-gray gases, including soot, can be handled in a similar manner.

For the overheat ratio of unity and a reference temperature of 555 K, which are considered in the present paper, radiation is found to alter significantly the characteristics of the flow and temperature fields. The heat transfer rate is increased to approximately four to six times the natural convection case. Due to the presence of the end walls, the field structure discloses a strong three-dimensionality when radiative transfer is accounted for.

A variety of thermal engineering systems can be modeled adequately by the present geometry. The obtained predictions serve as reference data for design of such systems.

Radiation may delay the flow to undergo transition to turbulence. Determination of the critical Rayleigh number in the presence of radiation can be considered as a future study.

NOMENCLATURE

- Bo Boltzmann number, $\rho_o c_{po} u_o / \sigma T_o^3$
- c_p specific heat at constant pressure
- Fr Froude number, $u_o^2 / g_o L_o$
- g gravitational acceleration
- L_o reference length (cavity height)
- J irradiance, $J = \int_{\Omega = 4\pi} I \, d\Omega$, where I is radiative intensity
- k thermal conductivity
- p pressure
- Pr Prandtl number, $c_{po} \mu_o / k_o$
- Ra Rayleigh number, $g_o \beta_o c_{po} \rho_o^2 L_o^3 (T_H - T_C) / \mu_o k_o$
- Re Reynolds number, $\rho_o u_o L_o / \mu_o$
- t time
- T temperature
- T_o reference temperature, $(T_H + T_C)/2$
- T_C, T_H cooled and heated side wall temperatures
- u_o reference velocity, $[g_o \beta_o L_o (T_H - T_C)]^{1/2}$
- u, v, w velocity components in the x, y and z directions

W_m weighting function for the m-th component gray gas
x, y, z Cartesian coordinates

Greek symbolds
β thermal expansion coefficient
δ overheat ratio, $(T_H - T_C) / T_o$
ε_w surface emissivity
κ absorption coefficient
μ viscosity
ρ density
σ Stefan-Boltzmann constant
τ optical thickness, κL

Subscripts
m m-th component gray gas
o reference quantities (dimensional)
w at the wall

Superscript
* dimensional quantities

REFERENCES

1. Yang, K. T. Numerical Modeling of Natural Convection-Radiation Interactions in Enclosures, vol. 1, pp. 131 - 140, *Proc. of 8th Int. Heat Transfer Conf.*, San Francisco, California, 1986.
2. Chang, L. C., Yang, K. T. and Lloyd, J. R. Radiation-Natural Convection Interaction in Two-Dimensional Complex Enclosures, *J. Heat Transfer*, Vol. 105, pp. 89 - 95, 1983.
3. Zhong, Z. Y., Yang, K. T. and Lloyd, J. R. Variable-Property Natural Convection in a Tilted Enclosure with Thermal Radiation, *Numerical Methods in Heat Transfer* (Ed. Lewis, R. W. and Morgan, K.), Vol. III, pp. 195 - 214, 1985.
4. Fusegi, T. and Farouk, B. Radiation-Convection Interactions of a Non-Gray Gas in a Square Enclosure, ASME-HTD Vol. 73, pp. 63 - 68, *Heat Transfer in Fire* (Ed. Kulkarni, A. K. and Jaluria, Y.), 1987.
5. Fusegi, T. and Farouk, B. Laminar and Turbulent Natural Convection-Radiation Interactions in a Square Enclosure Filled with a Nongray Gas, *Numerical Heat Transfer*, part A, Vol. 15, pp. 303 - 322, 1989.
6. Song., T. H. and Viskanta, R. Interaction of Radiation with Turbulence: Application to a Combustion System, *J. Thermophys. Heat Transfer*, Vol. 1, pp. 56 - 62, 1987.
7. Smith, T. F., Shen, Z. F. and Friedman, J. N. Evaluation of Coefficients for the Weighted Sum of Gray Gases Model, *J. Heat Transfer*, Vol. 104, pp. 602 - 608, 1982.
8. Armfield, S. W. Direct Simulation of Unsteady Natural Convection in a Cavity, pp. 305 - 310, *Proc. of Int. Symp. on Comput. Fluid Dynamics*, Nagoya, Japan, 1989 .
9. Paolucci, S. and Chenoweth, D. R. Transition to Chaos in a Differentially Heated Vertical Cavity, *J. Fluid Mech.*, Vol. 201, pp. 379 - 410, 1989.
10. Patankar, S. V. *Numerical Heat Transfer and Fluid Flow*, Chapter 6, Hemisphere, Washington, D. C., 1980.
11. Stone, H. L. Iterative Solution of Implicit Approximations of Multi-Dimensional Partial Differential Equations, *J. Numer. Anal.*, Vol. 5, pp. 530 - 558, 1968.
12. Freitas, C. J., Street, R. L., Findikakis, A. N. and Koseff, J. R. Numerical Simulation of Three-Dimensional Flow in a Cavity, *Int. J. Numer. Methods Fluids*, Vol. 5, pp. 561 - 575, 1985.
13. Shirayama, S. and Kuwahara, K. Patterns of Three-Dimensional Boundary Layer Separation, *25th Aerospace Sciences Meeting*, Reno, Nevada, AIAA Paper 87-0461, 1987.
14. Lankhorst, A. M. and Hoogendoorn, C. J. Three-Dimensional Numerical Calculations of High Rayleigh Number Natural Convective Flows in Enclosed Cavities, ASME HTD-96, Vol. 3, pp. 463 - 470, *Proc. of 1988 Nat. Heat Transfer Conf.*, Houston, Texas, 1988.
15. Markatos, N. C. and Pericleous, K. A. Laminar and Turbulent Natural Convection in an Enclosed Cavity, *Int. J. Heat Mass Transfer*, Vol. 27, pp. 755 - 772, 1984.

Approximate Matrix Inversion in the Zone Method for Radiative Heat Transfer in Nearly Black Enclosures

D.A. Lawson

Department of Mathematics, Coventry Polytechnic, Priory Street, Coventry CV1 5FB, England

ABSTRACT

When using the Hottel zone method for radiative heat transfer analyses in enclosures containing participating media it is necessary to compute the total exchange areas from the direct exchange areas. This requires the computation of the inverse of a matrix whose elements depend on the direct exchange areas and the surface zone areas and reflectivites. It is shown that when the reflectivities are small (in other words, when the enclosure is nearly black) a simple approximation of this matrix inverse can be found. Total exchange areas calculated using this approximate matrix inverse are compared with exact total exchange areas and found to be in good agreement.

INTRODUCTION

The Hottel zone method [1] is one of the most widely used methods of performing radiative heat transfer analyses in enclosures containing participating media. The basic idea of the method is to divide the bounding surfaces into a number of area zones and the enclosed gaseous medium into a number of volume zones. The temperature and thermal properties of each zone are assumed to be uniform. With this division made it is then necessary to define the direct exchange areas between zones and the total exchange areas between zones. It has become conventional to use lower case symbols such as s_is_j for direct exchange areas and upper case symbols such as S_iS_j for total exchange areas. These exchange areas are directed (in the sense that one zone is the emitting zone and the other the receiving zone) and the convention of Noble [2], that the first subscript refers to the receiving zone whilst the second refers to the emitting zone, is followed here.

The direct exchange areas are defined below; in these definitions A_j is the area of surface zone j, V_j is the volume of gas zone j and k is the absorption coefficient.

$s_i s_j / A_j$ = fraction of energy emitted by surface zone j which is directly incident on surface zone i

$g_i s_j / A_j$ = fraction of energy emitted by surface zone j which is directly absorbed by gas zone i

$g_i g_j / 4kV_j$ = fraction of energy emitted by gas zone j which is directly absorbed by gas zone i

$s_i g_j / 4kV_j$ = fraction of energy emitted by gas zone j which is directly incident on surface zone i

In these definitions the use of "directly" refers to events occurring without the radiation being reflected by any of the surfaces of the enclosure.

The direct exchange areas depend only on the geometry of the enclosure and the transmittance of the gas. They can be calculated from integral formulae given by many authors, for example Siegel and Howell [3]. These integrals are double area, double volume or area and volume integrals. Calculation of the direct exchange areas can therefore often be a laborious and computationally expensive task.

The direct exchange areas do not give a complete picture of the radiative heat transfer within an enclosure as radiation emitted from one zone may be reflected (possibly many times) before it is finally absorbed by another zone. It is therefore necessary to define total exchange areas as follows:

$S_i S_j / \epsilon_j A_j$ = fraction of radiation emitted by surface zone j which is absorbed by surface zone i

$G_i S_j / \epsilon_j A_j$ = fraction of radiation emitted by surface zone j which is absorbed by gas zone i

$G_i G_j / 4kV_j$ = fraction of radiation emitted by volume zone j which is absorbed by gas zone i

$S_i G_j / 4kV_j$ = fraction of radiation emitted by volume zone j which is absorbed by surface zone i

In these definitions ϵ_j is the emissivity of surface zone j.

From the definitions of the exchange areas it is clear that for a particular surface zone the sum of all the direct exchange areas from this zone is the zone area, whilst the sum of all the total exchange areas from this zone is the product of the zone's emissivity and area. For gas zones both these sums have the value of 4kV, where V is the volume of the particular zone considered.

There are no compact integral formulae for the total exchange areas. They can be calculated from the direct exchange areas by consideration of basic principles of radiative heat transfer. This relationship between direct exchange areas and total exchange areas can be given succinctly in matrix form. The next section which is based on the work of Noble [2] outlines this relationship.

RELATIONSHIP BETWEEN DIRECT AND TOTAL EXCHANGE AREAS

If there are n surface zones and m gas zones we can construct n×n matrices ss and SS of surface-to-surface exchange areas. Similarly gg and GG are m×m matrices of gas-to-gas exchange areas. All these matrices are symmetric. The gas-to-surface and surface-to-gas exchange areas can be assembled into matrices sg or SG (order n×m) and gs or GS (order m×n), respectively. These matrices satisfy the relations $sg = (gs)^t$ and $SG = (GS)^t$.

In this section we require the following notation. The vector of zone radiosities is represented by $\underset{\sim}{J}$, zone irradiations by $\underset{\sim}{H}$, blackbody emissive power by $\underset{\sim}{E}$ and net outward radiative flux by $\underset{\sim}{Q}$. A subscript s denotes quantities for surface zones and g for gas zones. Diagonal matrices are denoted by D_x, where the subscript x indicates the quantity which is found on the diagonal of the matrix. For example, D_ϵ is the matrix whose i,j entry is zero ($i \neq j$) and whose i,i entry is ϵ_i.

Irradiation upon a surface is made up of radiation coming from either a surface or a gas zone hence

$$D_A \underset{\sim}{H}_s = ss\underset{\sim}{J}_s + sg\underset{\sim}{J}_g \tag{1}$$

The radiosity of a surface has two components: emission and reflection and so

$$\underset{\sim}{J}_s = D_\epsilon \underset{\sim}{E}_s + D_\rho \underset{\sim}{H}_s \tag{2}$$

Here ρ denotes the reflectivity of a surface. The net outward radiative flux from a surface zone is found from the radiosity minus the irradiation giving

$$\underset{\sim}{Q}_s = D_A(\underset{\sim}{J}_s - \underset{\sim}{H}_s) = D_A D_\epsilon (\underset{\sim}{E}_s - \underset{\sim}{H}_s) \tag{3}$$

where we have used the identity $D_\epsilon + D_\rho \equiv I$ and equation (2).

Applying the same principles to gas zones leads to:

$$4kD_V \underset{\sim}{H}_g = gs\underset{\sim}{J}_s + gg\underset{\sim}{J}_g \tag{4}$$

$$\underset{\sim}{J}_g = \underset{\sim}{E}_g \tag{5}$$

$$\underset{\sim}{Q}_g = 4kD_V(\underset{\sim}{E}_g - \underset{\sim}{H}_g) \tag{6}$$

Equations (1) and (2) may be solved for $\underset{\sim}{H}_s$ to give

$$\underset{\sim}{H}_s = (D_A - ssD_\rho)^{-1}(ssD_\epsilon \underset{\sim}{E}_s + sg\underset{\sim}{E}_g) \tag{7}$$

The matrix $(D_A - ssD_\rho)^{-1}$ plays a crucial role in what follows so we define

$$R = (D_A - ssD_\rho)^{-1} \tag{8}$$

Substitution of equation (7) into equations (2) and (3) gives

$$\underset{\sim}{J}_s = (I + D_\rho Rss)D_\epsilon \underset{\sim}{E}_s + D_\rho Rsg\underset{\sim}{E}_g \tag{9}$$

$$\underset{\sim}{Q}_s = (D_\epsilon D_A - D_\epsilon D_A RssD_\epsilon)\underset{\sim}{E}_s - D_\epsilon D_A Rsg\underset{\sim}{E}_g \tag{10}$$

Equations (5) and (9) used in equation (4) produce

$$\underset{\sim}{H}_g = (1/4k)D_V^{-1}[gs(I + D_\rho Rss)D_\epsilon \underset{\sim}{E}_s + (gsD_\rho Rsg + gg)\underset{\sim}{E}_g] \tag{11}$$

and so equation (6) becomes

$$\underset{\sim}{Q}_g = [4kD_V - (gsD_\rho sg + gg)]\underset{\sim}{E}_g - gs(I + D_\rho Rss)D_\epsilon \underset{\sim}{E}_s \tag{12}$$

Alternative expressions for $\underset{\sim}{Q}_s$ and $\underset{\sim}{Q}_g$ can be found from the definitions of total exchange areas and the fact that net outward radiative flux is emission minus absorption. We have

$$\underset{\sim}{Q}_s = D_\epsilon D_A \underset{\sim}{E}_s - (SS\underset{\sim}{E}_s + SG\underset{\sim}{E}_g) \tag{13}$$

$$\underset{\sim}{Q}_g = 4kD_V\underset{\sim}{E}_g - (GG\underset{\sim}{E}_g + GS\underset{\sim}{E}_s) \tag{14}$$

Comparing equations (10) and (13) and equations (12) and (14) we find

$$SS = D_\epsilon D_A RssD_\epsilon \tag{15a}$$

$$SG = D_\epsilon D_A Rsg \tag{15b}$$

$$GG = gsD_\rho Rsg + gg \tag{15c}$$

$$GS = gs(I + D_\rho Rss)D_\epsilon \tag{15d}$$

The matrix R defined by equation (8) features in the formula for each of the four total exchange area matrices. In practice the matrix GS would not be computed as we know that $GS = (SG)^t$. (Although this is not immediately obvious from equations (15b) and (15d) it can be shown after a little algebraic manipulation.)

In order to find the net outward radiative fluxes the matrix R must be found. This means we must find the inverse of the matrix $(D_A - ssD_\rho)$. Assuming the direct exchange areas have been found then equations (15) can be used to find the total exchange areas and then equations (13) and (14) give the fluxes.

AN APPROXIMATE INVERSE

The calculation of a matrix inverse whilst not difficult is a relativley expensive computational process. In this section we examine a possible way of approximating the matrix R.

It is a well established result (see Isaacson and Keller [4]) that if $||C||<1$ (where $||.||$ is any natural matrix norm) then

$$I + C + C^2 + C^3 + \ldots + C^n \rightarrow (I-C)^{-1} \text{ as } n \rightarrow \infty$$

Now if we set $C = ssD_A^{-1}D_\rho$ then equation (8) gives

$$D_A R = (I-C)^{-1} \tag{16}$$

It is a relatively easy matter to establish that the l_1-norm of C is less than unity. The l_1-norm of a matrix is the maximum of the sums of the absolute values of the elements of a column. The i,j element of ssD_A^{-1} is $s_i s_j/A_j$ which by the definitions given earlier is the fraction of radiation emitted by zone j which is directly incident upon surface zone i. Since each of these fractions is non-negative the sum of the absolute values of the elements of column j of ssD_A^{-1} is the fraction of radiation emitted by surface j which is directly incident upon any of the surface zones. Clearly this fraction must be less than or equal to one (with equality if and only if the gas absorbs none of the radiation). Hence

$$||ssD_A^{-1}||_1 \leqslant 1 \tag{17}$$

Also, since D_ρ is a diagonal matrix

$$||D_\rho||_1 = \rho_{max} \tag{18}$$

where ρ_{max} is the greatest reflectivity of a surface zone.

By the properties of matrix norms

$$||C||_1 = ||ssD_A^{-1}D_\rho||_1 \leqslant ||ssD_A^{-1}||_1 ||D_\rho||_1 \leqslant 1 \times \rho_{max} \tag{19}$$

Assuming that no surface is purely reflecting $\rho_{max}<1$ and so $||C||_1<1$. It should be observed that inequality (19) will usually give a conservative upper bound for $||C||_1$. The first inequality will be a strict inequality unless the column of ssD_A^{-1} with the maximum sum corresponds to the surface with maximum reflectivity. Also, as observed earlier, the second inequality will be strict if the gas participates in the radiative heat transfer.

We may choose to approximate $(I-C)^{-1}$ by $I+C+C^2+\ldots+C^n$. If we define the error matrix X by

$$X = (I-C)^{-1} - (I+C+C^2+\ldots+C^n) \tag{20}$$

then

$$(I-C)X = I - (I-C^{n+1}) = C^{n+1} \tag{21}$$

giving

$$X = (I-C)^{-1}C^{n+1} \tag{22}$$

Hence

$$\|X\|_1 \leqslant \|(I-C)^{-1}\|_1 \|C^{n+1}\|_1 \tag{23}$$

Using standard results of numerical analysis (see Isaacson and Keller [4]) we have

$$\|X\|_1 \leqslant \frac{\|C\|_1^{n+1}}{1 - \|C\|_1} \tag{24}$$

This result clearly shows that $\|X\|_1 \to 0$ as $n \to \infty$ (since $\|C\|_1 < 1$).

The computation of a large number of powers of C can be more computationally expensive than computing $(I-C)^{-1}$ directly. In general, approximating the inverse by a truncated series of powers of C will not be a practical way to proceed. However, under certain conditions the norm of C may be so small that very few powers of C are needed in the truncated series for the approximation to be an accurate representation of the inverse.

If the enclosure under consideration is nearly black all the reflectivities will be small and so ρ_{max} will be small. Since ρ_{max} is an upper bound for $\|C\|_1$ this will also be small. Furthermore, if the gas in the enclosure is optically thick then the norm of the matrix ssD_A^{-1} will be considerably less than 1, as a significant proportion of the radiation emitted by any surface will be absorbed by the gas. This will also mean that $\|C\|_1$ is small. In such circumstances the bound on the norm of the error matrix given by inequality (24) may be acceptably small for small values of n such as n=2 or even n=1. The case n=2 means that $I+C+C^2$ is taken as the approximation for $D_AR=(I-C)^{-1}$. Whilst when n=1 $D_AR=(I-C)^{-1}$ is approximated by I+C. For both of these cases (and emphatically so for the case n=1) the construction of the approximation requires very little effort.

EXAMPLE CALCULATIONS

Exact values of direct exchange areas in the presence of a participating medium are available for only relatively few enclosure geometries. Siddall [5] has published the direct exchange areas for a cube containing an absorbing-emitting medium. These direct exchange areas will be used as the basis for some trial calculations to examine the accuracy of the approximations proposed in the previous section.

With a cube there is a natural division of the bounding surface into 6 zones (the faces of the cube). The symmetry of a cube means that there are essentially only three types of surface-to-surface exchange areas to consider: a) the receiving zone is opposite to the emitting zone; b) the receiving zone is adjacent to the emitting zone; and c) the receiving zone is the emitting zone. With the bounding surface so divided there is a single gas zone (the entire volume of the cube). Again symmetry means that each surface-to-gas (and gas-to-surface) exchange area is the same, whilst with only one gas zone there is only one gas-to-gas exchange area. If we consider a cube with edges of unit length containing a gas with unit absorption coefficient then the direct exchange areas given by Siddall [5] are:

Direct exchange areas where the emitting zone is a surface:

For opposite faces $ss = 0.066164$

For adjacent faces $ss = 0.121952$

For the same face $ss = 0.0$

For the gas zone $gs = 0.446028$

Direct exchange areas where the emitting zone is the gas:

For the gas zone $gg = 1.323836$

For a surface $sg = 0.446028$

It was stated in the second section of this paper that the sum of all the direct exchange areas with a particular zone as the emittor is equal to the zone area if it is a surface zone or the product of the zone volume and four times the absorption coefficient if it is a gas zone. That this is satisfied by these values is easily seen.

For a surface zone the zone area is 1 and the sum of the direct exchange areas is

$$0.066164 + 4\times0.121952 + 0.0 + 0.446028 = 1.000000$$

For a gas zone the product of the zone volume and four times the absorption coefficient is 4 and the sum of the direct exchange areas is

$$1.323836 + 6\times0.446028 = 4.000004$$

The slight error here is due to the rounding in giving the direct exchange areas to only 6 decimal places.

The total exchange areas depend on the emissivity of the surface zones. For the purpose of this validation it was assumed that all surfaces have the same emissivity. The approximation method of the previous section in no way requires this, it does however make the results easier to interpret.

Table 1 : Total Exchange Areas when ϵ=0.9, k=1

Type	Exact	Approx 1	Approx 2	Error 1	Error 2
opp SS	.05864	.05841	.05863	.229e-3	.137e-4
adj SS	.10275	.10250	.10274	.250e-3	.133e-4
same SS	.00540	.00517	.00539	.227e-3	.137e-4
GS=SG	.42497	.42366	.42489	.130e-2	.723e-4
GG	1.4502	1.4498	1.4502	.388e-3	.215e-4

In the above table approximation 1 refers to total exchange areas calculated by approximating $D_A R=(I-C)^{-1}$ by I+C whilst approximation 2 refers to use of $I+C+C^2$.

For both approximations the largest error occurs in the value of the surface-to-gas total exchange area. For approximation 1 this error represents 0.31% of the exact value of GS. For approximation 2 the percentage error is only 0.02%. In both cases it seems reasonable to assume that these errors will not have a very significant effect on the heat transfer calculation.

The largest percentage error for both approximations occurs in the values of SS for the same surface. This measures the amount of radiation emitted by a particular surface which is eventually absorbed by that same surface after at least one (and possibly many) surface reflections. For approximation 1 this percentage error is 4.2% whilst for approximation 2 it is 0.25%. Although these are relatively high when compared to the other percentage errors (being an order of magnitude greater) it should be observed that this total exchange area is very small and its effect on the overall radiative heat transfer within the enclosure will be very small.

The fact that the largest error occurs in the approximate value of GS is not unique to this value of ϵ, it seems to always be the case. In table 2 the effect upon this error of varying ϵ is presented. The results given in table 2 show that as is expected both approximations improve the more nearly black the enclosure is (ie the closer ϵ is to 1). In the case where ϵ=0.8 the actual size of the error is still very small, being 0.493e−2 for approximation 1 and 0.546e−3 for approximation 2 (the exact value of GS in this case is 0.40128).

Table 2 : Percentage Errors in GS for Different ϵ, k=1

ϵ	0.95	0.90	0.85	0.80
Approx 1	0.08	0.31	0.69	1.23
Approx 2	0.00	0.02	0.05	0.14

It was observed earlier that the definitions of total exchange area mean that the sum of all the total exchange areas which have a particular zone as the emitting zone is known (being the product of the zone area and emissivity for surface zones or the product of the zone volume and four times the absorption coefficient for gas zones.) Table 3 shows values of these sums when approximations 1 and 2 are used.

Table 3 : Sum of Total Exchange Areas from a Zone, k=1

ϵ		Surface	Gas
0.95	Approx 1	.949271	3.99795
	Approx 2	.949980	3.99995
0.90	Approx 1	.897238	3.99179
	Approx 2	.899847	3.99955
0.85	Approx 1	.844131	3.98153
	Approx 2	.849512	3.99847
0.80	Approx 1	.790180	3.96715
	Approx 2	.798912	3.99636

The exact value of the surface sum is ϵ and of the gas sum is 4. As can be seen both approximations always underestimate slightly this total exchange area sum. This is as expected because the approximations were based on truncating an infinite series all of whose terms are positive.

CONCLUSIONS

The results show that the two approximations based on truncating the infinite series expansions of D_AR after two or three terms produce results which agree very well with the exact values of the total exchange areas when the emissivities of the surface zones are close to 1. The closer the emissivitites to 1 the better the results.

It is also worth noting that for many enclosures containing a participating medium exact or very accurate values of the direct exchange areas are not available from the literature. These have to be calculated as part of the radiative heat transfer analysis. The direct exchange areas are frequently calculated using a Monte Carlo method (as described by Edwards [6]). There are then inevitably statistical errors in the values of the direct exchange areas used in the calculation of the total exchange areas. In such circumstances use of an approximation method which is quick and whose errors are only of the same size as those in the direct exchange areas is an attractive proposition.

REFERENCES

1. Hottel, H.C. and Cohen, E.S., Radiant Heat Transfer in a Gas-Filled Enclosure: Allowance for Non-Uniformity of Gas Temperature, AIChE J., Vol.4, pp. 3-14, 1958.
2. Noble, J., The Zone Mtethod: Explicit Matrix Relations for Total Exchange Areas, Int. J. Heat Mass Transfer, Vol.28, pp. 245-251, 1985.
3. Siegel, R. and Howell, J.R., Thermal Radiation Heat Transfer, Hemisphere, Washington D.C., 1981.
4. Isaacson, E. and Keller, H.B., Analysis of Numerical Methods, Wiley, New York, 1966.
5. Siddall, R.G., Accurate Evaluation of Radiative Direct Exchange Areas for Rectangular Geometries, in Heat Transfer 1986 (Ed. Tien, C.L., Carey, V.P. and Ferrell, J.K.), Vol.2, pp.751-756, Proceedings of the 8th Int. Conf. on Heat Transfer, San Francisco, CA, Hemisphere, Washington D.C., 1986.
6. Edwards, D.K., Hybrid Monte Carlo Matrix-Inversion Formulation of Radiation Heat Transfer with Volume Scattering, 23rd National Heat Transfer Conference, Denver, CO, Symposium on Heat Transfer in Fire and Combustion Systems, HTD-Vol. 45, pp. 273-278, August, 1985.

Radiative Transfer through a Nonhomogeneous Fly-Ash Cloud: Effects of Temperature and Wavelength Dependent Optical Properties

M.P. Mengüç, S. Subramaniam

Department of Mechanical Engineering, University of Kentucky, Lexington, KY 40506-0046, USA

ABSTRACT

The effects of spectral and nonhomogeneous radiative properties of fly-ash particles to radiative heat transfer through a planar layer near the walls of a combustion system are investigated. The spectral absorption and scattering coefficients as well as the scattering phase function of particles are obtained from the Lorenz-Mie theory as functions of temperature for three different particle diameters. These data are fitted to cubic spline polynomials and employed in the solution of the radiative transfer equation, which is modeled using a double spherical harmonics approximation. In the calculations, a linearly anisotropic scattering phase function is used. The results show that the effect of temperature dependent, spectral real properties of fly-ash particles on the divergence of radiative flux term, which enters into the energy equation as a source function, is significant.

INTRODUCTION

In large utility furnaces, radiation accounts for a large fraction of energy transfer from combustion products to the walls. If the fuel used is pulverized-coal, the importance of radiation is further increased, mainly because of the redistribution of radiant energy within the enclosure by absorbing, emitting and scattering particles, such as coal, char, fly-ash, and soot. Among these, the contribution of fly-ash particles on radiant energy balance is significant, since they exist everywhere in a furnace, whereas coal, char, and soot particles are usually confined in regions close to the flame.

It is very important to use accurate radiative properties of particles and corresponding volume fraction distributions for accurate modeling of heat transfer mechanism in furnaces. Also, in calculating radiation heat transfer for real systems, the spectral properties of the particles are to be used. Unfortunately, the radiative properties of most combustion products are not readily available, at least in spectral basis. An extensive review of radiation heat transfer in combustion chambers and the related literature on radiative properties of the combustion products, including fly-ash, is available [1].

The importance of fly-ash particles on radiative transfer in coal-fired furnaces has been discussed in several empirical and numerical studies [2-7]. Recently, a detailed experimental study has been conducted to obtain the spectral index of refraction data of fly-ash particles [8]. Using these data, a sensitivity analysis has been performed to determine the impact of the spectral fly-ash properties on the prediction of radiation blockage by a two-layer fly-ash cloud [9]. Although the effect of flame temperature on the spectral blockage of the cold layer was discussed in Ref. [9], the temperature dependence of the radiative properties was not considered.

In the present work, we will model a more realistic system, where a temperature profile for the plane-parallel medium will be assumed, and the radiative properties will be calculated as functions of local temperature and wavelength of radiation. Since we will consider nonhomogeneous distribution for the properties, the radiative transfer equation becomes a non-linear integro-differential equation. It will be solved using a hybrid double spherical harmonics (DP_1) approximation, which is capable of accounting for nonhomogenous medium properties.

ANALYSIS

The physical system considered in this work is a one-dimensional, plane-parallel, absorbing, emitting, and anisotropically-scattering medium with azimuthal symmetry. One side of the medium is exposed to diffuse radiation from flame inside the combustion chamber, and the other side is a black wall. Radiative properties are assumed to be nonuniform and wavelength dependent. The radiative transfer equation (RTE) for this system is written as

$$\mu\left[\partial I_\lambda(\tau_\lambda,\mu)/\partial\tau_\lambda\right] + I_\lambda(\tau_\lambda,\mu) = S_\lambda(\tau_\lambda,\mu)\,, \tag{1}$$

where the source function $S_\lambda(\tau_\lambda,\mu)$, the spectral optical thickness τ_λ, and the spectral single scattering albedo ω_λ are given as

$$S_\lambda(\tau_\lambda,\mu) = [1-\omega_\lambda(\tau_\lambda)]I_b[T(\tau_\lambda)] + [\omega_\lambda(\tau_\lambda)/4\pi]\int_{\phi'=0}^{2\pi}\int_{\mu'=-1}^{1} \Phi_\lambda(\tau_\lambda;\mu,\mu')I_\lambda(\tau_\lambda,\mu,\mu')d\mu'd\phi'\,, \tag{2}$$

$$\tau_\lambda(y) = \int_0^y \beta_\lambda(y)dy\,, \quad \omega_\lambda[\tau_\lambda(y)] = \sigma_\lambda(y)/\beta_\lambda(y)\,. \tag{3}$$

The wavelength and location dependency of the parameters will not be used from now on, but it is understood.

The double spherical harmonics (DP_1) approximation, which yields accurate predictions for radiative flux profiles in both homogeneous and nonhomogeneous layers [10] is used in this study to solve the RTE. The spherical space is divided into two domains and the intensity is represented in each domain separately using the first order spherical harmonics approximation, such as

$$M^+(\tau,\mu) = A_0 + A_1\mu\,, \quad 0 < \mu < 1\,, \tag{4}$$

$$M^-(\tau,\mu) = B_0 + B_1\mu\,, \quad -1 < \mu < 0 \tag{5}$$

We multiply Eq.(4) by 1 and μ and integrate over Ω_1 (positive hemisphere) to obtain

$$M_0^+ = \int_{\phi=0}^{2\pi}\int_{\mu=0}^{1} M^+ d\Omega = 2\pi A_0 + \pi A_1\,, \tag{6}$$

$$M_1^+ = \int_{\phi=0}^{2\pi}\int_{\mu=0}^{1} \mu M^+ d\Omega = \pi A_0 + (2\pi/3)A_1\,, \tag{7}$$

where M_0^+ and M_1^+ are the zeroth and first moments of intensity, respectively. By solving Eqs. (6) and (7) for A_0 and A_1 and substituting back into Eq. (4), we obtain

$$M^+(\tau,\mu) = \frac{1}{\pi}\left[(2M_0^+-3M_1^+) + (-3M_0^++6M_1^+)\mu\right]. \tag{8}$$

We repeat the same procedure for the second domain (negative hemisphere), which yields

$$M^-(\tau,\mu) = \frac{1}{\pi}\left[(2M_0^- + 3M_1^-) + (3M_0^- + 6M_1^-)\mu\right] \tag{9}$$

Using a linearly anisotropic scattering phase function approximation such as

$$\Phi_\lambda(\tau_\lambda;\mu,\mu') = 1 + a_{1,\lambda}(\tau_\lambda)\mu\mu' , \tag{10}$$

and after a lengthy algebraic manipulation, the RTE in each domain is expressed as [10]

$$\frac{\mu}{\tau_0}\frac{\partial M^+}{\partial \tau} + M^+ = (1-\omega)I_b(T) + \frac{\omega}{4\pi}\left[(M_0^+ + M_0^-) + a_1\mu(M_1^+ + M_1^-)\right], \tag{11}$$

$$\frac{\mu}{\tau_0}\frac{\partial M^-}{\partial \tau} + M^- = (1-\omega)I_b(T) + \frac{\omega}{4\pi}\left[(M_0^+ + M_0^-) + a_1\mu(M_1^+ + M_1^-)\right]. \tag{12}$$

We multiply each of these equations by 1 and μ and integrate over the corresponding hemispheres to obtain

$$\frac{1}{\tau_0}\frac{\partial M_1^+}{\partial \tau} = 2\pi(1-\omega)I_b + (\frac{\omega}{2}-1)M_0^+ + \frac{\omega}{2}M_0^- + \frac{\omega a_1}{4}M_1^+ + \frac{\omega a_1}{4}M_1^- \tag{13}$$

$$\frac{1}{\tau_0}\frac{\partial M_0^+}{\partial \tau} + M^- = 6\pi(1-\omega)I_b + (\frac{3\omega}{2}-6)M_0^+ + \frac{3\omega}{2}M_0^- + (\frac{\omega a_1}{2}+6)M_1^+ + \frac{\omega a_1}{2}M_1^- \tag{14}$$

$$\frac{1}{\tau_0}\frac{\partial M_1^-}{\partial \tau} = 2\pi(1-\omega)I_b + \frac{\omega}{2}M_0^+ + (\frac{\omega}{2}-1)M_0^- - \frac{\omega a_1}{4}M_1^+ - \frac{\omega a_1}{4}M_1^- \tag{15}$$

$$\frac{1}{\tau_0}\frac{\partial M_0^-}{\partial \tau} = -6\pi(1-\omega)I_b - \frac{3\omega}{2}M_0^+ + (6-\frac{3\omega}{2})M_0^- + \frac{\omega a_1}{2}M_1^+ + (6+\frac{\omega a_1}{2})M_1^- \tag{16}$$

These equations are to be solved simultaneously with appropriate boundary conditions in terms of the moments. These moments are then employed to evaluate the coefficients of intensity in each domain, as well as the heat fluxes.

Boundary Conditions

We impose Marshak's boundary relations on both of the boundary surfaces, which reads

$$\int_{\Omega_j} I(\tau,\mu)\ell d\Omega_j = \int_{\Omega_j} h_w \ell d\Omega_j \qquad (j = 1,2) , \tag{17}$$

where ℓ is the direction cosine. The h_w functions for diffusely emitting and reflecting surfaces are given as

$$h_w = \epsilon_w I_b(T_w) + (\rho_w^d/\pi) \int_{\Omega'=2\pi} I(\tau,\mu')\ell d\Omega' , \tag{18}$$

which can be written for this problem as

$$h_1 = I_b(T_1) \qquad h_2 = I_b(T_2)\,. \tag{19}$$

We substitute for h_1 and h_2 into Eqs. (19) and integrate over each domain after multiplying by $\ell = 1$ and μ to get the boundary conditions

$$M_0^+ = 2\pi h_1\,, \quad M_1^+ = \pi h_1\,, \quad M_0^- = 2\pi h_2\,, \quad M_1^- = -\pi h_2\,. \tag{20}$$

Radiation Heat Flux

The radiative heat flux at any given point in the medium is expressed as

$$q^r = \int_{4\pi} I\mu d\Omega = \int_{\phi=0}^{2\pi} \int_{\mu=-1}^{1} I\mu d\Omega = M_1^+ + M_1^-\,. \tag{21}$$

In order to couple radiation with convection or conduction heat transfer, the divergence of radiative heat flux distribution in the medium is required. The divergence term enters into the energy equation as the source term, and is written as

$$\partial q^r(\tau)/\partial\tau = \partial M_1{}^+/\partial\tau + \partial M_1{}^-/\partial\tau\,. \tag{22}$$

Once the M moments are determined, the flux and divergence of flux are calculated.

Solution Method

Equations (13)-(16) are solved along with the boundary conditions Eqs. (20) using the software package DISPL2 [11] which was developed at the Argonne National Laboratory. The package is capable of handling non-linear systems of second-order partial differential equations. Also, non-linear boundary conditions can be incorporated in the computer code.

DISPL2 is used by specifying the number of grid points, and by using a polynomial approximation between the grid points. These two parameters determine the accuracy of the solution. Once these are specified, the various array sizes required by the program are calculated. The equations and the boundary conditions are then input into the built-in routines.

RESULTS AND DISCUSSION

In this section, we will present results to show the effect of temperature dependent optical properties of fly-ash particles (i.e., complex index of refraction) on the spectral radiative properties (efficiency factors, albedo) and radiative heat transfer parameters (radiative flux and its divergence). Spectral complex index of refraction data for several different fly-ash samples were obtained by Goodwin [8]. Here, we will consider only the sample categorized as SD05 [8]. We chose to use this slag, because it has about the same iron content as natural slags. Note that absorption of radiation by fly-ash particles at wavelengths less than 4μm is mainly due to iron content.

Fly-ash complex index of refraction displays a strong wavelength dependency, as shown in Fig. 1 [8]. The real part of the refractive index is a weak function of temperature, and its variation is in the order of 10^{-5}/K. The imaginary part, however, has strong temperature dependence at larger wavelengths, as shown in Fig. 2.

Radiative properties for fly-ash particles, i.e., absorption, extinction, and scattering efficiency factors and scattering phase function are calculated spectrally using the corresponding index of refraction data in the Lorenz-Mie theory [12]. The scattering phase function is expressed in terms of Legendre polynomials; however, in this study only the first term is considered (see Eq.(10)). It was shown that this approximation yields very accurate predictions for highly-forward scattering particles when used with the DP_1-approximation [10]. Here, we have to indicate that in most radiative transfer

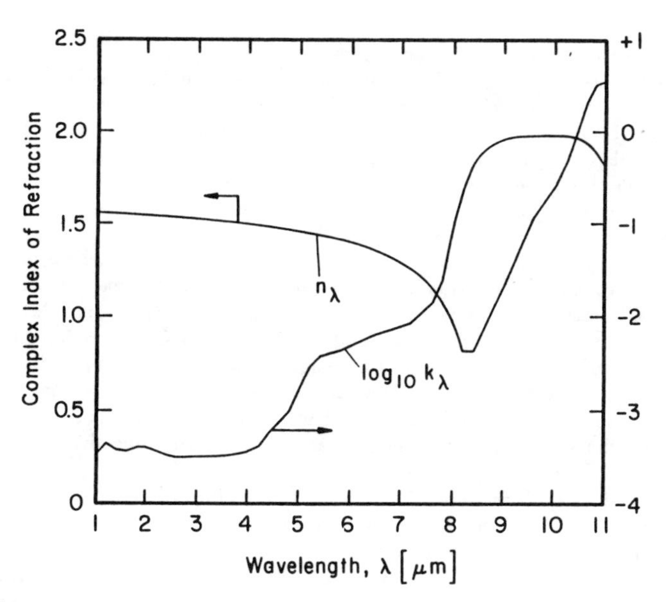

Fig. 1
Spectral variation of the real and imaginary parts of the complex index of refraction of SD05 slag [8].

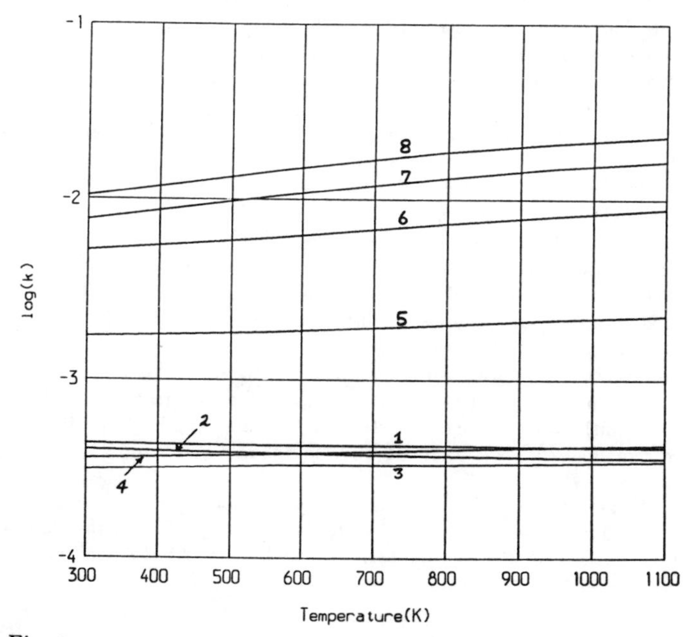

Fig. 2
Temperature dependence of the imaginary part of the complex index of refraction of SD05 slag [8].

models, a similar simple phase function approximation is employed [1].

In combustion systems the size of the fly-ash particles is not known accurately; although, there are several theories concerned with order-of-magnitude of the particle sizes [13]. It has been conclusively established that the particles are polydisperse, rather than monodisperse; however, the size distribution of fly-ash particles is not readily available. In this paper, we assume that particles are monodisperse, and three different effective diameters for the fly-ash particles in the cloud are considered. This assumption helps to identify size effects more clearly.

Extinction efficiency factor ($Q_{e,\lambda}$) is a measure of how effectively a spherical particle removes energy from incident radiation beam. Extinction (or attenuation) coefficient of a cloud of monodisperse particles is calculated by multiplying $Q_{e,\lambda}$, particle cross-section and number density. In Figs. 3, 4, and 5, variation of $Q_{e,\lambda}$ as a function of temperature is shown for several different wavelengths, for 1, 5, and 10 μm diameter particles, respectively. The temperature dependence of Q_e is negligible for the entire range considered; although, it is a strong function of λ. Note that, $Q_{e,\lambda}$ is approaching 2 at small wavelengths with increasing particle diameter (i.e., increasing size parameter $x = \pi D/\lambda$).

In Figs. 6, 7, and 8, the variation of single scattering albedo (ω_λ) as a function of temperature is shown for 1, 5, and 10 μm particles, respectively. Albedo is defined by Eq. (3), and for monodisperse particle cloud it is equal to the ratio of the scattering efficiency factor to the extinction efficiency factor. For larger particles, scattering is more pronounced than that for smaller particles, except at 8 μm wavelength. The temperature dependence is not significant at wavelengths less than 6 μm; however, it is noticeble for λ= 6 and 7 μm. From Wien's law, one can deduce that if the temperature of incident radiation is less than 500 K, then the scattering of radiation by fly-ash particles will be more sensitive to the temperature variations.

In an earlier study, it was shown that spectral variation of $\hat{n}$ affects predictions significantly if the volume fraction distribution for fly ash particles is known accurately [9]. Here, we want to assess how critical is the use of temperature dependent complex index of refraction data to predict radiative heat transfer through a cloud of fly-ash particles.

Because of the insufficient data available in the literature, many researchers have assumed that fly-ash properties can be approximated by a single, "constant" index of refraction value. In the earlier studies, values of $\hat{n}$=1.50-0.012i [6], 1.50-0.005i, 1.50-0.05i [7], and 1.50-0.02i [15-17] were used for fly-ash particles. Here, we want to compare the profiles for the radiative flux and divergence of flux obtained from the spectral complex index of refraction data to those predicted using constant values.

In order to determine the effects of the temperature dependent ω_λ and $Q_{e,\lambda}$ on radiative transfer predictions, β_λ and ω_λ were calculated as functions of location within the one-dimensional cloud of fly-ash particles. It was assumed that the walls of the furnace were at 500 K and black. The outer boundary was assumed to receive diffuse radiation from the flame at temperature of 1000 K. As expected, for wavelengths at which ω_λ did not vary significantly with temperature, the single scattering albedo profile within the fly-ash cloud remained constant. Only at wavelengths of 6 and 7 μm a slight variation in ω_λ was observed.

Using ω_λ and β_λ values calculated as a function of location within the medium, we determined the radiative flux and divergence of radiative flux profiles in the slab. The volume fraction of fly-ash particles was assumed to be $f_v = 2.5\text{x}10^{-6}$ m^3/m^3. The extinction coefficient was calculated from

$$\beta_\lambda = \frac{3}{2}\frac{Q_{e,\lambda} f_v}{D} \tag{23}$$

Fig. 9 depicts the variation in q^r for three different particles. The results were obtained using either a temperature dependent set of properties or a set of average properties. It is apparent from these results that the temperature dependent properties do

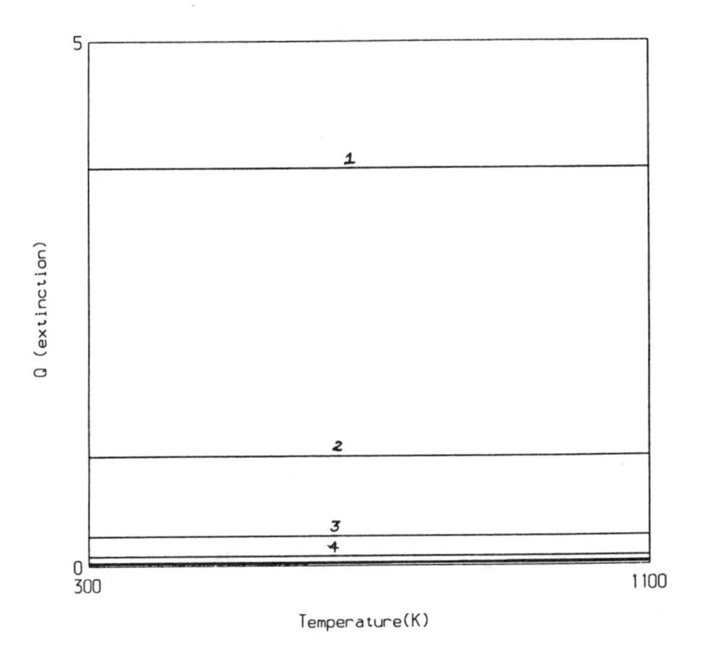

Fig. 3
Temperature and wavelength dependence of the extinction efficiency factor, $Q_{e,\lambda}$, for 1 μm diameter particles. λ in μm.

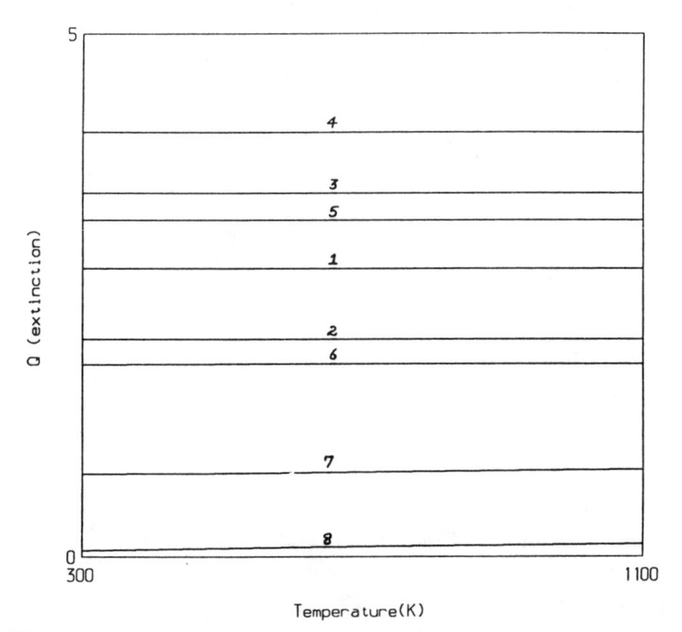

Fig. 4
Temperature and wavelength dependence of the extinction efficiency factor, $Q_{e,\lambda}$, for 5 μm diameter particles. λ in μm.

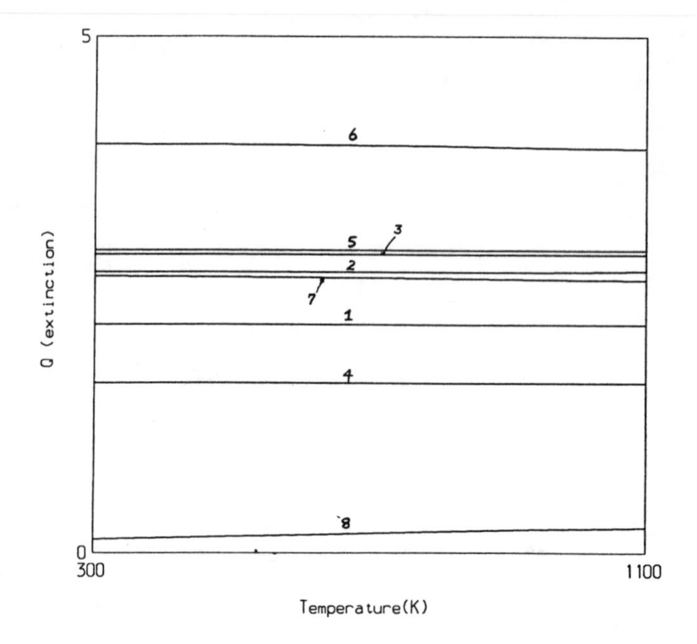

Fig. 5
Temperature and wavelength dependence of the extinction efficiency factor, $Q_{e,\lambda}$, for 10 μm diameter particles. λ in μm.

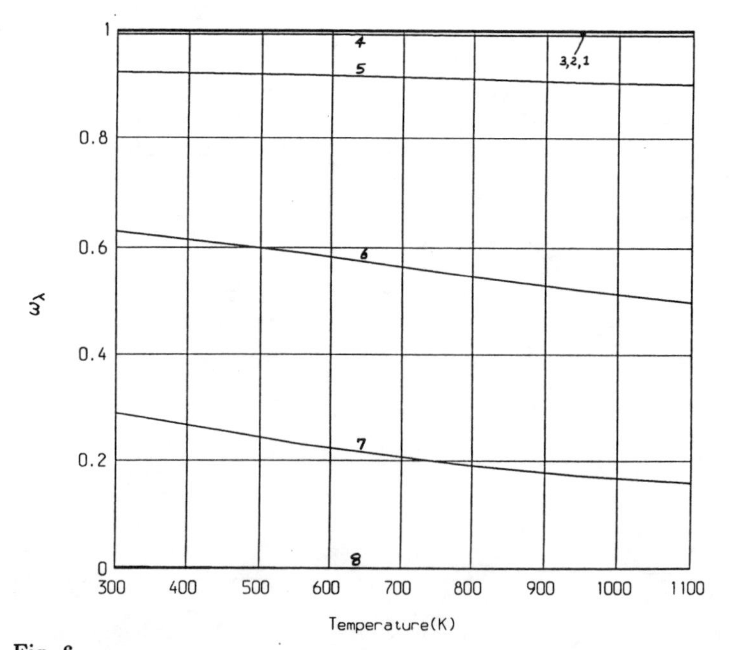

Fig. 6
Temperature and wavelength dependence of the single scattering albedo, ω_λ, for 1 μm diameter particles. λ in μm.

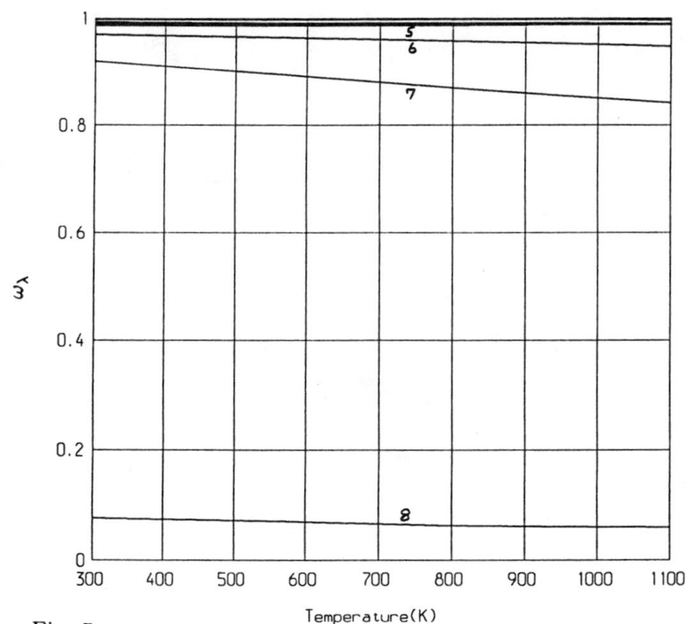

Fig. 7
Temperature and wavelength dependence of the single scattering albedo, ω_λ, for 5 μm diameter particles. λ in μm.

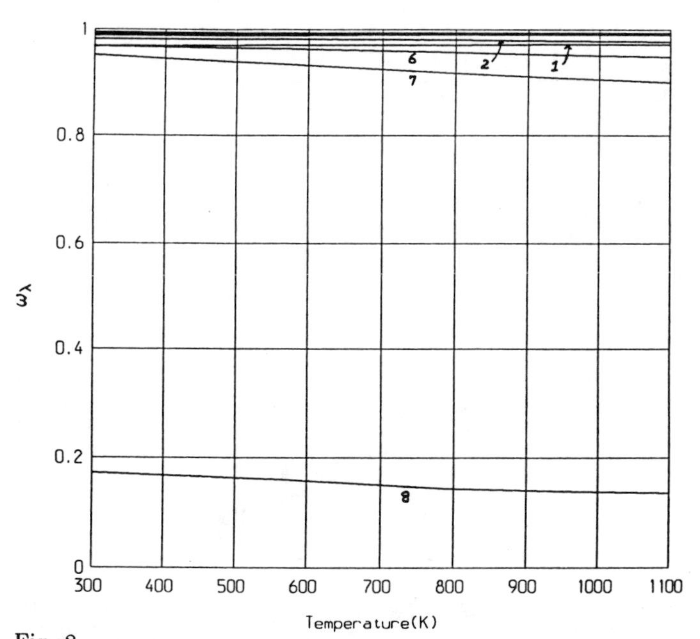

Fig. 8
Temperature and wavelength dependence of the single scattering albedo, ω_λ, for 10 μm diameter particles. λ in μm.

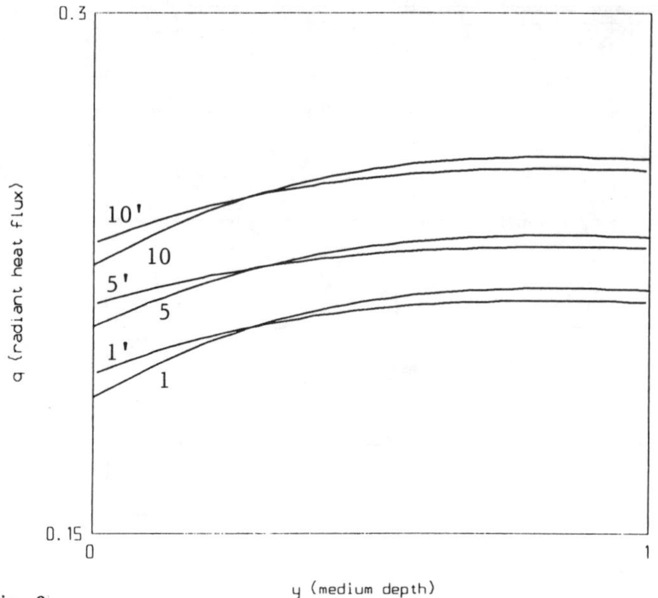

Fig. 9
Comparison of radiative flux distributions within the fly-ash cloud calculated using temperature dependent and constant radiative properties (denoted with prime). For 1, 5, 10 μm diameter particles.

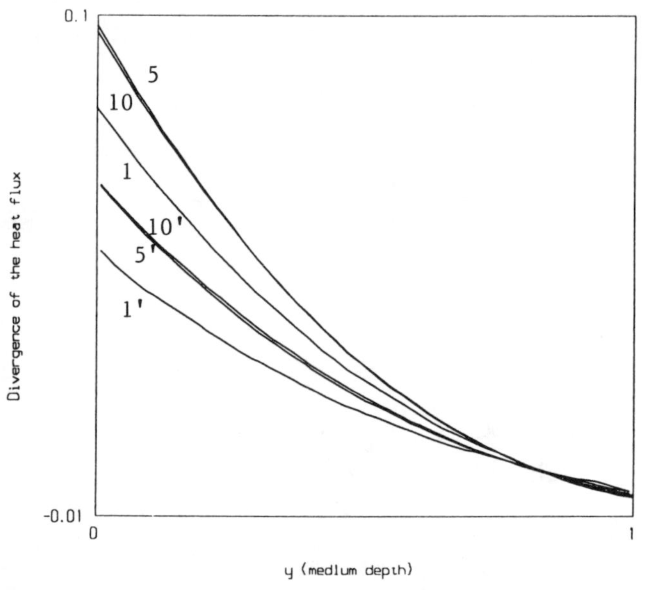

Fig. 10
Comparison of the divergence of radiative flux distributions within the fly-ash cloud calculated using temperature dependent and constant radiative properties (denoted with prime). For 1, 5, 10 μm diameter particles.

not yield results much different than those obtained from the average values.

Fig. 10 depicts the divergence of radiative flux profiles calculated for average and temperature dependent radiative properties. As seen from this figure, there are significant variation in predictions especially near the cold walls. The main reason for this is that with decreasing temperature, the effect of properties at longer wavelengths become stronger. It is worth to note that the divergence term $\partial q^r/\partial \tau$ is more important than the radiative flux, because it enters into energy equation to account for radiative heating/cooling. Without an accurate prediction for $\partial q^r/\partial \tau$ profile, it is not possible to calculate radiation-convection or radiation-conduction interaction accurately.

CONCLUSIONS

In this paper, the effect of temperature dependent optical properties of fly-ash particles on radiative properties as well as radiative transfer calculations through a a layer of fly-ash particles near the walls of a coal-fired furnace was discussed. The layer was exposed to diffuse radiation which originates from hot, optically thick flame region. The calculations were performed on spectral basis using spectral complex index of refraction data of fly-ash.

It was shown that fly-ash radiative properties were temperature dependent only for wavelengths larger than 6 μm. This means that if the medium temperature (or, the temperature of the incident radiation) is high (greater than about 600 K) and uniform, then temperature dependence of fly-ash particles can be safely neglected. However, if there is a temperature profile in the medium and if the local temperatures drops to about 400 K, then the temperature dependent radiative properties should be considered. The wavelength dependence of the radiative properties are also very important, as shown here as well as elsewhere [8,9].

REFERENCES

1. Viskanta, R. and Menguc, M.P., "Radiation Heat Transfer in Combustion Systems", *Progress in Energy and Combustion Sciences, Vol.13, pp. 97-160,* 1987.
2. Lowe, A., Wall, T.F., and Stewart, I.McC., "Combustion Kinetics in the Modeling of Large, Pulverized-Fuel Furnaces: A Numerical Experiment in Sensitivity", *AIChE Journal,* Vol.23, pp. 440-448, 1977.
3. Macek, A., "Coal Combustion in Boilers: A Mature Technology Facing New Constraints", in *Seventeenth Symposium (International) on Combustion,* The Combustion Institute, Pittsburgh, PA, pp. 65-74, 1979.
4. Gupta, R.P., Wall, T.F., and Truelove, J.S., "Radiation Scatter by Fly-Ash in Pulverized-Coal Fired Furnaces: Applications of the Monte Carlo Method to Anisotropic Scatterer", *International Journal of Heat and Mass Transfer,* Vol. 26, pp. 1649-1660, 1983.
5. Boothroyd, S.A. and Jones, A.R., "Radiative Transfer Scattering Data Relevant to Fly-Ash", *Journal of Physics D: Applied Physics,* Vol.17, pp. 1107-1114, 1984.
6. Gupta, R.P. and Wall, T.F., "The Optical Properties of Fly-Ash in Coal Fired Furnaces", *Combustion and Flame,* Vol. 61, pp. 145-151, 1985.
7. Wall, T.F., Phelan, W.J., and Bortz, S., "Coal Burnout in the IFRF No. 1 Furnace", *Combustion and Flame, Vol. 66, pp. 137-150, 1986.*
8. Goodwin, D.G., *Infrared Optical Constants of Coal Slags,* Technical Report T-255, Stanford University, CA, 1986.
9. Menguc, M.P. and Viskanta, R., "Effect of Fly-Ash Particles on Spectral and Total Radiation Blockage", *Combustion Science and Technology,* Vol.60, pp. 97-115, 1988.
10. Menguc, M.P. and Iyer, R.K., "Modeling of Radiation Heat Transfer Using Multiple Spherical Harmonics Approximations", *J. Quantitative Spectroscopy and*

Radiative Transfer, Vol.39, pp. 445-461, 1988.

11. Leaf, G.K. and Minkoff, M., "DISPL2-A Software Package for One and Two Spatially Dimensioned Kinetics-Diffusion Problems", ANL-84-56, Argonne National Laboratory, Argonne, IL, 1984.
12. Bohren, F. and Huffman, D.R., *Absorption and Scattering of Light by Small Particles,* John Wiley, New York, 1983.
13. Flagan, R.C. and Friedlander, S.K., "Particle Formation in Pulverized Coal Combustion - A Review ", *Recent Developments in Aerosol Science,* edited by D.T. Shaw, pp.25-59, Wiley, New York, 1978.
14. Viskanta, R., Ungan, A., and Menguc, M.P., "Predictions of Radiative Properties of Pulverized Coal and Fly-Ash Polydispersions", ASME Paper No: 81-HT-24, 1981.
15. Menguc, M.P. and Viskanta, R., "A Sensitivity Analysis for Radiative Heat Transfer in a Pulverized-Coal Fired Furnace", *Combustion Science and Technology,* Vol. 51, p. 51, 1986.
16. Varma, K.R. and Menguc, M.P., "Effects of Particulate Concentrations on Temperature and Heat Flux Distributions in a Pulverized-Coal Furnace", *Int. J. Energy Research,* Vol. 13, pp. 555-572, 1989.

Numerical Experimentation of the Sn-Approximation for the Radiative Transfer Equation

L. Castellano(*), S. Pasini(**)

(*) MATEC, Modelli Matematici, 20146 Milano, Italy

(**) ENEL, Centro Ricerche Termiche e Nucleari, 56100 Pisa, Italy

ABSTRACT

The Discrete Ordinates method (S_n-Approximation) has been applied for modeling radiative heat transfer in two-dimensional enclosures with participating media. The main purpose of the study was to investigate the ray effects for non-rectangular geometries and non-isothermal bounding walls. The numerical experiments performed show that: a) unacceptable anomalies in the scalar flux distribution can arise also in the case of scattering gas and diffusely-reflecting walls; b) the qualitative improvements due to the use of finite difference weighting factor greater than 0.5 are consequences of the numerical diffusion related to the truncation error. It was also observed that the S_n-Approximation always converges; but sometimes the obtained solution suffers from unacceptable error in the overall balance for radiative energy. On the other hand, unphysical solutions which properly satisfy the overall balance have been also observed.

INTRODUCTION

The major obstacle confronting the researcher in the mathematical simulation of systems such as combustion chambers, furnaces and industrial boilers, is the solution of coupled equations for fluid flow, heat transfer and chemical reactions.

As far as computer memory requirements and computational efforts are concerned, the key point is the degree of compatibility between the available approximations for the Radiative Transfer Equation (RTE) and the Finite Differences (FD) or Finite Elements (FE) models required for flow-field calculations.

From this point of view, the comparison of the main radiative heat transfer models suggests that the better choice should consist in using the Discrete Ordinates Method (S_n-Approximation).
This conclusion is not new. In fact, a number of computer codes for nuclear reactor safety analysis combine the FD approximation of Balance Equations for Multi-Phase, Multi-Component flows with the S_n-Approximation of the Neutron Transport Equation, e.g. Bell [1].
Unfortunately the S_n-Approximation, which also makes it easy to describe complicated geometries, e.g. Castellano [2], Yucel [3] and sophisticated optical models, e.g. Fiveland [4], suffers from the so-called "ray effects", as was demonstrated by Lathrop [5], which may cause generation of significantly inaccurate and physically inacceptable results. These effects are especially pronounced when localized radiation sources (e.g. Viskanta [6]), or non-rectangular enclosures and non-isothermal bounding walls are considered.
This paper deals with heuristic analysis of the above-mentioned unwanted results. Our goal is to investigate the extent of ray effects on the prediction of radiative flux distribution on the walls of irregular-shaped geometries. In order to isolate the geometric effects from those due to medium-nonhomogeneities, we consider uniform medium properties. We investigate both absorbing as well as absorbing-scattering media. It must be noted that the intention of the present paper is not a criticism of the S_n-Approximations for the solution of RTE. This account should be interpreted as a contribution to the correct use of the method.

GOVERNING EQUATIONS

The present study considers a single emitting-absorbing and scattering gray gas in radiative equilibrium. The surface bounding the participating medium is assumed to be gray and emits and reflects diffusely. All the thermophysical and optical properties are assumed to be constant.
With the above assumptions the governing equations are

$$\nabla \cdot \underline{q}_R = K\,(4\pi I_b - \int_{\omega=4\pi} I d\omega) = 0 \tag{1}$$

$$\underline{q}_R = \int_{\omega=4\pi} I \underline{s} d\omega \tag{2}$$

$$(\underline{s}\cdot\nabla)I = -\beta I + \beta(1-\gamma)I_b + \frac{\beta\gamma}{4\pi}\int_{\omega'=4\pi} \varphi(\underline{s}'\!\to\!\underline{s}) I'(\underline{x},\underline{s}')d\omega' \tag{3}$$

with the boundary conditions

$$T(\underline{x}) = T_w \quad \underline{x} \in S_w \tag{4}$$

$$I(\underline{x},\underline{s}) = \varepsilon_w I_b(T_w) + \frac{\rho_w}{4\cdot\pi}\int_{\underline{n}'\cdot\underline{s}<0} |\underline{n}\cdot\underline{s}|\ I(\underline{x},\underline{s}')d\omega' \quad \underline{x}\varepsilon S_W \tag{5}$$

S_n-APPROXIMATION

Background

The S_n-Approximation of RTE has already been documented in numerous papers; e.g. Carlson [7], Hyde [8], Fiveland [9,10,12], Truelove [11,13].

We will not discuss the details here, interest readers should refer to the previous publications. In the present study the S_n angular quadrature scheme proposed by J.S. Truelove [11] has been adopted. The numerical model is based on the FD approximation described by W.A. Fiveland [12]. The computational cell for 2D plane problems is shown in Fig.1.

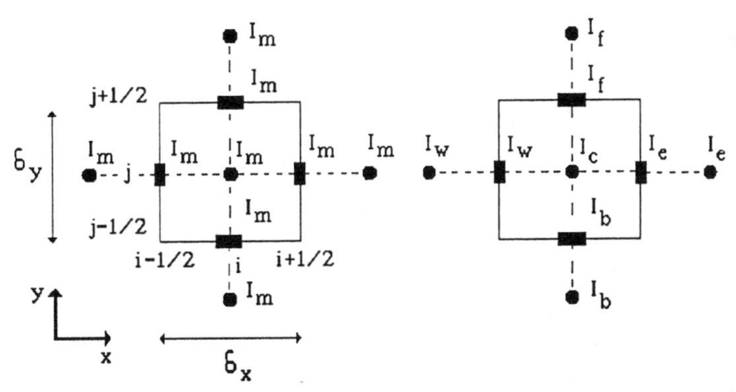

Fig. 1 - Computational cell for 2D approximation

The computational procedure consists of a point-by-point solution of the algebraic equations given for radiation intensity:

$$(I_C)_m = \frac{|\xi_m|A_{yc}(I_{xi})_m+|\eta_m|A_{xc}(I_{yi})_m+\alpha V_c(S_{1c}+S_{2c})}{|\xi_m|A_{yc}+|\eta_m|A_{xc}+\alpha\beta V_c} \tag{6}$$

$$(I_{xu})_m = \frac{(I_C)_m-(1-\alpha)(I_{xi})_m}{\alpha}$$

$$(I_{yu})_m = \frac{(I_C)_m-(1-\alpha)(I_{yi})_m}{\alpha} \tag{7}$$

where

$$A_{xc} = \delta x_C \tag{8}$$

$$A_{yc} = \delta y_C \tag{9}$$

$$I_{xi} = \begin{cases} I_w & \text{if } \xi_m > 0 \\ I_e & \text{if } \xi_m < 0 \end{cases} \tag{10}$$

$$I_{yi} = \begin{cases} I_b & \text{if } \eta_m > 0 \\ I_f & \text{if } \eta_m < 0 \end{cases} \tag{11}$$

$$I_{xu} = \begin{cases} I_e & \text{if } \xi_m > 0 \\ I_w & \text{if } \xi_m < 0 \end{cases} \tag{12}$$

$$I_{yu} = \begin{cases} I_f & \text{if } \eta_m > 0 \\ I_b & \text{if } \eta_m < 0 \end{cases} \tag{13}$$

$$S_{1c} = \beta\ (1-\gamma)\, \sigma\, T^4/\pi \tag{14}$$

$$S_{2c} = \frac{\beta\gamma}{4\pi} \sum_{m'} \varphi_{m'm} (I_C)_{m'} \tag{15}$$

$$V_c = \delta x_C \delta y_C \tag{16}$$

for each of the discrete directions $\underline{s}_m(\xi_m, \eta_m, \zeta_m)$.
According to the assumption of linear anisotropic scattering, the shape parameters $\varphi_{m'm}$ are

$$\varphi_{m'm} = 1.+a_o(\xi_m\xi_{m'}+\eta_m\eta_{m'}+\zeta_m\zeta_{m'}) \tag{17}$$

Equations (6), (7) are the results of the substitutions

$$\left(\frac{\partial I_m}{\partial x}\right)_C \cong \frac{(I_e)_m-(I_w)_m}{\delta x_C} \tag{18}$$

$$\left(\frac{\partial I_m}{\partial y}\right)_C \cong \frac{(I_f)_m-(I_b)_m}{\delta y_C} \tag{19}$$

$$(I_C)_m = \alpha\ (I_{xu})_m + (1-\alpha)(I_{xi})_m \tag{20}$$

$$(I_C)_m = \alpha\ (I_{yu})_m + (1-\alpha)(I_{yi})_m \tag{21}$$

in eq. (3).

The scheme of the solution procedure for (ξ_m>0, η_m>0) is shown in Fig. 2; the extensions to cases (ξ_m>0, η_m<0), (ξ_m<0, η_m>0), (ξ_m<0, η_m<0) are straightforward.

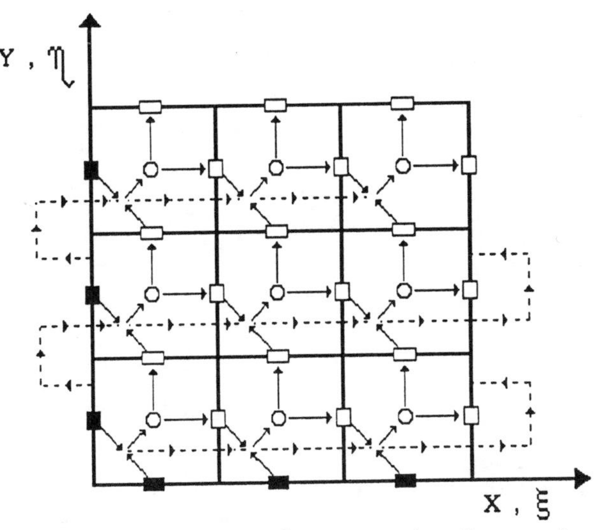

Fig. 2 - Scheme of computational procedure

Because the temperature of the gas is unknown and in-scattering source terms and boundary conditions depend on incoming intensity, an iterative procedure is needed. Iterations are continued until the net radiant heat fluxes at the walls satisfy the following condition

$$\frac{|q_{Rn}^{new} - q_{R1}^{old}|}{q_{Rn}^{old}} \times 100 < \varepsilon_{conv} \qquad (22)$$

where ε_{conv} is a preselected tolerance.

Numerical Experiments

The implemented computer code (STRODI) reproduced quite well [2], all the numerical experiments described by Truelove [11]. It has been also observed that the use of the S_4 angular quadrature scheme proposed by Truelove [11], improves the solutions described by Fiveland [9]. Interesting results have been also obtained for the complex enclosures, e.g. [2].
The objective of the present study is to focus on the ray effects, due to the solution scheme. Thus, in order to bring out these effects, the complicated irregular geometry sketched in Fig. 3 is examined.
All the numerical studies were performed by dividing the

enclosure into 20 uniformly spaced grid points in each direction. The iterative procedures were stopped when the condition (22) was satisfied with ε_{conv}=0.01.

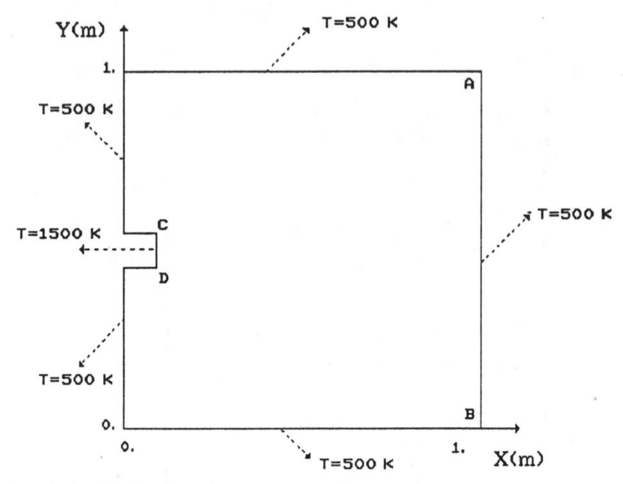

Fig. 3 - First 2D enclosure analyzed in the present study

Figure 4 shows the non-dimensional net radiative flux profile at cold wall AB, when a non-scattering gas and black walls are considered, and the finite-difference weighting factor is equal 0.5 (Lathrop diamond scheme). Two different absorption coefficients have been considered, which yield optical thickness of 0.1 and 1.0 for 1mx1m enclosure. The results show that on the AB wall there is a significant drop of the heat flux at the center.

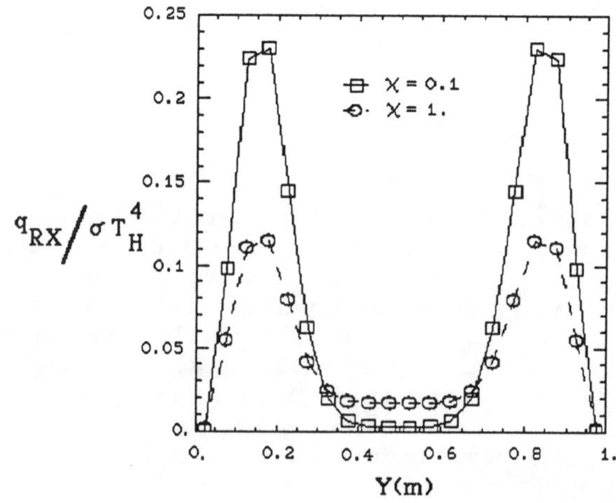

Fig. 4 - Nondimensional net radiative flux at wall AB; $\bar{\sigma}$=0. (m^{-1}), ε_w=0., ϱ_w=0., T_H=1500 K, α=0.5

The reason for these unphysical results can be found in Fig. 5 where the computational cells invested by the radiative energy outgoing from the hot wall CD are marked out at the first iteration step for each discrete direction (a,b,c in the figure).

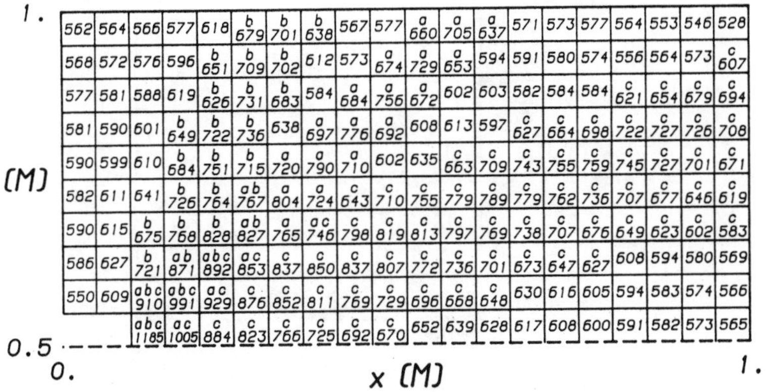

1.

y [M]

562	564	566	577	618	b 679	b 701	b 638	567	577	a 660	a 705	a 637	571	573	577	564	553	546	528
568	572	576	596	b 651	b 709	b 702	612	573	a 674	a 729	a 653	594	591	580	574	556	564	573	c 607
577	581	588	619	b 626	b 731	b 683	584	a 684	a 756	a 672	602	603	582	584	584	c 621	c 654	c 679	c 694
581	590	601	b 649	b 722	b 736	638	a 697	a 776	a 692	608	613	597	c 627	c 664	c 698	c 722	c 727	c 726	c 708
590	599	610	b 684	b 751	b 715	a 720	a 790	a 710	602	635	c 663	c 709	c 743	c 755	c 759	c 745	c 727	c 701	c 671
582	611	641	b 726	b 764	ab 767	a 804	a 724	c 643	c 710	c 755	c 779	c 789	c 779	c 762	c 736	c 707	c 677	c 646	c 619
590	615	b 675	b 768	b 828	ab 827	a 765	ac 746	c 798	c 819	c 813	c 797	c 769	c 738	c 707	c 676	c 649	c 623	c 602	c 583
586	627	b 721	ab 871	abc 892	ac 853	c 837	c 850	c 837	c 807	c 772	c 736	c 701	c 673	c 647	c 627	608	594	580	569
550	609	abc 910	abc 991	ac 929	c 876	c 852	c 811	c 769	c 729	c 696	c 668	c 648	630	616	605	594	583	574	566
		abc 1185	ac 1005	c 884	c 823	c 766	c 725	c 692	c 670	652	639	628	617	608	600	591	582	573	565

0.5

0. x [M] 1.

Fig. 5 - Computational cells invested by radiative energy outgoing from wall CD (iteration 1) and temperature distribution at end of the overall computation cycles; $\alpha=0.5$, $K=1.$ (m^{-1}); a) $\xi=.29588$ $\zeta=.90825$; b) $\xi=.29588$ $\eta=.90825$ $\zeta=.29588$; c) $\xi=.90825$ $\eta=.29588$ $\zeta=.29588$.

In spite of the iterative procedure, the effects of the preferencial paths drawn in Fig. 5 still persist, as proved by the spatial oscillations in temperature distribution calculated at the end of the overall computational cycle.
It must be also noted that the finite-difference weighting factor $\alpha=0.5$ satisfies the first stability condition proposed by Fiveland [12]

$$\left.\begin{aligned} \delta x_c &< |\xi_m|/\beta(1-\alpha) \\ \delta y_c &< |\eta_m|/\beta(1-\alpha) \end{aligned}\right\} \quad \forall\ m \tag{23}$$

Figures 6, 7 show the results obtained when the parameter α is selected in such a way as to satisfy the constraints suggested by Fiveland [10,12]:

$$\left.\begin{aligned} \delta x_c &< |\xi_m|\ \psi/\beta(1-\alpha) \\ \delta y_c &< |\eta_m|\ \psi/\beta(1-\alpha) \\ \psi &= (\alpha^3+(1-\alpha)^2(2-5\alpha))/\alpha \end{aligned}\right\} \tag{24}$$

and equal 1.0 respectively.
Although the results are still unsatisfactory, a progressive

improvement can be observed. This improvement, confirmed by a minor error in the overall balance in the radiative energy (see TABLE 1) is a consequence of the truncation error connected with the use of a first order scheme.

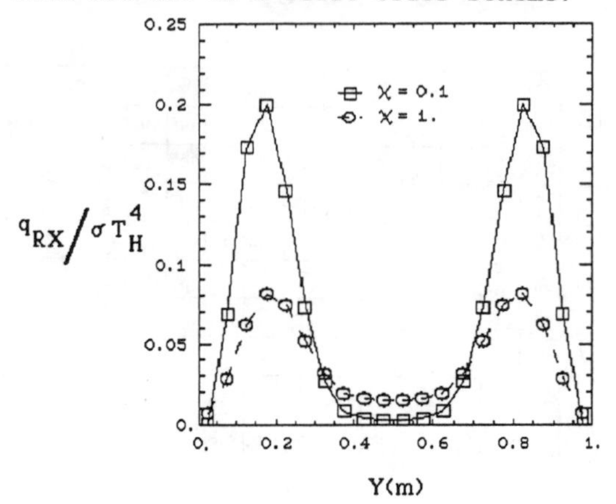

Fig. 6 - Nondimensional net radiative flux at wall AB; $\bar{\sigma}$=0.(m^{-1}), ε_w=0., T_H=1500 K, α according to stability conditions eq. (24)

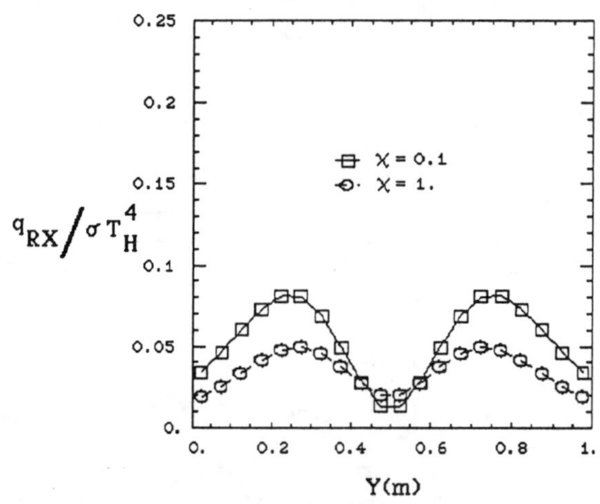

Fig. 7 - Nondimensional net radiative flux at wall AB; $\bar{\sigma}$=0. (m^{-1}), ε_w=0., T_H=1500 K, α=1.0

Indeed, from the substitution

$$I_e = I_C + (1-\alpha)\delta x_C \left(\frac{\partial I}{\partial x}\right)_C + (1-\alpha)^2 \frac{\delta x_C^2}{2!} \left(\frac{\partial^2 I}{\partial x^2}\right)_C \ldots$$

$$I_w = I_c - \alpha\delta x_c\left(\frac{\partial I}{\partial x}\right)_c + \alpha^2\frac{\delta x^2_c}{2!}\left(\frac{\partial^2 I}{\partial x^2}\right)_c + \ldots \tag{25}$$

into (18), it follows that

$$\xi\frac{I_e - I_w}{\delta x_c} = \left(\xi\frac{\partial I}{\partial x}\right)_c + \xi\frac{1-2\alpha}{2!}\delta x_c\left(\frac{\partial^2 I}{\partial x^2}\right)_c + \ldots \tag{26}$$

The diffusion-type effects of the truncation error are shown in Figs. 8 and 9.

Y(m): 1. (top) – 0.5 (bottom); X(m): 0. – 1.

				b	b	b	b		a	a	a	a	a						
			b	b	b	b		a	a	a	a	a	a						c
			b	b	b	b	a	a	a	a	a	a				c	c	c	c
			b	b	b	ab	a	a	a	a	a		c	c	c	c	c	c	c
		b	b	b	ab	a	a	a	a	a	c	c	c	c	c	c	c	c	c
		b	b	ab	ab	a	a	ac	ac	c	c	c	c	c	c	c	c	c	c
		b	ab	ab	a	ac	ac	ac	c	c	c	c	c	c	c	c	c	c	c
		ab	ab	abc	ac	ac	ac	c	c	c	c	c	c	c	c	c	c	c	
		abc	abc	ac	ac	ac	c	c	c	c	c	c	c	c	c				
		abc	abc	ac	ac	c	c	c	c	c	c	c							

Fig. 8 - Computational cells invested by radiative energy outgoing from wall CD (iteration 1); α=0.59, K=1. (m^{-1}); a,b,c as in Fig. 5

Y(m): 1. (top) – 0.5 (bottom); X(m): 0. – 1.

			b	b	b	b	ab	ab	a	a	a	a	a						c
			b	b	b	ab	ab	ab	a	a	a	a	a		c	c	c	c	c
		b	b	b	ab	ab	ab	ab	a	a	a	ac	ac	c	c	c	c	c	c
		b	b	ab	ab	ab	ab	ab	ac	ac	ac	ac	c	c	c	c	c	c	c
		b	ab	ab	ab	ab	abc	ac	ac	ac	ac	ac	c	c	c	c	c	c	c
		b	ab	ab	abc	abc	abc	ac	ac	ac	ac	ac	c	c	c	c	c	c	c
		ab	abc	abc	abc	abc	abc	ac	ac	ac	ac	c	c	c	c	c	c	c	c
		abc	abc	abc	abc	abc	ac	ac	ac	ac	ac	c	c	c	c	c	c	c	c
		abc	abc	abc	abc	ac	ac	ac	ac	c	c	c	c	c	c	c	c	c	c
		abc	abc	abc	ac	ac	ac	ac	c	c	c	c	c	c	c	c			

Fig. 9 - Computational cells invested by radiative energy outgoing from wall CD (iteration 1); α=1.0, K=1. (m^{-1}); a,b,c as in Fig. 5

Figure 10 and TABLE 1 show a further reduction of the difference between the minimum and maximum heat fluxes caused by the scattering phenomena and reflecting walls. However, these phenomena do not remove the unphysical minimum at the center of the wall AB.

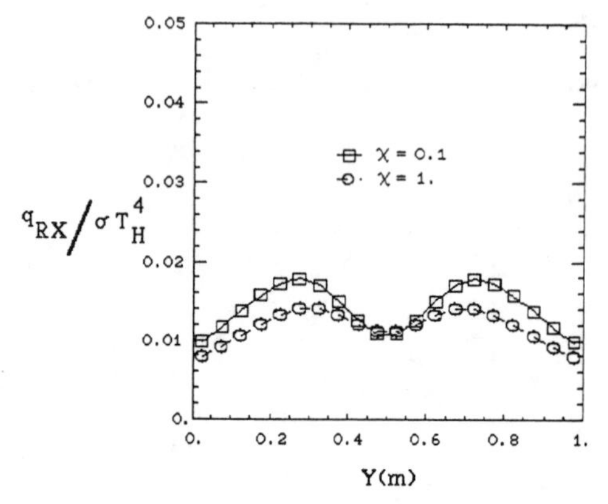

Fig. 10 - Nondimensional net radiative flux at wall AB; $\bar{\sigma}$=1.0 (m^{-1}), ε_w=0.5, ϱ_w=0.5, T_H=1500 K, α=1.0

Table 1 - Error in the overall balance of radiative energy

Case		Error (%)	Iterations
Fig. 4	χ = 0.1	27.39	8
	χ = 1.0	23.7	15
Fig. 6	χ = 0.1	17.5	7
	χ = 1.0	5.66	16
Fig. 7	χ = 0.1	$9.7\ 10^{-7}$	7
	χ = 1.0	$1.5\ 10^{-4}$	17
Fig. 10	χ = 0.1	$3.95\ 10^{-2}$	21
	χ = 1.0	$9.9\ 10^{-4}$	35
Fig. 12	χ = 0.1	$8.6\ 10^{-4}$	24

The error also persists for the system outlined in Fig. 11 where the radiation streams through a small aperture localized at the center of the enclosure and aligned with the hot surface element. The obtained result is shown in Fig. 12. Better solutions could be obtained, of course, by using denser meshes and higher order quadrature sets; but the precision is expected to increase very slowly (e.g. Yücel [3]).

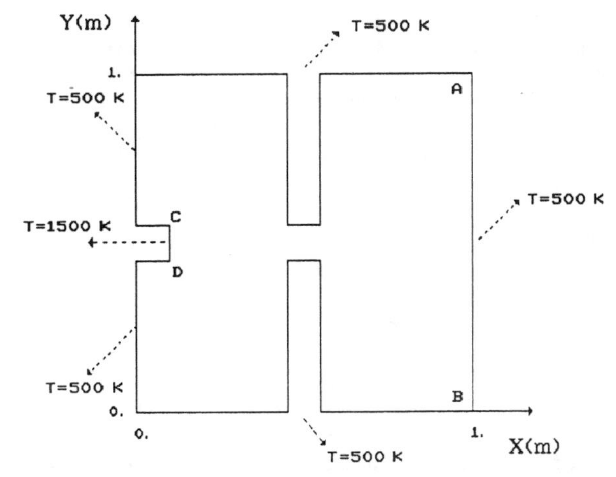

Fig. 11 - Second 2D enclosure analyzed in the present study

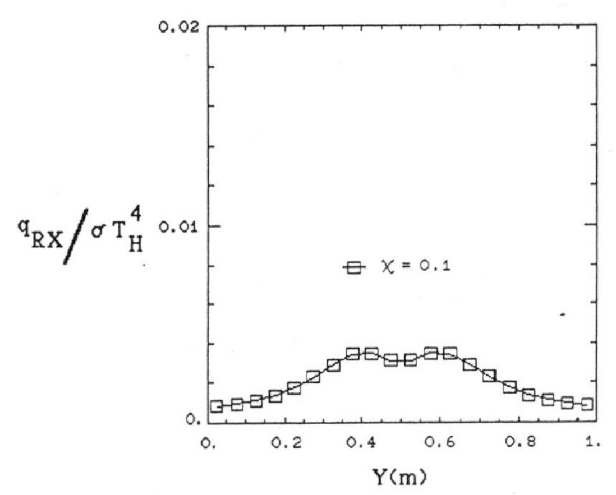

Fig. 12 - Small aperture; nondimensional net radiative flux at wall AB; $\bar{\sigma}$=1.0 (m^{-1}), ε_w=0.5, ϱ_w=0.5, T_H=1500 K, α=1.0

CONCLUSIONS

The ray effects in S_n-approximations of the Boltzmann transport equation have been known ever since 1965 (e.g. Carlson [7]).
These undesired effects have been discussed by Viskanta and Menguc [6] in the S_n solutions of RTE, too, when localized radiation sources have to be taken into account.
The numerical experiments described in the present paper prove the validity of this forecast.
Although the enclosures considered have been appropriately selected in order to exasperate the undesired effects present

in the method, the results obtained suggest that the S_n solutions should always be viewed with caution.
It should also be noted that the correctness of the radiative energy balance is not sufficient to guarantee an acceptable solution.

ACKNOWLEDGEMENTS

The authors are indebted to Mr., Glauco Mazzucchi of MATEC Modelli Matematici Srl for the preparation of the graphic support.

REFERENCES

1. Bell, C.R., Bleiweis, P.B., Boudreau, J.E., Parker, F.R. and Smith, L.L. SIMMER-I: An Sn, Implicit, Multifield, Multicomponent, Eulerian, Recritically Code for LMFBR Disrupted Core Analysis, LA-NUREG-6467-MS, Los Alamos, New Mexico, 1977.

2. Castellano, L. Programmi di Calcolo per lo Scambio Termico Radiativo, MATEC MTC/RD3/89, Milano, 1989.

3. Yücel, A. Heat Transfer Phenomena in Radiation, Combustion and Fires (Ed. Shah, R.K.), HTD-Vol. 106, p. 35, ASME, 1989.

4. Fiveland, W.A. and Jamaluddin, A.S. Heat Transfer Phenomena in Radiation, Combustion and Fires (Ed. Shah, R.K.), HTD-Vol. 106, p. 43, ASME, 1989.

5. Lathrop, K.D. Ray effects in Discrete Ordinates Equations, Nucl. Sci. Engng. 32, 357, 1968.

6. Viskanta, R. and Mengüç, M.P. Radiation Heat Transfer in Combustion Systems, Prog. Energy Combust. Sci. 13, 97, 1987.

7. Carlson, B.G. and Lathrop, K.D. Computing Methods in Reactor Physics (Eds. H. Greenspan, C.N. Kelber and D. Okrent), p. 65, Gordon and Breach, 1968.

8. Hyde, D.J. and Truelove, J.S. HTFS RS 189: The Discrete Ordinates Approximation for Multidimensional Radiant Heat Transfer in Furnaces, UKAEA AERE-R8502, 1977.

9. Fiveland, W.A. Discrete-Ordinates Solutions of the Radiative Transport Equation for Rectangular Enclosures, J. Heat Transfer (Trans. ASME) 106, 699, 1984.

10. Fiveland, W.A. Radiation Heat Transfer (Eds. B.F. Arenaly and A.F. Emery), HTD-Vol. 49, p. 42, ASME, 1985.

11. Truelove, J.S. Discrete-Ordinates Solutions of Radiation Transport Equations, J. Heat Transfer (Trans. ASME) 109, 1048, 1987

12. Fiveland, W.A. Fundamentals and Applications of Radiation Heat Transfer (Eds. A.M. Smith and T.F. Smith), HTD-Vol. 72, p. 9, ASME, 1987.

13. Truelove, J.S. Three-Dimensional Radiation in Absorbing-Emitting-Scattering Medium Using Discrete-Ordinates Approximations, J.Q.S.R.T. 39(1), 27, 1988.

NOMENCLATURE

a_o	asymmetric factor for scattering
I	radiant intensity ($W/m^2/Sr$)
$\underline{n}$	unit normal
$\underline{q}_R$	radiation flux vector (W/m^2)
s_m	outgoing direction of radiation
S_w	bounding surface
T	temperature (K)
$\underline{x} \equiv (x,y,z)$	position vector (m)
w_m	weight factor in m-direction
α	finite difference weighting factor
$\beta = K + \bar{\sigma}$	extinction coefficient (m^{-1})
$\gamma = \bar{\sigma}/\beta$	scattering albedo
δx_C	x-dimension of computational cell
δy_C	y-dimension of computational cell
ε_{conv}	tolerance on the iterative procedure (%)
ε_w	surface emittance
K	absorption coefficient (m^{-1})
(ξ_m, η_m, ζ_m)	direction cosines of s_m
ρ_w	surface reflectance
$\sigma = 5.669\ 10^{-8}$	Boltzmann's constant ($W/m^2/K^4$)
$\bar{\sigma}$	scattering coefficient
Φ	phase function
$\phi_{m'm}$	shape parameter for the phase function
ω	solid angle

Subscripts

b	backbody
C	control volume center
m	outgoing ordinate direction
m'	incoming ordinate direction
R	radiative
w	value at bounding surface

Solving 3-D Heat Radiation Problems in Cavities Filled by a Participating Non-Gray Medium using BEM

R. Białecki

Institute of Thermal Technology, Silesian Technical University, Konarskiego 22, 44-101 Gliwice, Poland

ABSTRACT

Heat radiation is governed by integral equations therefore the BEM numerical procedures and codes can be readily adopted to solve heat radiation problems. Two alternative representations of the integral equations of radiative heat transfer are derived. The first representation contains volume integrals while the other uses solely surface integrals. It is demonstrated that the known Hottel's zoning technique is a specific case of the weighted residual solution of the first representation. The second, boundary only, formulation is discretized by employing standard Boundary Element methodology i.e. collocation and locally based interpolating functions defining both the geometry and the variation of the unknown functions. A numerical solution for a 5 band radiating gas filling a nongray rectangular enclosure is discussed.

INTRODUCTION

Heat radiation plays a dominant role in energy transfer at elevated temperatures and in rarefied gases. The calculation of heat radiation fluxes constitutes the crucial portion of the thermal analysis carried out in many important branches of science and industry to name only industrial combustion chambers and furnaces, spacecrafts, solar energy utilization.

Radiation is one of the few phenomena governed by integral (integrodifferential) equations. This feature is a source of both conceptual and computational difficulties to most of the engineers whose mathematical background bases on differential equations. Additional difficulties inherent in heat radiation computations are due to the severe nonlinearity of the problem and very complex characteristic of the material properties appearing in the radiation transport equations.

Equations of heat radiation have been solved by many approximate methods described in fundamental literature of heat radiation [1-3]. Typical examples of such methods are exponential kernel approximation, flux method, discrete ordinate, differential approximation, Monte Carlo and Hottel's zoning method [4]. The idea of using the power and versatility of the Finite Element Method (FEM) to attack radiation problems have been exploited by several authors for some relatively simple physical situations [6],[7].

The Boundary Element Method (BEM) is an efficient numerical technique of similar field of application as FEM. Both techniques are developed to deal with differential equations and base on weighted residuals [8] [9]. BEM consists in transforming the boundary value problem into an equivalent integral equation which is then discretized and solved. As the governing equations of heat radiation are integral ones the idea of using BEM for the solution of radiation problems arises in a quite natural way.

The paper is a continuation of earlier work on this topic published by the present author [10-13]. The model considered here has been extended by including the allowance for nonisothermal, participating nongray gas radiation within the cavity.

As a byproduct of the analysis a novel mathematical interpretation of the classical zoning method derived usually by physical reasoning has been found.

FORMULATION OF THE PROBLEM. TWO SETS OF GOVERNING EQUATIONS

An enclosure composed of diffuse nonisothermal surfaces having wavelength dependent emissivities is considered. The participating medium (usually gas) filling the enclosure is a nonisothermal emitting - absorbing one with wavelength dependent absorption coefficient. It is assumed that scattering is negligible.

Due to the variation of the material properties within the spectrum the governing equation will be derived at a single wavelength. Index λ designates throughout the paper the spectral quantities. The equations of radiation transfer will be derived employing the notion of spectral intensity defined as

$$I_\lambda = \frac{d^4E}{d\Omega\ dS\ d\lambda\ d\tau} \tag{1}$$

where:

- E – radiant energy; J
- $d\Omega$ – element of a solid angle centered around a direction $\vec{\Omega}$; sr
- dS – element of surface normal to $\vec{\Omega}$; m^2
- $d\lambda$ – unit wavelength centered around λ; m
- $d\tau$ – infinitesimal time increment; s

Consider a pencil of rays starting at a point having vector coordinate **r**. The intensity of radiation at an observation point **p** can be calculated upon integration of the differential equation of radiation transfer along the line connecting **r** and **p** [1].

$$I_\lambda(\mathbf{p}) = I_\lambda(\mathbf{r})\exp[-\int_{L_{\mathbf{rp}}} a_\lambda(\mathbf{r}')dL(\mathbf{r}')] +$$

$$+\int_{L_{\mathbf{rp}}} a_\lambda(\mathbf{r}')I_{\lambda b}[T_g(\mathbf{r}')]\{\exp[-\int_{L_{\mathbf{r}'\mathbf{p}}} a_\lambda(\mathbf{r}'')dL(\mathbf{r}'')]\}\ dL(\mathbf{r}') \tag{2}$$

where:

- a_λ – monochromatic absorption coefficient
- T_g – temperature of the intervening medium (gas) $T_g = T_g(\mathbf{r}')$
- $I_{\lambda b}$ – monochromatic intensity of radiation of a blackbody
- $\mathbf{r}', \mathbf{r}''$ vector coordinates of points lying on the line connecting **r** and **p** (cf Fig.1)
- dL – differential element of line

symbol $\int_{L_{\mathbf{rp}}} (\ )dL(\mathbf{r}')$ designates integration along a straight line connecting points **r** and **p**

Assuming the source of radiation (point **r**) is placed on the wall of the enclosure the initial condition in Eq.2 can be expressed in terms of the net outgoing radiation (radiosity) of the wall at point **r**. Similarly the intensity of the blackbody radiation can be substituted by the emissive power. Taking into account the diffusive behaviour of the radiation Eq.2 acquires a form

$$I_\lambda(\mathbf{p})=\frac{b_\lambda(\mathbf{r})\tau_\lambda(\mathbf{r},\mathbf{p})}{\pi}+\int_{L_{\mathbf{rp}}}\frac{e_{\lambda b}(\mathbf{r}')a_\lambda(\mathbf{r}')\tau_\lambda(\mathbf{r}',\mathbf{p})}{\pi}dL(\mathbf{r}') \quad (3)$$

where:

b_λ - radiosity being a sum of self emission and radiation reflected from a surface

$e_{\lambda b}$ - blackbody emissive power (Planck's function)

τ_λ - transmissivity defined as

$$\tau_\lambda(\mathbf{r},\mathbf{p})=\exp[-\int_{L_{\mathbf{rp}}}a_\lambda(\mathbf{r}')dL(\mathbf{r}')] \quad (4)$$

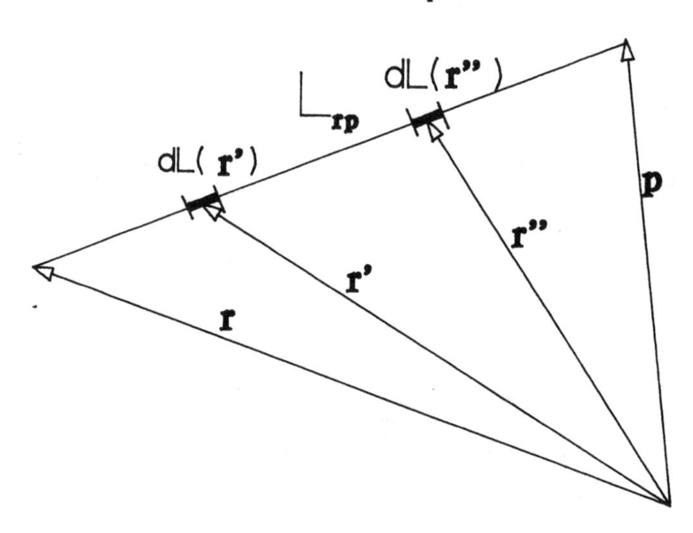

Fig. 1. *Nomenclature used when integrating the equation of radiation transport along a line*

Two equations of radiant transfer can be derived. The first one is the heat balance for the unit wall surface. The equation can be obtained upon placing the observation point **p** on the surface of the enclosure. Eq. 4 is then multiplied by the cosine of an angle between the line connecting **r** and **p** and the outward normal at point **p**. Integration of the result over the solid angle 2π yields the radiant energy impinging the unit surface dS(**p**). The net heat flux gained by this surface by radiation is a difference between the radiation absorbed and emitted by the surface

$$q_\lambda^r(\mathbf{p}) = \varepsilon_\lambda(\mathbf{p}) \int_{2\pi} \{I_\lambda(\mathbf{p}) - I_{\lambda b}[T(\mathbf{p})]\} \cos\varphi_p \, d\Omega \tag{5}$$

where:

- q_λ^r - net incoming radiant heat flux
- ε_λ - emissivity of the wall
- T - temperature of the wall
- φ_p - angle that makes the normal at point **p** and the line connecting **r** and **p**

The solid angle can be expressed as

$$d\Omega = dS(\mathbf{r}) \cos\varphi_r / |\mathbf{r}-\mathbf{p}|^2 \tag{6}$$

where:

$|\mathbf{r}-\mathbf{p}|$ distance between points **r** and **p**

Substitution of Eq.3 and 6 into 5 yields

$$q_\lambda^r(\mathbf{p})/\varepsilon_\lambda(\mathbf{p}) + e_{\lambda b}[T(\mathbf{r})] = \int_S b_\lambda(\mathbf{r}) \tau_\lambda(\mathbf{r},\mathbf{p}) K(\mathbf{r},\mathbf{p}) dS(\mathbf{r}) +$$

$$+ \int_S \left\{ \int_{L_{\mathbf{rp}}} a_\lambda(\mathbf{r}') e_{\lambda b}[T_g(\mathbf{r}')] \tau_\lambda(\mathbf{r}',\mathbf{p}) dL(\mathbf{r}') \right\} K(\mathbf{r},\mathbf{p}) dS(\mathbf{r}) \tag{7}$$

where:

$$K(\mathbf{r},\mathbf{p}) = \cos\varphi_r \cos\varphi_p / (\pi |\mathbf{r}-\mathbf{p}|^2) \tag{8}$$

The second integral in Eq.7 can be written in an alternative form noting that

$$d\Omega \; dL(\mathbf{r}') = dV(\mathbf{r}') / |\mathbf{r}'-\mathbf{p}| \tag{9}$$

where:

$|\mathbf{r}'-\mathbf{p}|$ - distance between points **r'** and **p**

Taking into account Eq.9, Eq.8 can be rewritten as

$$q_\lambda^r(\mathbf{p})/\varepsilon_\lambda(\mathbf{p}) + e_{\lambda b}[T(\mathbf{p})] = \int_S b_\lambda(\mathbf{r}) \tau_\lambda(\mathbf{r},\mathbf{p}) K(\mathbf{r},\mathbf{p}) dS(\mathbf{r}) +$$

$$+ \int_V a_\lambda(\mathbf{r}') e_{\lambda b}[T_g(\mathbf{r}')] K_p(\mathbf{r}',\mathbf{p}) \tau_\lambda(\mathbf{r}',\mathbf{p}) dV(\mathbf{r}') \tag{10}$$

where:

$$K_p(\mathbf{r}',\mathbf{p})=\cos\varphi_p/(\pi|\mathbf{r}'-\mathbf{p}|) \tag{11}$$

The second equation of heat radiation is a heat balance for a unit volume dV of the gas. The equation is obtained upon placing the observation point **p** within the participating medium and performing appropriate integration of the incoming and outgoing intensities

$$q^r_{\lambda v}(\mathbf{p})= a_\lambda\int_{4\pi}\{I_\lambda(\mathbf{p})-I_{\lambda b}[T_g(\mathbf{p})]\}d\Omega \tag{13}$$

where:

$q^r_{\lambda v}(\mathbf{p})$ - generation rate of the net radiative heat source i.e. heat gained by a volume dV due to radiative heat exchange.

$$q^r_{\lambda v}(\mathbf{p})= -\mathrm{div}[q^r_\lambda(\mathbf{p})] \tag{14}$$

By virtue of Eq.4, Eq 13 can be written in a form

$$q^r_{\lambda v}(\mathbf{p})/a_\lambda(\mathbf{p})+4e_{\lambda b}[T_g(\mathbf{p})]=\int_S b_\lambda(\mathbf{r})\tau_\lambda(\mathbf{r},\mathbf{p})K_r(\mathbf{r},\mathbf{p})dS(\mathbf{r}) +$$

$$+\int_S\left\{\int_{L_{\mathbf{rp}}}a_\lambda(\mathbf{r}')e_{\lambda b}[T_g(\mathbf{r}')]\tau_\lambda(\mathbf{r}',\mathbf{p})dL(\mathbf{r}')\right\}K_r(\mathbf{r},\mathbf{p})dS(\mathbf{r}) \tag{15}$$

where:

$$K_r(\mathbf{r},\mathbf{p}) = \cos\varphi_r/(\pi\ |\mathbf{r}-\mathbf{p}|^2) \tag{16}$$

Taking into account Eq.9,Eq.15 can be rewritten in an alternative form containing volume integral

$$q^r_{\lambda v}(\mathbf{p})/a_\lambda(\mathbf{p})+4e_{\lambda b}[T_g(\mathbf{p})]=\int_S b_\lambda(\mathbf{r})\tau_\lambda(\mathbf{r},\mathbf{p})K_r(\mathbf{r},\mathbf{p})dS(\mathbf{r}) +$$

$$+\int_V a_\lambda(\mathbf{r}')e_{\lambda b}[T_g(\mathbf{r}')]K_o(\mathbf{r}',\mathbf{p})\tau_\lambda(\mathbf{r}',\mathbf{p})dV(\mathbf{r}') \tag{17}$$

where:

$$K_o(\mathbf{r}', \mathbf{p}) = 1/(\pi|\mathbf{r}' - \mathbf{p}|) \tag{18}$$

Eqs.7,15 and Eqs.10,17 constitute two equivalent sets of integral equations. Both sets require integration along the path of the ray within the medium. Representations 7 and 15 however, are numerically much easier to handle as they do not demand volume integration.

Each of these two sets link four unknowns:

- -temperature of the wall; T
- -temperature of the gas: T
- -spectral radiative heat flux; q^r_λ
- -spectral radiative heat sources; $q^r_{\lambda v}$

Additional equations to be solved in conjunction with the heat radiation ones are:

i) equation of heat conduction within the walls forming the enclosure

ii) equation of convective and conductive energy transfer within the gas

Radiative heat flux enters the first equation giving contribution to the boundary conditions. Radiative heat sources constitute a nonhomogeneous term in the differential equation of energy transfer within the medium filling the enclosure. Both of these radiative heats are total quantities and are integrals of the spectral ones over the whole spectrum.

$$q^r(\mathbf{p}) = \int_0^\infty q^r_\lambda(\mathbf{p})d\lambda \tag{19}$$

$$q^r_v(\mathbf{p}) = \int_0^\infty q^r_{\lambda v}(\mathbf{p})d\lambda \tag{20}$$

Integration of Eqs.19,20 can be performed numerically. This can be accomplished upon forming equation of radiative transfer for a number of chosen wavelength and solving this equation for spectral radiative quantities. Then, formulae 19,20 are approximated by appropriate quadratures.
There are specific cases when the changes of material properties ε_λ and a_λ within the spectrum can be assumed stepwise. In these instances integration over the intervals of the spectrum where the material properties remain constant can be carried out semi analytically. In these cases unknown emissive powers in the equations of radiative transfer can be substituted by the fraction of blackbody radiation emitted within the mentioned intervals. Similarly the spec-

tral radiative heats q_λ^r and $q_{\lambda v}^r$ can be substituted by heats transmitted within these intervals. More details of this procedure can be found in [12,13].

DISCRETIZATION

Transmissivity. Line integrals

To discretize the line integral one have to approximate the variation of both temperature of the gas and the absorption coefficient within the enclosure. Many possibilities of such approximations exist. The simplest one is to divide the entire volume of the gas into a finite number of cells and assume constant values of both temperature and absorptivity within each cell. Under these assumptions transmissivities present in Eqs.7,10,15,17 can be calculated as

$$\tau_\lambda(\mathbf{r},\mathbf{p}) \cong \tau_\lambda(1,I) = \exp\left(-\sum_{i=1}^{I} a_{\lambda i} d_i\right) \tag{21}$$

where:

$a_{\lambda i}$ - absorption coefficient within i-th cell intersected by a ray going from point **r** to **p**
d_i - length of ray within i-th cell
I - number of cell intersected by the ray
$\tau_\lambda(1,I)$ - approximate value of transmissivity between points **r** and **p**

The line integral from Eqs.7, 15 can be computed as

$$\int_{L_{\mathbf{rp}}} a_\lambda(\mathbf{r}')e_{\lambda b}[T_g(\mathbf{r}')]\tau_\lambda(\mathbf{r}',\mathbf{p})dL(\mathbf{r}') \cong$$

$$\cong \sum_{i=1}^{I} e_{\lambda b}(T_{gi})A_{\lambda i}\tau_\lambda(i+1,I) \tag{22}$$

where:

T_{gi} - temperature of gas in the i-th cell
$A_{\lambda i}$ - absorptivity of the i-th cell

$$A_{\lambda i} = 1 - \exp(-a_{\lambda i} d_i) \tag{23}$$

$\tau_\lambda(i+1,I)$ - approximate transmissivity

$$\tau_\lambda(i+1,I)=\exp\left(-\sum_{j=i+1}^{I} a_{\lambda j}d_j\right) \qquad (24)$$

Surface integrals

Surface integrals are discretized following the standard BEM methodology [13] i.e. employing weighted residuals. The first step of this approach is to divide the entire surface S bounding the enclosure into a finite number of subsurfaces (boundary elements) ΔS_l ; $l=1,2..L$. Then one has to choose a set of interpolating functions approximating the variation of the sought for functions over surface S. The simplest possible choice is a set of locally based functions (shape functions) $N^l(r)$ satisfying a simple relationship

$$N^l(r)=\begin{cases} 1 & \text{if} \quad r\in\Delta S_l \\ 0 & \text{if} \quad r\notin\Delta S_l \end{cases} \qquad (25)$$

Appropriate interpolation formulae have the form

$$e_{\lambda b}[T(r)]\cong\sum_{l=1}^{L} e_{\lambda b}[T(r_l)]N^l(r) \qquad (26)$$

$$q_\lambda^r(r)\cong\sum_{l=1}^{L} q_\lambda^r(r_l)N^l(r) \qquad (27)$$

where:

r_l - nodal point $r_l\in \Delta S_l$

Substituting Eqs.26,27 into integral equations 7,15 produces residuals. To minimize them another sequence of functions $w_k(p_k)$ $k=1..K$ termed weighting function is introduced. The simplest possible choice is to define $w(p)$ as a sequence of Dirac's distributions acting at nodal points $p_k=r_k$. Upon multiplication of the residuals by subsequent weighting function and integration of the result over the entire surface S one arrives at two sets of algebraic equations.

$$\mathbf{A}_\lambda\ e_{\lambda b}(T)=\mathbf{B}_\lambda\ q_\lambda^r + \mathbf{C}_\lambda\ e_{\lambda b}(T_g) \qquad (28)$$

$$q_{\lambda v}^r = \mathbf{D}_\lambda\ e_{\lambda b}(T) + \mathbf{E}_\lambda\ q_\lambda^r + \mathbf{F}_\lambda\ e_{\lambda b}(T_g) \qquad (29)$$

When deriving the above equations the radiosity has been eliminated employing the relationship [1]

$$b_\lambda(p)=e_{\lambda b}[T(p)] - [1-\varepsilon_\lambda(p)]q_\lambda^r(p)/\varepsilon(p) \tag{30}$$

The entries of the matrices in Eqs. 28, 29 are defined as

$$\{e_{\lambda b}(T)\}_k = e_{\lambda b}[T(p_k)] \tag{31}$$

$$\{q_\lambda^r\}_k = q_\lambda^r(p_k) \tag{32}$$

$$\{q_{\lambda v}^r\}_k = q_\lambda^r(p_k) \tag{33}$$

$$\{e_{\lambda b}(T_g)\}_i = e_{\lambda b}[T_g(r_i')] \tag{34}$$

$$\{A_\lambda\}_{kl} = \delta_{kl} - \int_{\Delta S_l} K(r,p_k)\tau_\lambda(1,I)dS(r) \tag{35}$$

$$\{B_\lambda\}_{kl} = -\delta_{kl}/\varepsilon_\lambda(p_k) - \int_{\Delta S_l} [1-\varepsilon_\lambda(r)]K(r,p_k)\tau_\lambda(1,I)/\varepsilon_\lambda(r)dS(r) \tag{36}$$

$$\{C_\lambda\}_{ki} = \sum_{l=1}^{L} \int_{\Delta S_l} A_\lambda\tau_\lambda(i+1,I)K(r,p_k)dS(r) \tag{37}$$

$$\{D_\lambda\}_{kl} = -4a_\lambda(p_k)\delta_{kl} + a_\lambda\int_{\Delta S_l} K_r(r,p_k)\ \tau_\lambda(1,I)dS(r) \tag{38}$$

$$\{E_\lambda\}_{kl} = -\int_{\Delta S_l} [1-\varepsilon_\lambda(r)]K_r(r,p_k)\tau_\lambda(1,I)/\varepsilon_\lambda(r)dS(r) \tag{39}$$

$$\{F_\lambda\}_{ki} = \sum_{l=1}^{L} \int_{\Delta S_l} A_\lambda\tau_\lambda(i+1,I)K_r(r,p_k)dS(r) \tag{40}$$

where:

δ_{kl} - Kronecker's symbol

Integration over surface elements is performed in local coordinates following standard BEM approach. The described above version of the weighted residual technique employed to solve the integral equations is the collocation method frequently used in BEM. Details of numerical procedures used to discretize the surface integrals are described in [13].

It should be stressed that Dirac's distributions chosen as weighting functions are not the only possible sequence that can be employed when using the residual technique. Another simple choice is to take the weighted functions as equal to the interpolating functions. This version of the residual technique is referred to as the Galerkin approach. This technique is frequently used in FEM context. As shown in Refs.-[11,12] the standard view factor method of solving radiative heat exchange problems in enclosures filled with transparent gas is equivalent to the weighted residuals solution of Eq.7 with both weighting and interpolating functions chosen as constant within boundary elements (i.e. defined by Eq. 25).

HOTTEL'S ZONING METHOD AS A WEIGHTING RESIDUAL TECHNIQUE

Consider the set of integral equations containing volume integrals (Eqs. 10, 17). To discretize this set using weighted residuals the surface of the enclosure is divided into a finite number of boundary elements ΔS_l exactly as in the previous paragraph. Similarly, subdivision of the volume into a number of finite (volume) elements ΔV_m $m=1,2..M$ is introduced. Interpolating functions associated with this subdivision are the step functions both for surface and volume elements. These sets of functions are defined as (cf Eq.25)

$$N^l(\mathbf{r})=\begin{cases}1 & \text{if} \quad \mathbf{r}\in\Delta S_l \\ 0 & \text{if} \quad \mathbf{r}\notin\Delta S_l\end{cases} \tag{41}$$

$$M^m(\mathbf{r})=\begin{cases}1 & \text{if} \quad \mathbf{r}\in\Delta V_m \\ 0 & \text{if} \quad \mathbf{r}\notin\Delta V_m\end{cases} \tag{42}$$

Introducing approximating formulae spanned on nodal values of unknown functions and being linear combinations of interpolation sets 42 and 43 into the integral equations 10, 17 gives a residuum. The residuum of Eq. 10 is made orthogonal to functions (41) and the residuum of Eq. 17 is orthogonal to set (42). In terms of the weighted residual technique Eqs.10 and 17 are solved by the Galerkin approach.

As a result of the described above Galerkin discretization one obtains two sets of algebraic equations. The discretized version of Eq.10 reads

H

$$\left[q^r_{\lambda l}/\varepsilon_{\lambda l}+e_{\lambda bl}\right]\Delta S_l = \sum_{k=1}^{L} b_{\lambda k}\, \overline{s_k s_l} + \sum_{i=1}^{M} e_{\lambda bi}\, \overline{g_i s_l} \quad (43)$$

where:

$$\overline{s_k s_l} = \int_{\Delta S_k}\int_{\Delta S_l} K(r,p)\tau_\lambda(r,p)dS(r)dS(p) \quad (44)$$

$$\overline{g_i s_l} = a_{\lambda i} \int_{\Delta V_i}\int_{\Delta S_l} K_p(r',p)\tau_\lambda(r',p)dV(r')dS(p) \quad (45)$$

Discretization of Eq. 17 yields:

$$(q^r_{\lambda vl}+4a_\lambda e_{\lambda bl})\Delta V_l = \sum_{k=1}^{L} b_{\lambda k}\overline{s_k g_l} + \sum_{i=1}^{M} e_{\lambda bi}\, \overline{g_i g_k} \quad (46)$$

$$\overline{s_k g_l} = \int_{\Delta S_k}\int_{\Delta V_l} K_r(r,p)\tau_\lambda(r,p)dS(r)dV(p) \quad (47)$$

$$\overline{g_i g_l} = a_{\lambda i} a_{\lambda l} \int_{\Delta V_i}\int_{\Delta V_l} K_o(r',p)\tau_\lambda(r',p)dV(r')dV(p) \quad (48)$$

Integrals 44, 45 and 47, 48 are referred to as *surface-surface, gas-surface, surface-gas, and gas-gas direct exchange areas* respectively . Equations 43 and 46 are identical with those of Hottel's zoning method [5]. Thus, Hottel's method can be interpreted as a Galerkin's weighted residual solution with weighting functions being constant in subregions. The discretization technique of the zoning method is therefore similar to that used in the Finite Element Method. The classical Hottel's equations are derived from the physical interpretation using the simplest possible zeroth order elements. Knowing the mathematical background of the zoning approach it is possible to gain higher accuracy upon employing higher order elements, as it is done in the Finite Element Method.

NUMERICAL EXAMPLE

To test the proposed algorithm and to gain some numerical experience several sample problems of different complexity have been solved. The example to be discussed is an open rectangular 0.7 x 0.7 x 0.4m cavity problem . The enclosure is filled by a participating gas having known nonuniform temperature. The walls of the enclosure are subdivided into 210 rectangular elements (0.1 x 0.1 m) of known temperatures. The base of the cavity is maintained at temperature 700K whereas the top of the cavity is open to the environment of known temperature 300K. The temperature of other walls vary linearly in height from 700K at the bottom to 300K at the top of the cavity.

The spectrum was divided into 5 bands

1 band- 0.0 to 0.260e-5; m
2 band- 0.26e-5 to 0.286e-5; m
3 band- 0.286e-5 to 0.126e-4; m
4 band- 0.126e-4 to 0.185e-4; m
5 band- 0.185e-4 to ∞

Within each band both emissivities and absorption coefficient was assumed constant. The base of the cavity was divided into 3 subregions having different emissivities $\varepsilon_1, \varepsilon_2, \varepsilon_3$ as shown in Fig.2a. Material properties are shown in Table 1. ε_5 designates the emissivity of the environment, ε_4 stands for the emissivity of the side walls.

The volume of the enclosure has been divided into 27 cubic isothermal cells consisting of 3 layers each of 9 cells. The height of the bottom and middle layer was 0.1m whereas height of the top layer was 0.2m. The temperature of the gas layers is shown in Fig. 2b,c,d.

Table 1. *Numerical values of the material properties used in computations.*

Band No	ε_1	ε_2	ε_3	ε_4	ε_5	a m^{-1}
1	0.9	0.1	0.1	0.5	1.0	0.0
2	0.7	0.3	0.2	0.5	1.0	1.0
3	0.5	0.5	0.3	0.5	1.0	0.0
4	0.3	0.7	0.2	0.5	1.0	0.1
5	0.1	0.9	0.1	0.5	1.0	0.1

Figures 3 - 7 show the obtained results (radiative heat fluxes in kW/m^2) at the base of the cavity. The isolines are constant values of the incoming radiative heat fluxes. Except for Fig. 4b the fluxes are negative. This means that only within band 2 the heat recieved by the base from the gas is larger than that send to the side walls and the environment. The global heat flux absorbed within the gas filling the cavity was found to be 989 W. Total fluxes (Fig. 3) and nontransparent bands i.e. band 2 (large absorption coefficient, Fig. 4) and band 4 (medium absorption coefficient Fig. 5) are shown together with the values of radiative fluxes for a situation when the gas is transparent.

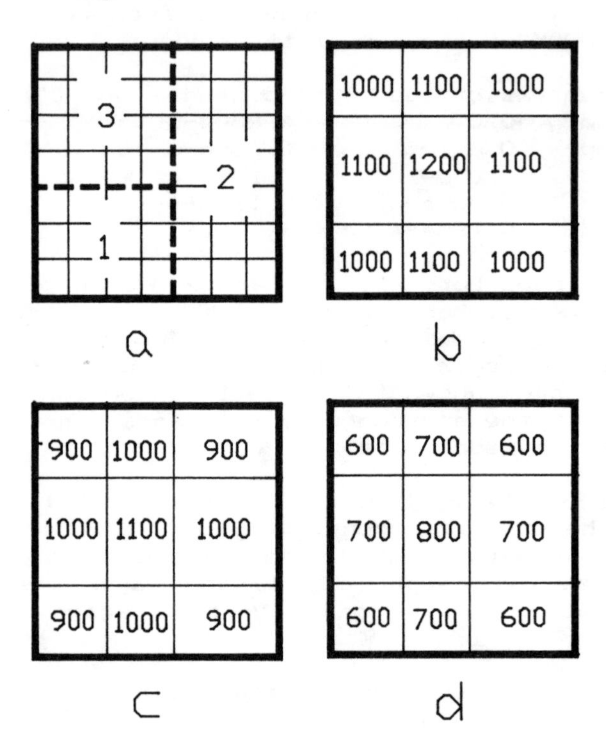

Fig. 2. *Numerical grid at the base of the cavity and subdivision into three fields of different emissivity (a).*

Geometry and temperatures (Kelvins) in gas cells in the bottom (b) middle (c) and upper layer of the gas

CONCLUSIONS

The proposed method enables one to deal with nonisothermal nongray gases filling an enclosure formed by nongray, nonisothermal walls. Forming the appropriate algebraic equations demands numerical integration over the boundary only and can be easily automated upon employing the BEM methodology.

The proposed numerical method of solving complex heat radiation problems bases on a novel formulation of the integral equations of heat radiation transfer. The standard formulation employed in the zoning method demands two types of integration to be carried out: the integration over the boundary and the integration over the volume. The integral equations derived in the present paper do not demand volume integration. Thus, as it is in BEM the dimensionality of the problem is reduced by one.

Moreover, though there are two boundary integrals in one integral equation, the cost of forming the set of algebraic equations corresponding to one integral equation is equivalent to performing single numerical integration over the boundary. This is due to the fact that the integrand of these two integrals can be expressed as a product of transmissivity and a kernel function identical for both integrals. As the zoning method demands one surface plus one volume integration, the time needed for forming matrices is in this method much longer.

The numerical treatment of the equations bases on the well established BEM techniques and consists of local interpolation and collocation. Therefore many routines from existing BEM codes can be employed in programs implementing the proposed approach (input of geometry, parameterization of both geometry and function over the boundary, numerical integration,and assembling the matrices etc.)

Mathematical interpretation of the zoning method as a specific weighed residuals technique gives some insight into the numerical methods used in that classical approach. It enables one also to use higher order approximation and adopting some of the FEM routines in the zoning method codes.

ACKNOWLEDGMENTS

The financial support of the Central Plan of Fundamental Research CPBP 02-22/02-17 coordinated by the Technical University of Poznań is gratefully acknowledged.

REFERENCES

1 Siegel, R. and Howell, J.R. Thermal Radiation Heat Transfer, Hemisphere Publishing Corporation, Washington, 1981, 2nd ed.
2 Özişik, M.N. Radiative Transfer and Interactions with Conduction and Radiation, J. Wiley & Sons,

New York, London 1973.

3 Sparrow, E.M. & Cess, R.D. Radiative Heat Transfer. Hemisphere, Washington and London 1971, 2nd Edition

4 Sarofim, A.F. Radiative heat transfer in combustion: friend of foe. Twenty-first Symposium (International) on Combustion. The Combustion Institute, 1986, pp 1-23

5 Hottel, H.C. & Sarofim, A.F. Radiative Transfer, McGraw Hill, New York 1967

6 Mason, W.E. Finite Element analysis of coupled heat conduction and enclosure radiation. International Conference on Numerical Methods in Thermal Problems, Swansea UK, 1979.

7 Chen, T.J. Integral and Integro-differential Systems, Chapter 14 in Handbook of Numerical Heat Transfer pp. 579-624 (Eds. Minkowycz, W.J., Sparrow E.M., Schneider, G.E., Pletcher), Wiley Interscience, N. York, 1988

8 Zienkiewicz, O.C. The Finite Element Method. McGraw Hill London 1977.

9 Brebbia, C.A., Telles J.F.C., Wrobel L.C. Boundary Element Techniques, Springer Verlag, Berlin and New York, 1984.

10 Białecki, R., Nahlik, R., Nowak, A.J. Temperature field in a solid forming an enclosure where heat transfer by convection and radiation is taking place, in Proceedings of the 1st National UK Heat Transfer Conference, pp. 989-1000, Leeds, UK, 1984. Pergamon Press, London, 1984.

11 Białecki, R. Applying BEM to calculations of temperature field in bodies containing radiating enclosures, in Boundary Elements VII, (Eds. Brebbia, C.A. and Maier, G.), Vol. 1, pp. 2-35 to 2-50, Proceedings of the 7th International Conference on Boundary Elements , Como, Italy, 1985. Springer-Verlag, Berlin and New York, 1985.

12 Białecki, R. Radiative heat transfer in cavities. BEM solution, in Boundary Elements X, (Ed. Brebbia, C. A.) Vol. 2, pp. 246-256, Proceedings of the 10th International Conference on Boundary Elements, Southampton, UK, 1988. Springer Verlag, Berlin and New York, 1988.

13 Białecki, R. Modelling 3D band thernmal radiation in cavities using BEM, in Advances in Boundary Elements (Eds. Brebbia, C.A. & Connor J.J.) vol 2 Field and Flow Solution pp 116-135, Proceedings of the 11th International Conference on Boundary Elements, Cambridge (USA), 1989, Springer Verlag Berlin and New York, 1989.

INCOMING RADIATIVE HEAT FLUXES in kW/m^2

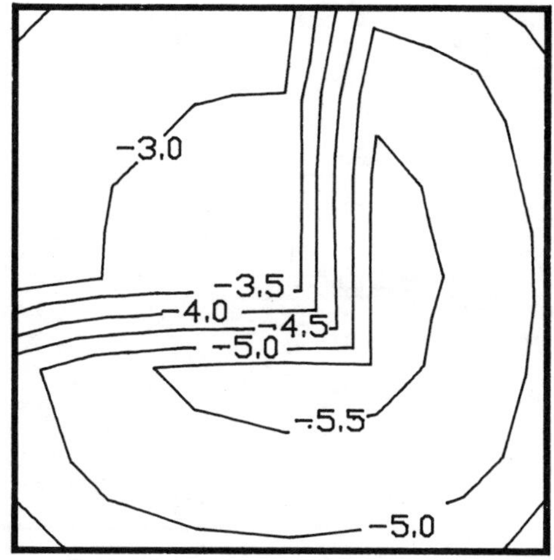

total flux, transparent gas

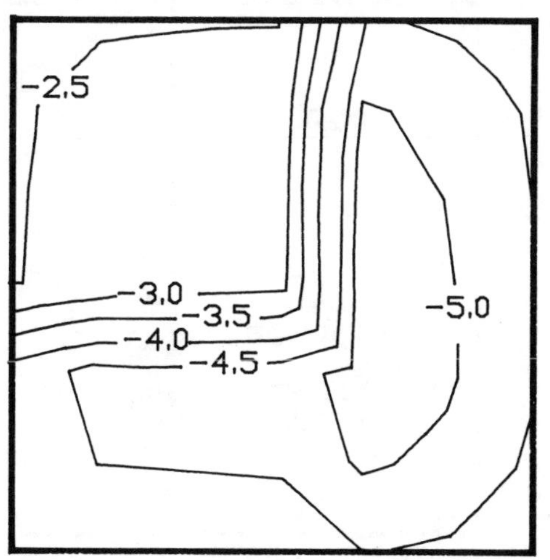

total flux, participating gas

Fig. 3 *Total radiative heat fluxes (incoming) at the base of the rectangular cavity for transparent (a) and participating (b) medium filling the cavity*

INCOMING RADIATIVE HEAT FLUXES in kW/m^2

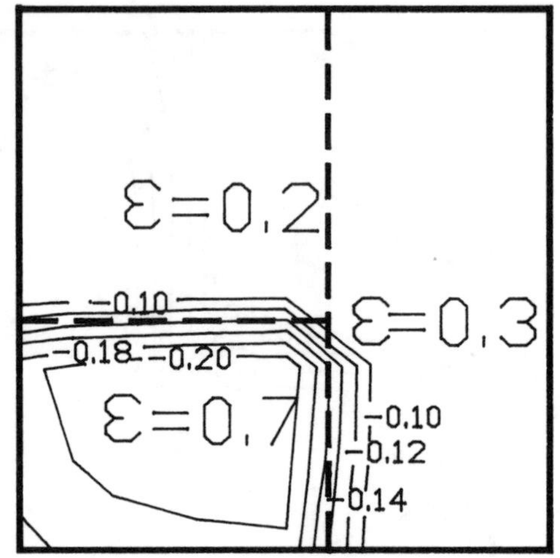

band 2, transparent gas $a=0m^{-1}$

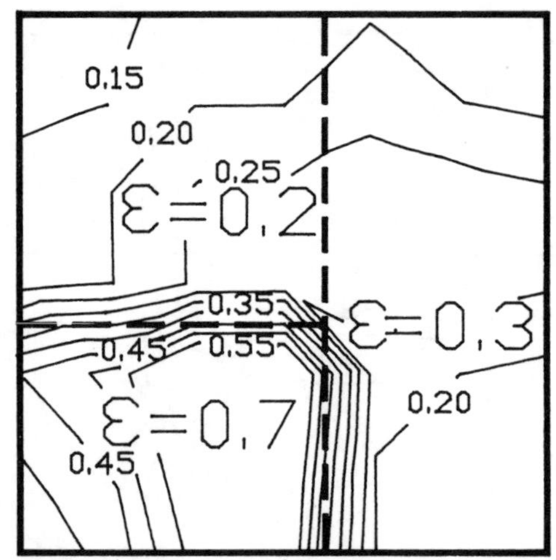

band 2, participating gas $a=1m^{-1}$

Fig. 4 *Radiative heat fluxes (incoming) transmitted at the base of the cavity within the second spectral band, transparent (a) and participating gas (b) cases*

INCOMING RADIATIVE HEAT FLUXES in kW/m^2

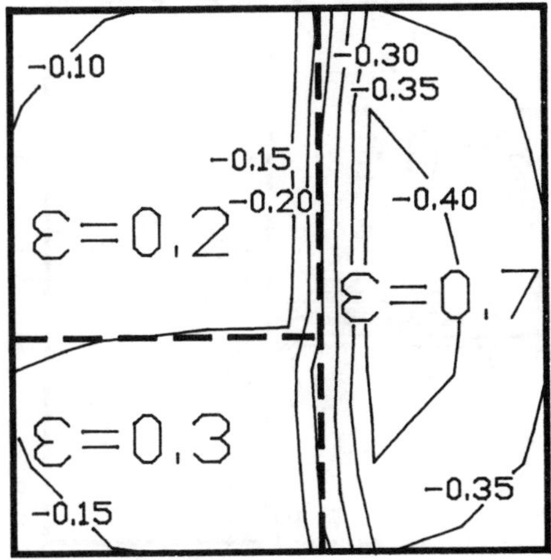

band 4, transparent gas $a=0.0m^{-1}$

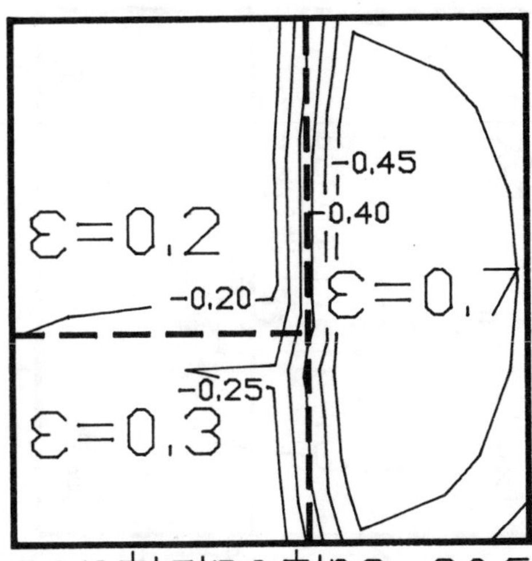

band 4, participating gas $a=0.1m^{-1}$

Fig. 5 *Radiative heat fluxes (incoming) transmitted at the base of the cavity within the fourth spectral band, transparent (a) and participating gas (b) cases*

INCOMING RADIATIVE HEAT FLUXES in kW/m²

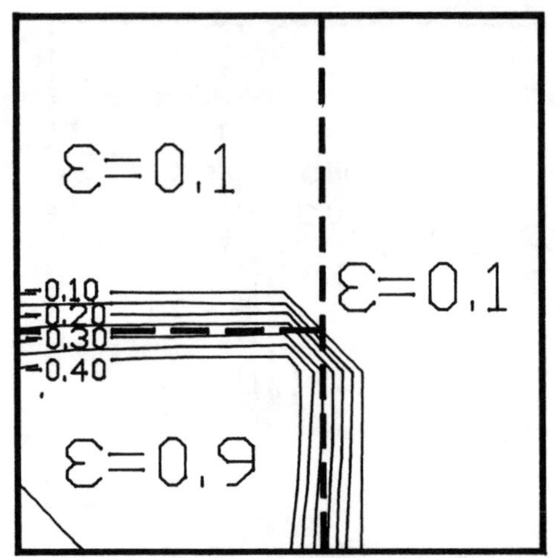

band 1, transparent gas

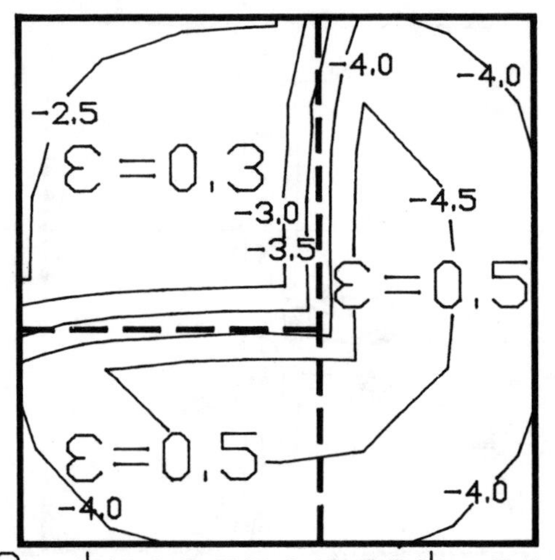

band 3, transparent gas

Fig. 6 *Radiative heat fluxes (incoming) transmitted at the base of the cavity within the first (a) and third (b) spectral bands where the gas is transparent tp radiation*

INCOMING RADIATIVE HEAT FLUXES in kW/m^2

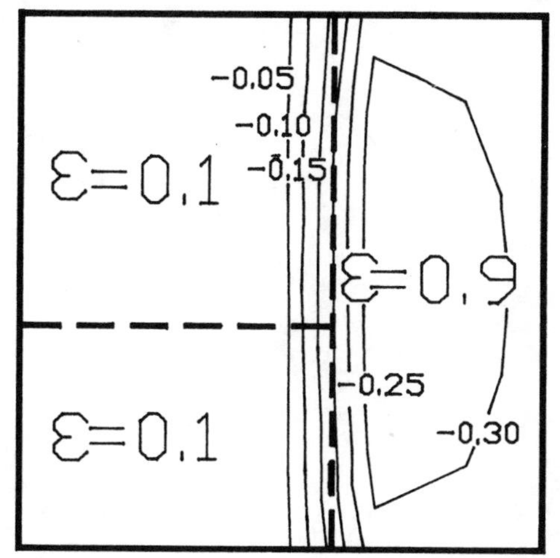

Fig. 7 *Radiative heat fluxes (incoming) transmitted at the base of the cavity within the fifth spectral bandwhere the gas is transparent to radiation*

SECTION 3: CONVECTION-DIFFUSION

A Boundary Element Analysis of Convective Heat Diffusion Problems

D.B. De Figueiredo(*), L.C. Wrobel
Computational Mechanics Institute, Wessex Institute of Technology, Ashurst Lodge, Ashurst, Southampton SO4 2AA, England
(*) On leave from CNPq, Brazilian Council for Scientific and Technological Development, Brazil

ABSTRACT

This paper presents a boundary element formulation for steady-state convection-diffusion problems employing the fundamental solution of the equation with constant velocity. Thus, the mathematical problem can be described in terms of boundary values only. Numerical results show that the formulation does not present oscillations or damping of the wave front as appear in other numerical techniques.

INTRODUCTION

Convective effects are of importance in many practical physical problems. When the medium where heat transfer is taking place is fluid, the complete mathematical description of the problem is given by the Navier-Stokes equations incorporating a buoyancy term, coupled with the Fourier heat conduction equation with convective terms included. In some situations where the medium is solid and moves as a whole at a constant velocity, the heat transfer mechanism can be described by heat conduction with a constant convective velocity, and the mathematical model reduces to the convection-diffusion equation.

The numerical solution of the convection-diffusion equation is no easy task because of the nature of the equation, which includes first-order and second-order partial derivatives in space. According to the value of the Péclet number, the equation becomes parabolic (for diffusion-dominated processes) or hyperbolic (for convection-dominated processes). Traditional finite difference and finite element algorithms are generally accurate for solving the former but not the latter, in which case oscillations and smoothing of the wave front are introduced. This can be interpreted as an "artificial diffusion" intrinsic to these methods [1],[2].

This work presents a boundary element formulation for two-dimensional steady-state convection-diffusion problems, employing the proper fundamental solution of the equation. Thus, the problem is described in terms of boundary values only, consequently reducing its dimensionality by one. The paper is a follow-up of earlier work of Ikeuchi and Onishi [3] and Okamoto [4] using more refined numerical algorithms. The formulation is at present restricted to constant velocity fields; however, its extension to problems with variable velocity is possible through the use of the Dual Reciprocity Method [5],[6].

Results of several analyses are presented and compared to analytical solutions. They show that the boundary element formulation does not display any artificial diffusion or oscillatory behaviour even for relatively large values of the Péclet number, thus precluding the need for "upwind" or other algorithms common to finite element analysis.

FORMULATION OF THE PROBLEM

The two-dimensional steady-state convection-diffusion equation can be written in the form

$$D\nabla^2\phi - v_x\frac{\partial\phi}{\partial x} - v_y\frac{\partial\phi}{\partial y} = f(x,y,\phi) \tag{1}$$

where ϕ is the temperature, v_x and v_y the components of the velocity vector $\mathbf{v}$, D is the thermal conductivity (assuming the medium is homogeneous and isotropic) and f represents a heat source. Herein, sources whose intensity are a linear function of the temperature will be considered, in which case equation (1) becomes

$$D\nabla^2\phi - v_x\frac{\partial\phi}{\partial x} - v_y\frac{\partial\phi}{\partial y} - k\phi = 0 \tag{2}$$

where k is a proportionality factor. The mathematical description of the problem is complemented by boundary conditions of the Dirichlet, Neumann or Robin (mixed) types. Non-linear conditions of the radiative type can be included in the formulation as explained in [7].

The above differential equation can be transformed into an equivalent integral equation by applying a weighted residual technique. Starting with the weighted residual statement

$$\int_\Omega \left(D\nabla^2\phi - v_x\frac{\partial\phi}{\partial x} - v_y\frac{\partial\phi}{\partial y} - k\phi\right)\phi^* d\Omega = 0 \tag{3}$$

and integrating by parts twice the Laplacian and once the first-order derivatives, the following equation is obtained

$$\phi(\xi) = D\int_\Gamma \phi^*\frac{\partial\phi}{\partial n}d\Gamma - D\int_\Gamma \phi\frac{\partial\phi^*}{\partial n}d\Gamma - \int_\Gamma \phi\phi^* v_n d\Gamma \tag{4}$$

where $v_n = \mathbf{v}\cdot\mathbf{n}$, $\mathbf{n}$ is the unit outward normal vector and the dot stands for scalar product.

In the above equation, ϕ^* is the fundamental solution of equation (2), *i.e.* the solution of

$$D\nabla^2\phi^* + v_x\frac{\partial\phi^*}{\partial x} + v_y\frac{\partial\phi^*}{\partial y} - k\phi^* = -\delta(\xi,\eta) \tag{5}$$

in which ξ and η are the source and field points, respectively. It can be noticed that the sign of the first-derivative terms is reversed in (2) and (5), since this operator is not self-adjoint. For two-dimensional problems, ϕ^* is of the form

$$\phi^*(\xi,\eta) = \frac{1}{2\pi D} e^{-\frac{\mathbf{v}\cdot\mathbf{r}}{2D}} K_0(\mu r) \tag{6}$$

where

$$\mu = \left[\left(\frac{|\mathbf{v}|}{2D}\right)^2 + \frac{k}{D}\right]^{\frac{1}{2}} \tag{7}$$

and r is the modulus of $\mathbf{r}$, the distance vector between the source and field points. The derivative of the fundamental solution with respect to the outward normal direction is given by

$$\frac{\partial \phi^*}{\partial n} = \frac{1}{2\pi D} e^{-\frac{\mathbf{v}\cdot\mathbf{r}}{2D}} \left[-\mu K_1(\mu r)\frac{\partial r}{\partial n} - \frac{v_n}{2D} K_0(\mu r)\right] \tag{8}$$

In the above, K_0 and K_1 are Bessel functions of second kind, of orders zero and one, respectively.

Equation (4) permits calculating the value of ϕ at any internal point once the boundary values of ϕ and $\partial\phi/\partial n$ are all known. In order to obtain a boundary integral equation, the source point ξ is taken to the boundary and a limit analysis carried out due to the jump of $\partial\phi^*/\partial n$. The result is the equation

$$c(\xi)\phi(\xi) = D\int_\Gamma \phi^* \frac{\partial \phi}{\partial n} d\Gamma - D\int_\Gamma \phi \frac{\partial \phi^*}{\partial n} d\Gamma - \int_\Gamma \phi\phi^* v_n d\Gamma \tag{9}$$

in which $c(\xi)$ is a function of the internal angle the boundary Γ makes at point ξ [8].

NUMERICAL SOLUTION

For the numerical solution of the problem, equation (9) is written in a discretized form in which the integrals over the boundary are approximated by a summation of integrals over individual boundary elements, *i.e.*

$$c_i\phi_i = D\sum_{j=1}^{N}\int_{\Gamma_j} \phi^* \frac{\partial \phi}{\partial n} d\Gamma - D\sum_{j=1}^{N}\int_{\Gamma_j} \left(\frac{\partial \phi^*}{\partial n} + \frac{v_n}{D}\phi^*\right)\phi d\Gamma \tag{10}$$

where the index i stands for values at the source point ξ and N elements have been employed. In the above equation, it can be seen that

$$\frac{\partial \phi^*}{\partial n} + \frac{v_n}{D}\phi^* = \frac{1}{2\pi D} e^{-\frac{\mathbf{v}\cdot\mathbf{r}}{2D}} \left[-\mu K_1(\mu r)\frac{\partial r}{\partial n} + \frac{v_n}{2D} K_0(\mu r)\right] \tag{11}$$

Next, the variation of functions ϕ and $\partial\phi/\partial n$ within each element are approximated by interpolating from the values at the element nodes. Herein, linear elements are used, for which the expressions are

$$\phi = \Phi_1\phi_1 + \Phi_2\phi_2$$

$$\frac{\partial\phi}{\partial n} = q = \Phi_1 q_1 + \Phi_2 q_2$$

where Φ_1 and Φ_2 are linear interpolation functions. Substituting the above into equation (10), the following expression is obtained

$$c_i\phi_i = \sum_{j=1}^{N}\left(g_{ij}^1 q_1 + g_{ij}^2 q_2 - h_{ij}^1\phi_1 - h_{ij}^2\phi_2\right) \tag{12}$$

Note that the indexes 1 and 2 refer to the nodal (extreme) points of each element, and

$$g_{ij}^k = D\int_{\Gamma_j}\Phi_k\phi^* d\Gamma$$

$$h_{ij}^k = D\int_{\Gamma_j}\Phi_k\left(\frac{\partial\phi^*}{\partial n} + \frac{v_n}{D}\phi^*\right)d\Gamma$$

Adding up the contributions of adjoining elements to each nodal point, equation (12) can be rewritten as

$$c_i\phi_i = \sum_{j=1}^{N}(G_{ij}q_j - H_{ij}\phi_j) \tag{13}$$

The above equation involves N values of ϕ and N values of q, half of which are prescribed as boundary conditions. In order to calculate the remaining unknowns, it is necessary to generate N equations. This can be done by using a simple collocation technique, *i.e.* by making the equation be satisfied at the N nodal points. The result is a system of equations of the form

$$\mathbf{H}\boldsymbol{\phi} = \mathbf{G}\mathbf{q} \tag{14}$$

where the c values have been incorporated into the diagonal coefficients of matrix **H**. After introducing the boundary conditions, the system is reordered and solved by a direct method, *e.g.* Gauss elimination.

Evaluation of the coefficients of matrices **H** and **G** is carried out numerically. For the off-diagonal terms, a selective Gaussian integration with number of integration points as a function of the distance between source point and field element is employed, as described in [8]. The diagonal coefficients of matrix **G**

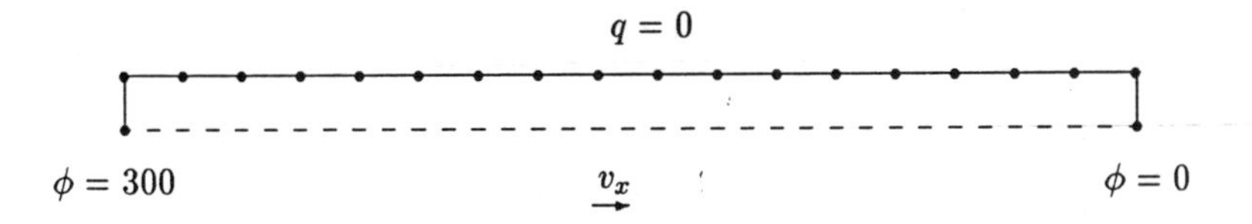

Figure 1: Moving bar: geometry, discretization and boundary conditions

have a weak singularity of the logarithmic type, and are calculated using the self-adaptive scheme of Telles [9]. The terms H_{ii} can be calculated, in the absence of heat generation, by noting that a consistent solution for a prescribed uniform temperature along the boundary can only be obtained if matrix $\mathbf{H}$ is singular, *i.e.*

$$H_{ii} = -\sum_{j=1}^{N} H_{ij} \quad (i \neq j) \tag{15}$$

However, when $k \neq 0$, there is heat flux even if a uniform temperature is applied. In this case, the coefficients H_{ii} have to be evaluated explicitly. These terms are composed of two parts, one being a sum of integrals of the form h_{ij}^k and the other the free term c_i. The former possesses a logarithmic plus a Cauchy Principal Value singularity (see expression 8); for the case of linear elements, the CPV term (embodied in the Bessel function K_1) vanishes since $\partial r/\partial n$ is identically zero; the logarithmic singular integral is also calculated using Telles' scheme [9]. The free terms c_i depend solely on geometry, and have the same values as for Laplace's equation [8].

APPLICATIONS

To test the validity of the present boundary element scheme, two numerical applications were studied. The first is the one-dimensional problem of a long bar moving at a constant velocity; the second is a plate with a sinusoidal temperature distribution along one face, with internal heat generation.

The moving bar

The algorithm was initially tested with the problem of a moving bar with constant velocity v_x and specified temperature at the edges, *i.e.* $\phi = 300$ at $x = 0$ and $\phi = 0$ at $x = L$. Other values adopted were $D = 1$, $k = 0$. The problem was analyzed as two-dimensional with cross-section 6.0×0.7, with two insulated edges parallel to the x-axis. Symmetry was taken into account by reflection and condensation as described in [8], thus only the upper half of the region needed be considered. A sketch of the problem is shown in figure 1.

The discretization employed 17 elements on the longer side and 1 element

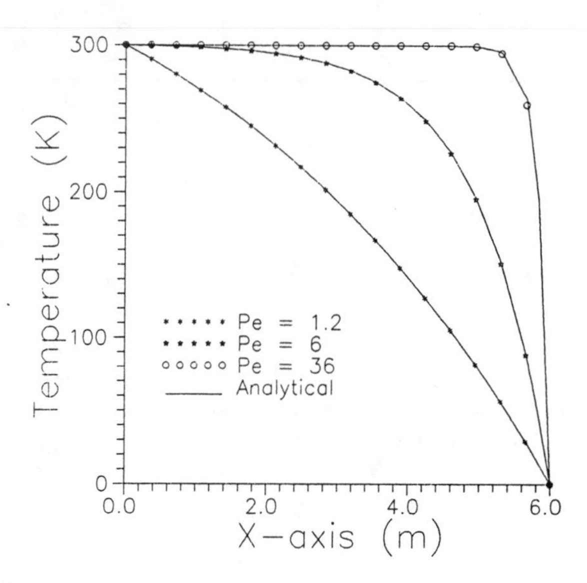

Figure 2: Results for positive velocity

on each of the smaller ones, making up a total of 19 linear elements and 22 nodes, for double nodes were used in the corners to allow for the discontinuity of the normal at these points [8].

Results are plotted in figures 2 and 3 for several velocity values, compared with the analytical solution:

$$\phi = 300 e^{\frac{v_x}{2}x} \frac{\sinh\left[\frac{v_x}{2}(L-x)\right]}{\sinh\left[\frac{v_x}{2}L\right]}$$

Besides the perfect agreement between the two solutions, it is important to notice that, unlike finite difference and finite element solutions, the boundary element results do not display any kind of oscillation or smoothing of the wave front even for moderately high values of the Péclet number ($Pe = 6v_x$ in the present case).

Plate with internal heat generation

The temperature distribution in a plate with the geometry and boundary conditions shown in figure 4 was studied next for a range of values of the Péclet number (from 10^{-6} to 8.33) and the parameter k (from 0 to 13.88). The discretization employed 20 linear elements and 23 nodes (with double nodes at corners), taking symmetry into account. A unit value was assumed for coefficient D.

The results obtained with the present boundary element scheme are plotted in figures 5 to 7 for the temperature along the centre line, compared with the following analytical solution:

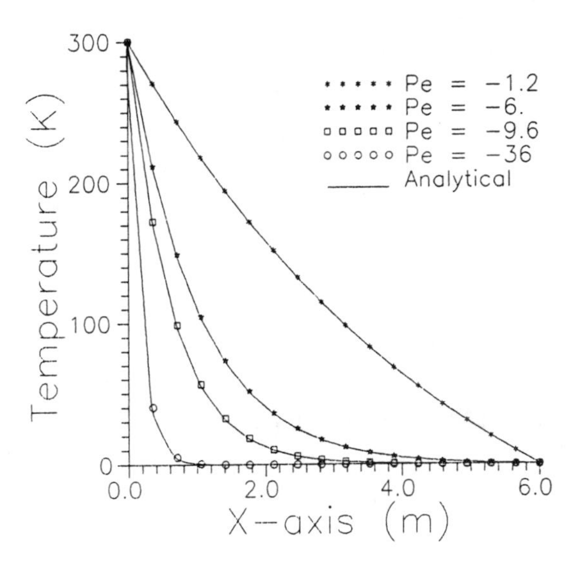

Figure 3: Results for negative velocity

$$\phi = \frac{e^{m_1(L-x)} - e^{m_2(L-x)}}{e^{m_1 L} - e^{m_2 L}} sin\frac{\pi}{l}y$$

where

$$m_i = \frac{1}{2}\left[-v_x \pm \sqrt{v_x^2 + 4\left(\frac{\pi^2}{L^2} + k\right)}\right] \quad (i = 1, 2)$$

and L and l are the dimensions in the x and y directions, respectively 6.0 and 4.0 in the present case.

It can be seen in the figures that the results compare very well with the analytical solution, showing again no oscillations or damping.

CONCLUSIONS

This paper reported an application of the boundary element method to two-dimensional steady-state convection-diffusion problems, employing the fundamental solution for the equation with a constant velocity field. Results of applications have shown that the solutions do not display numerical problems of oscillations and damping of the wave front, common in finite difference and finite element formulations.

The present research is now directed to extending the formulation to deal with transient problems and/or problems with variable coefficients, with the help

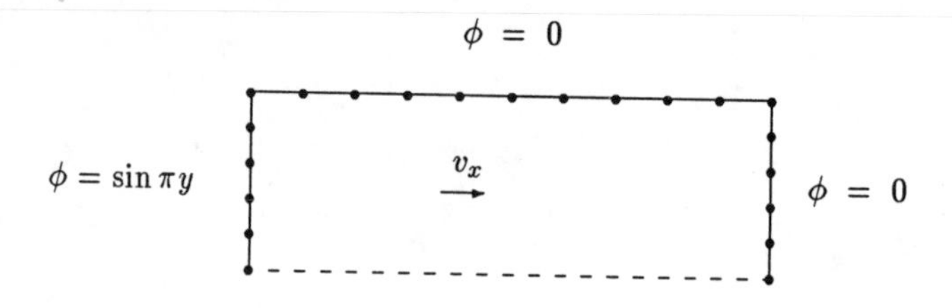

Figure 4: Rectangular plate: geometry, discretization and boundary conditions

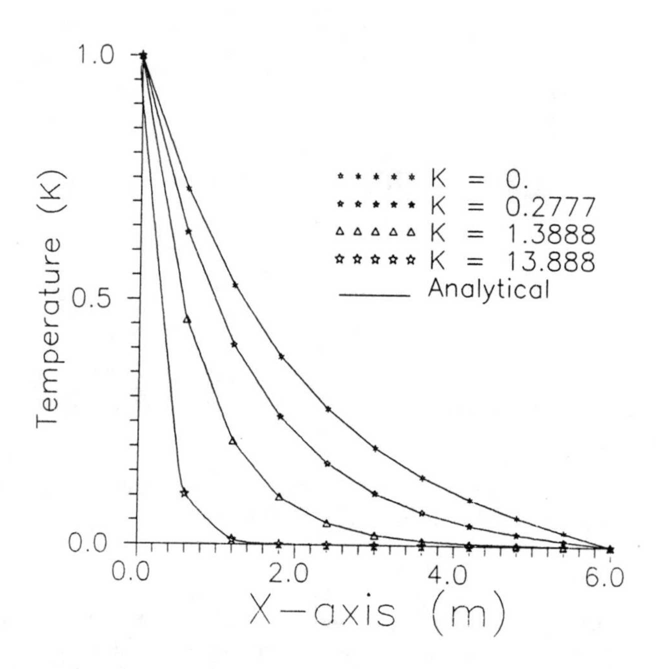

Figure 5: Results for $Pe = 1.6 \times 10^{-6}$

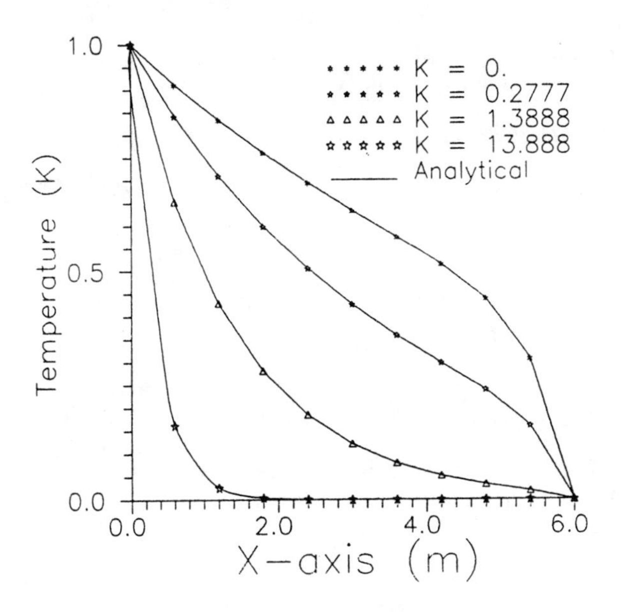

Figure 6: Results for $Pe = 1.66$

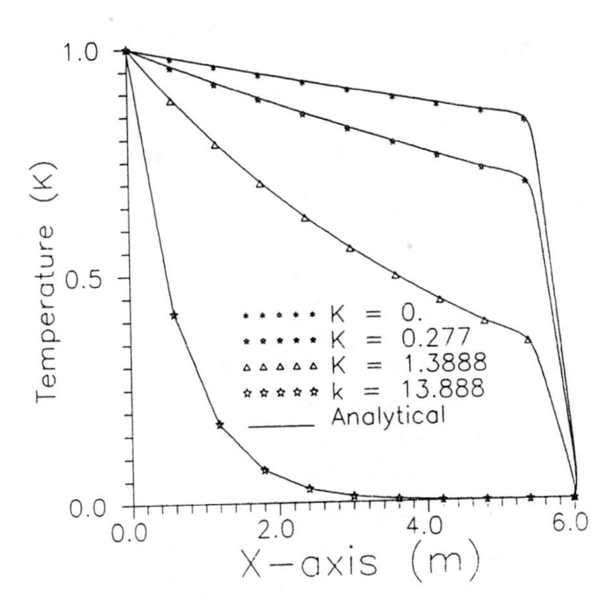

Figure 7: Results for $Pe = 8.33$

of the Dual Reciprocity Method [5],[6]. Results of these developments will be published in a forthcoming paper [10].

ACKNOWLEDGEMENTS

The first author would like to acknowledge the financial support of CNPq, Brazil.

REFERENCES

1. Roache, P.J., Computational Fluid Dynamics, Hermosa Publishers, Albuquerque, New Mexico, USA, 1972.

2. Hughes, T.J.R. (Ed.), Finite Element Methods for Convection Dominated Flows, ASME AMD-Vol.34, New York, USA, 1979.

3. Ikeuchi, M. and Onishi, K., Boundary Element Solutions to Steady Convective Diffusion Equations, Appl. Math. Mod., Vol. 7, pp. 115-118, 1983.

4. Okamoto, N., Boundary Element Method for Chemical Reaction System in Convective Diffusion, Proc. IV Int. Conf. on Numerical Methods in Laminar and Turbulent Flow, Pineridge Press, Swansea, UK, 1985.

5. Partridge, P.W., Brebbia, C.A. and Wrobel, L.C., The Dual Reciprocity Boundary Element Method, Computational Mechanics Publications, Southampton, 1990.

6. DeFigueiredo, D., Boundary Element Analysis of Convection-Diffusion Problems, Ph.D. Thesis, Wessex Institute of Technology, Southampton, UK (in preparation).

7. Azevedo, J.P.S. and Wrobel, L.C., Non-linear Heat Conduction in Composite Bodies: A Boundary Element Formulation, Int. J. Num. Meth. Engng, Vol. 26, pp. 19-38, 1988.

8. Brebbia, C.A., Telles, J.C.F. and Wrobel, L.C., Boundary Element Techniques, Springer-Verlag, Berlin and New York, 1984.

9. Telles, J.C.F., A Self-Adaptive Coordinate Transformation for Efficient Numerical Evaluation of General Boundary Element Integrals, Int. J. Num. Meth. Engng, Vol. 24, pp. 959-973, 1987.

10. DeFigueiredo, D. and Wrobel, L.C., A Boundary Element Analysis of Transient Convection-Diffusion Problems, to be presented at the XII Int. Conf. on Boundary Element Methods, Sapporo, Japan, 1990.

Integral Transform Solution of a Class of Transient Convection-Diffusion Problems

R.M. Cotta, R. Serfaty, R.O.C. Guedes(*)
Programa de Engenharia Mecânica, COPPE/UFRJ, Universidade Federal do Rio de Janeiro, Cidade Universitária, Cx. Postal 68503, Rio de Janeiro, RJ 21945, Brasil
(*) Present address: Mechanical & Aerospace Engineering Dept., North Carolina State University, USA

ABSTRACT

The generalized integral transform technique is further extended to allow for the analytical solution of a class of transient convection-diffusion problems, that can not be directly handled by the classical approaches. Besides the complete solution from the resulting coupled system of ordinary differential equations, approximate explicit solutions are presented for practical purposes. An application is considered related to the one- dimensional Burgers equation so as to illustrate convergence behavior and relative accuracy of the proposed solutions.

INTRODUCTION

In the last few years, a hybrid numerical-analytical approach to the solution of diffusion-type problems has been gradually advanced, based on formal exact solutions available for certain classes of problems [1]. This so-called generalized integral transform technique [2-10] resulted from various extensions of the ideas in the classical approach [1], as applied to a priori non-transformable problems, such as in the cases of problems with variable equation and/or boundary coefficients, problems with moving or irregular boundaries, problems that involve difficult auxiliary eigensystems, and nonlinear problems. For the definitive establishment of this recently developed computational tool in the heat and mass transfer field, an important class of problems yet to be dealt with is that of transient convection- diffusion, which models various physical situations of interest. The present work advances the generalized integral transform technique

(G.I.T.T.) to handle this class of problems, by offering a complete solution based on the numerical evaluation of an infinite system of coupled ordinary differential equations for the transformed potentials. In the realm of applications, approximate explicit solutions are also desirable, and the G.I.T.T. offers a quite straightforward lowest order solution and its once analytically iterated version as simple alternative expressions for practical estimations. In order to illustrate the formalism here advanced, the one-dimensional linearized Burgers equation [11] is more closely considered, allowing for a critical inspection of convergence characteristics of the complete solution as well as of the relative accuracy of approximate solutions.

ANALYSIS

For the sake of simplicity in the analysis that follows, we take a sufficiently general transient linear convection-diffusion problem in the form

$$w(\tilde{x}) \frac{\partial T(\tilde{x},t)}{\partial t} + \tilde{u}(\tilde{x},t) \cdot \nabla T(\tilde{x},t) = \nabla \cdot K(\tilde{x}) \nabla T(\tilde{x},t) - d(\tilde{x})\, T(\tilde{x},t) + P(\tilde{x},t) \quad , \quad \tilde{x} \in V,\ t > 0 \qquad (1.a)$$

with initial condition

$$T(\tilde{x},0) = f(\tilde{x}) \quad , \quad \tilde{x} \in V \qquad (1.b)$$

and boundary conditions

$$\left[\alpha(\tilde{x}) + \beta(\tilde{x})\, K(\tilde{x}) \frac{\partial}{\partial \tilde{n}} \right] T(\tilde{x},t) = \phi(\tilde{x},t) \quad , \quad \tilde{x} \in S \quad , \quad t > 0 \qquad (1.c)$$

The results from the classical integral transform technique [1] are not directly applicable to this class of problems, due to the appearance of the convection terms,

$\underset{\sim}{u} \cdot \underset{\sim}{\nabla}T$, in eq. (1.a). However, the ideas in the so-called generalized integral transform technique [2-10] can be extended as now demonstrated. By following the formalism in [4], the auxiliary problem is chosen as:

$$\underset{\sim}{\nabla} \cdot K(\underset{\sim}{x}) \underset{\sim}{\nabla} \psi(\mu_i,\underset{\sim}{x}) + \left[\mu_i^2 w(\underset{\sim}{x}) - d(\underset{\sim}{x})\right] \psi(\mu_i,\underset{\sim}{x}) = 0 \quad , \quad \underset{\sim}{x} \in V \tag{2.a}$$

with boundary conditions

$$\left[\alpha(\underset{\sim}{x}) + \beta(\underset{\sim}{x}) K(\underset{\sim}{x}) \frac{\partial}{\partial \underset{\sim}{n}}\right] \psi(\mu_i,\underset{\sim}{x}) = 0 \quad , \quad \underset{\sim}{x} \in S \tag{2.b}$$

whose solution is assumed to be known at this point.

The appropriate integral transform pair is readily derived from the eigenvalue problem (2) as

$$\bar{T}_i(t) = \int_V w(\underset{\sim}{x}) \frac{\psi_i(\underset{\sim}{x})}{N_i^{1/2}} T(\underset{\sim}{x},t)\, dv \quad , \quad \text{transform} \tag{3.a}$$

$$T(\underset{\sim}{x},t) = \sum_{i=1}^{\infty} \frac{1}{N_i^{1/2}} \psi_i(\underset{\sim}{x}) \bar{T}_i(t) \quad , \quad \text{inversion} \tag{3.b}$$

We now operate on eq. (1.a) with $\int_V \frac{\psi_i(\underset{\sim}{x})}{N_i^{1/2}} dv$, to obtain

$$\frac{d\bar{T}_i(t)}{dt} + \int_V \frac{\psi_i(\underset{\sim}{x})}{N_i^{1/2}} \left[\underset{\sim}{u}(\underset{\sim}{x},t) \cdot \underset{\sim}{\nabla} T(\underset{\sim}{x},t)\right] dv = -\mu_i^2 \bar{T}_i(t) +$$

$$+ \bar{g}_i(t) \quad , \quad t > 0 \quad , \quad i = 1,2,... \tag{4.a}$$

where the known source term $\bar{g}i(t)$ is given by

$$\bar{g}_i(t) = \int_V P(\tilde{x},t) \frac{\psi_i(\tilde{x})}{N_i^{1/2}} dv + \frac{1}{N_i^{1/2}} \int_S K(\tilde{x}) \left[\psi_i(\tilde{x}) \frac{\partial T}{\partial \tilde{n}} - \right.$$

$$\left. - T(\tilde{x},t) \frac{\partial \psi_i(\tilde{x})}{\partial \tilde{n}} \right] ds , \tag{4.b}$$

The integration in eq. (4.a) can be performed by making use of the inversion formula, eq. (3.b), to yield

$$\frac{d \bar{T}_i(t)}{dt} + \mu_i^2 \bar{T}_i(t) + \sum_{j=1}^{\infty} A_{ij}^*(t) \bar{T}_j(t) = \bar{g}_i(t), \; i=1,2,... \tag{5.a}$$

where,

$$A_{ij}^*(t) = \frac{1}{N_i^{1/2} N_j^{1/2}} \int_V \psi_i(\tilde{x}) \left[\tilde{u}(\tilde{x},t) \cdot \tilde{\nabla} \psi_j(\tilde{x}) \right] dv \tag{5.b}$$

Also, the initial condition, eq. (1.b), is transformed according to the operator $\int_V w(\tilde{x}) \frac{\psi_i(\tilde{x})}{N_i^{1/2}} dv$ that provides

$$\bar{T}_i(0) = \bar{f}_i \equiv \int_V w(\tilde{x}) \frac{\psi_i(\tilde{x})}{N_i^{1/2}} f(\tilde{x}) \, dv \quad , \quad i = 1,2,... \tag{5.c}$$

Equations (5.a) are, in fact, a system of infinetly many coupled ordinary differential equations for the transformed potentials, $\bar{T}_i$'s, and once system (5) has been solved, the inversion formula (3.b) is recalled to produce the desired complete potential, $T(\underset{\sim}{x},t)$. From the computational point of view, system (5) is actually truncated at the $N^{\underline{th}}$ row and column, with N sufficiently large for the required accuracy, and the finite system is then readily handled by standard numerical approaches available in well-established scientific subroutines packages, such as subroutine DGEAR from the IMSL library. The formal aspects of convergence analysis and error bounds for the truncated system have been previously investigated for such infinite system of O.D.E.'s [12], and for practical purposes it suffices to observe the solution convergence by increasing the order of truncation N. In matrix form, the truncated version of system (5) becomes

$$\underset{\sim}{y}'(t) + A(t)\,\underset{\sim}{y}(t) = \underset{\sim}{g}(t) \quad , \quad t > 0 \tag{6.a}$$

$$\underset{\sim}{y}(0) = \underset{\sim}{f} \tag{6.b}$$

where,

$$\underset{\sim}{y} = \left\{ \bar{T}_1(t),\ \bar{T}_2(t),\ \ldots,\ \bar{T}_N(t) \right\}^T \tag{6.c}$$

$$A = \left\{A_{ij}\right\} \ , \ A_{ij} = \delta_{ij}\,\mu_i^2 + A_{ij}^{*}(t) \ , \ i,j = 1,2,\ldots,N \tag{6.d}$$

$$\underset{\sim}{g} = \left\{ \bar{g}_1(t),\ \bar{g}_2(t),\ \ldots,\ \bar{g}_N(t) \right\} \tag{6.e}$$

$$\underset{\sim}{f} = \left\{ \bar{f}_1,\ \bar{f}_2,\ \ldots,\ \bar{f}_N \right\}^T \tag{6.f}$$

An interesting special case occurs for t-independent function vector, $\underset{\sim}{u}(\underset{\sim}{x},t) \equiv \underset{\sim}{u}(\underset{\sim}{x})$, when the matrix of coefficients, A, becomes constant, and the solution of system (6) can be explicitly written in terms of the exponential matrix, after variation of parameters has been applied to this nonhomogeneous system to yield

$$\underset{\sim}{y}(t) = \exp(At)\,\underset{\sim}{f} + \int_0^t \exp\left[A(t-t')\right]\underset{\sim}{g}\,dt' \qquad (7.a)$$

while the exponential matrix can be computed once eigenvalues and eigenvectors of A have been determined through solving the algebraic problem

$$(A - \lambda I)\,\underset{\sim}{\xi} = 0 \qquad (7.b)$$

Again, scientific subroutine libraries are readily available to accurately accomplish this task [12].

Approximate Solutions

In the realm of applications, however, the establishment of approximate explicit solutions might be of interest. Through the same arguments employed in [4], a lowest order solution can be derived provided the non-diagonal elements of the coefficients matrix A are not so significant with respect to diagonal ones, which would then approximately correspond to a decoupled system; therefore, by retaining only the diagonal elements of A, such an approximate solution is obtained from:

$$y'_{\ell,i}(t) + A_{ii}(t)\,y_{\ell,i}(t) = \bar{g}_i(t)\ ,\ t > 0\ ,\ i = 1,2,\ldots \qquad (8.a)$$

$$y_{\ell,i}(0) = \bar{f}_i \qquad (8.b)$$

which is readily solved for each mode i, in the explicit form

$$y_{\ell,i}(t) = \bar{f}_i \exp\left[-\int_0^t A_{ii}(t')\,dt'\right] +$$

$$+ \int_0^t \bar{g}_i(t') \exp\left[- \int_{t'}^t A_{ii}(t'')\, dt'' \right] dt' \qquad (8.c)$$

In addition, this explicit solution can be used to correct, though approximately, for less negligible non-diagonal elements, by substituting it into the terms of the infinite summation that correspond to non-diagonal elements of *A*, so as to provide an analytically iterated lowest order solution, obtained from the solution of the still decoupled system given below

$$y'_{h,i}(t) + A_{ii}(t) y_{h,i}(t) = \bar{g}_i(t) + G_i(t) \ , \ t > 0 \ , \ i = 1,2,\ldots \qquad (9.a)$$

$$y_{h,i}(0) = \bar{f}_i \qquad (9.b)$$

where,

$$G_i(t) = - \sum_{\substack{j=1 \\ j \neq i}}^{\infty} A_{ij}(t) y_{\ell,j}(t) \qquad (9.c)$$

This higher order solution can then be written in terms of a correction term to the lowest order solution as

$$y_{h,i}(t) = y_{\ell,i}(t) + y_{c,i}(t) \qquad (10.a)$$

where the correction is given by

$$y_{c,i}(t) = \int_0^t G_i(t') \exp\left[- \int_{t'}^t a_{ii}(t'')\, dt'' \right] dt' \qquad (10.b)$$

Again, error bounds for the approximate solutions have been established elsewhere [12], in terms of norms of the matrix formed by the non-diagonal elements only, and there is no need to repeat such analysis here.

APPLICATION AND DISCUSSION

The one-dimensional linearized Burgers equation [11], which provides the simplest mathematical model for the convection-diffusion class of problems, is here considered for illustration purposes, according to the following formulation:

$$\frac{\partial T(x,t)}{\partial t} + u_o \frac{\partial T(x,t)}{\partial x} = \nu \frac{\partial^2 T}{\partial x^2} \quad , \quad 0 < x < 1 \quad , \quad t > 0 \qquad (11.a)$$

with initial and boundary conditions

$$T(x,0) = 1 \quad , \quad 0 \le x \le 1 \qquad (11.b)$$

$$T(0,t) = 1 \quad , \quad T(1,t) = 0 \quad , \quad t > 0 \qquad (11.c,d)$$

The boundary conditions can be made homogeneous for best computational performance by extracting the steady-state solution as:

$$T(x,t) = T_\infty(x) + T^*(x,t) \qquad (12.a)$$

where,

$$T_\infty(x) = \frac{1 - \exp\left[\lambda(x-1)\right]}{1 - \exp(-\lambda)} \qquad (12.b)$$

so that the problem for $T^*(x,t)$ becomes

$$\frac{\partial T^*(x,t)}{\partial t} + u_o \frac{\partial T^*(x,t)}{\partial x} = \nu \frac{\partial^2 T^*(x,t)}{\partial x^2} \quad , \quad 0 < x < 1 \quad , \quad t > 0 \qquad (13.a)$$

$$T^*(x,0) = f(x) \equiv 1 - T_\infty(x) \quad , \quad 0 \le x \le 1 \qquad (13.b)$$

$$T^{*}(0,t) = T^{*}(1,t) = 0 \quad , \quad t > 0 \qquad (13.c,d)$$

and the expressions previously developed are directly applied to the problem formulated by eqs. (13).

In order to report a few numerical results, the complete solution of this problem was obtained for $N \leq 30$, which was sufficient for convergence within the range of t considered, and for different combinations of the parameters that govern the relative magnitudes of convection (u_o) and diffusion (ν). Table I indicates, as usual in eigenfunction expansion techniques, that a larger number of terms is required for convergence when time decreases; it also demonstrates that the complete solution convergence behavior is not markedly affected by the relative magnitude of the non-transformable convection term (u_o=10, 2, 1 or 0.1), where results for N=30 are fully converged for all the cases considered. Also shown are results for the lowest order solution, which confirm the reasonably good accuracy in general terms, specially for less important non-diagonal elements (smaller u_o), when the coupling of the infinite system of O.D.E.'s becomes weaker. These results can be further improved by considering the iterated lowest order solution of the previous section, although under most practical situations the relative accuracy of such simple expressions might suffice. When accuracy is at a premium, the complete solution stil provides a faster and more reliable alternative to purely numerical approaches. The last column in Table I illustrates the excellent agreement with numerical results from the well-established subroutine DMOLCH of the IMSL library, that employs the method of lines and a collocation scheme with cubic Hermite polynomials.

TABLE I - Convergence of complete solution, T(x,t), and accuracy of lowest order solution, $T_\ell(x,t)$

	t = 0.1		(u_o = .1	; ν =1)		Nume-
x/N	5	10	15	30	L.O.S.	rical*
0.1	0.97112	0.97106	0.97107	0.97107	0.97269	0.97105
0.3	0.89006	0.89006	0.89006	0.89006	0.89267	0.88999
0.5	0.74378	0.74377	0.74377	0.74377	0.74493	0.74366
0.7	0.50536	0.50534	0.50534	0.50534	0.50449	0.50524
0.9	0.18105	0.18110	0.18110	0.18109	0.18003	0.18105

	t = 0.5		(u_o =.1	; ν = 1)		Nume-
x/N	5	10	15	30	L.O.S.	rical*
0.1	0.90579	0.90579	0.90579	0.90579	0.90583	0.90579
0.3	0.71400	0.71400	0.71400	0.71400	0.71406	0.71400
0.5	0.51696	0.51696	0.51696	0.51696	0.51698	0.51695
0.7	0.31421	0.31421	0.31421	0.31421	0.31420	0.31421
0.9	0.10596	0.10597	0.10597	0.10597	0.10595	0.10596

	t = 0.1		(u_o =1.0	; ν = 1)		Nume-
x/N	5	10	15	30	L.O.S.	rical*
0.1	0.98145	0.98101	0.98104	0.98105	0.99402	0.98103
0.3	0.92105	0.92107	0.92109	0.92108	0.94444	0.92103
0.5	0.79842	0.79823	0.79821	0.79821	0.81247	0.79813
0.7	0.57225	0.57211	0.57207	0.57206	0.56947	0.57198
0.9	0.21993	0.22033	0.22026	0.22024	0.21281	0.22020

	t = 0.5		(u_o =1.0	; ν = 1)		Nume-
x/N	5	10	15	30	L.O.S.	rical*
0.1	0.93958	0.93957	0.93957	0.93957	0.93993	0.93957
0.3	0.79864	0.79864	0.79864	0.79864	0.79936	0.79864
0.5	0.62553	0.62553	0.62553	0.62553	0.62613	0.62553
0.7	0.41277	0.41276	0.41276	0.41276	0.41299	0.41276
0.9	0.15170	0.15171	0.15170	0.15170	0.15168	0.15170

(*) IMSL routine DMOLCH (31 grid points)

TABLE I (Cont.) - Convergence of complete solution, T(x,t), and accuracy of lowest order solution, $T_{\ell}(x,t)$

t = 0.1			(u_o = 2.	;	ν = 1)	Nume-
x/N	5	10	15	30	L.O.S.	rical*
0.1	0.98917	0.98851	0.98854	0.98855	1.0076	0.98854
0.3	0.94705	0.94709	0.94710	0.94709	0.98530	0.94705
0.5	0.84976	0.84933	0.84928	0.84928	0.87845	0.84922
0.7	0.64281	0.64256	0.64248	0.64247	0.64639	0.64240
0.9	0.26667	0.26723	0.26708	0.26706	0.25850	0.26702

t = 0.5			(u_o = 2.	;	ν = 1)	Nume-
x/N	5	10	15	30	L.O.S.	rical*
0.1	0.96567	0.96566	0.96566	0.96566	0.96623	0.96566
0.3	0.87234	0.87234	0.87234	0.87234	0.87363	0.87234
0.5	0.73260	0.73259	0.73259	0.73259	0.73390	0.73259
0.7	0.52333	0.52332	0.52332	0.52332	0.52411	0.52332
0.9	0.21034	0.21035	0.21035	0.21035	0.21052	0.21035

t = 0.1			(u_o = 10.	;	ν = 1)	Nume-
x/N	5	10	15	30	L.O.S.	rical*
0.1	1.0002	0.99993	0.99994	0.99994	1.0055	0.99994
0.3	0.99953	0.99923	0.99927	0.99926	1.0147	0.99926
0.5	0.99470	0.99370	0.99377	0.99376	1.0146	0.99376
0.7	0.95264	0.95122	0.95135	0.95132	0.96914	0.95131
0.9	0.63447	0.63322	0.63332	0.63329	0.63976	0.63329

t = 0.5			(u_o = 10.	;	ν = 1)	Nume-
x/N	5	10	15	30	L.O.S.	rical*
0.1	0.99992	0.99992	0.99992	0.99992	1.0000	0.99992
0.3	0.99913	0.99913	0.99913	0.99913	0.99947	0.99913
0.5	0.99331	0.99331	0.99331	0.99331	0.99372	0.99331
0.7	0.95026	0.95026	0.95026	0.95026	0.95059	0.95025
0.9	0.63215	0.63215	0.63215	0.63215	0.63228	0.63215

(*) IMSL routine DMOLCH (31 grid points)

REFERENCES

1. Mikhailov, M.D. and M.N. Özisik, "Unified Analysis and Solutions of Heat and Mass Diffusion", John Wiley, New York, 1984.

2. Özisik, M.N. and R.L. Murray, "On the Solution of Linear Diffusion Problems with Variable Boundary Condition Parameters", J. Heat Transfer, V. 96, pp. 48-51, 1974.

3. Mikhailov, M.D., "On the Solution of the Heat Equation with Time Dependent Coefficient", Int. J. Heat & Mass Transfer, V. 18, pp. 344-345, 1975.

4. Cotta, R.M. and M.N. Özisik, "Diffusion Problems with General Time-Dependent Coefficients", Braz. J. Mech. Sciences, RBCM, V. 9, no. 4, pp. 269-292, 1987.

5. Cotta, R.M. and M.N. Özisik, "Laminar Forced Convection in Ducts with Periodic Variation of Inlet Temperature", Int. J. Heat & Mass Transfer, V. 29, no. 10, pp. 1495-1501, 1986.

6. Cotta, R.M., "Steady-State Diffusion with Space-Dependent Boundary Condition Parameters", Proc. of the 1st National Meeting on Thermal Sciences - ENCIT 86, pp. 163-166, Rio de Janeiro, Brasil, 1986.

7. Cotta, R.M., "Diffusion in Media with Prescribed Moving Boundaries: - Application to Metals Oxidation at High Temperatures", II Latin American Congress of Heat & Mass Transfer, V. 1, pp. 502-513, São Paulo, Brasil, 1986.

8. Aparecido, J.B., R.M.Cotta, and M.N. Özisik, "Analytical Solutions to Two-Dimensional Diffusion Type Problems in Irregular Geometries", J. Franklin Inst., V. 326, pp.421-434, 1989.

9. Cotta, R.M., "On the Solution of Periodic Multidimensional Diffusion Problems", Int. Comm. Heat & Mass Transfer, V. 16, no. 4, pp. 569-579, 1989.

10. Cotta, R.M., "Hybrid Numerical-Analytical Approach to Nonlinear Diffusion Problems", Num. Heat Transfer, in press.

11. Benton, E.R. and G.W. Platzman, "A Table of Solutions of the One-Dimensional Burgers Equation", Quart. of Appl. Math., pp. 195-212, July, 1972.

12. Cotta, R.M., "Hybrid Numerical-Analytical Approach to Diffusion/Convection Problems", Proc. of the XV National Summer School (Invited Lecture), Institute of Applied Mahematics & Computer Science, Bulgaria, August 1989.

Comparison of Control Volume- and Weighted Residual- FEMs for Transient Convection-Diffusion Problems

J. Banaszek

Institute of Heat Engineering, Warsaw Technical University, 00-665 Warsaw, Poland

ABSTRACT

The weighted residual- and control-volume based FEMs are compared for the convection-diffusion problems. The latter method is inferior to the former one in terms of the mass balance and dispersion errors. Therefore, some comments are given on how to improve the performance of the CVFEM over two-dimensional FE grids.

INTRODUCTION

Many papers published over the past decade have shown that the FEM can be successfully applied in the heat transfer and fluid flow problems but its utility and efficiency are still being investigated in order to study the possibilities offered by the FE methodology and to contrast the method with alternative ones, at present well established. However, still another question seems to be equally important in the subject,namely the one of whether the FE numerical analogue properly reflects major physical features of the phenomena considered. For engineers and physicists such appraisal is the most convincing way for verifying a correctness of the discrete model used.

In this context, some comments are given on the conservative property of the FE models and on a degradation in their accuracy,when the solenoidal velocity field is interpolated with the functions which do not satisfy the continuity requirement. Furthermore, using the Fourier wave analysis the dissipation and phase errors are examined for both the weighted residual FEM and the Control Volume FEM.

The latter method becomes more and more popular due to both its direct physical background and the retention of the FE discretization procedure versatility. Over the past decade the CVFEM has been successfully used for transient and steady-state convection-diffusion problems (e.g. [1], [2], [3]). Nevertheless, its features and potential advantages over the weighted residual FEM have not been the subject of a comprehensive study. The aim of this paper is therefore to present some results drawn from the comparative analysis of both methods when they are applied to discretize the convection-diffusion problems.

CONSERVATIVE PROPERTY AND MASS BALANCE ERROR

The conservation principle for a scalar quantity Φ, transferred by convection and diffusion in a control-volume Ω_k confined by the surface Γ_k, has the following integral form:

$$\int_{\Omega_k} \frac{\partial}{\partial t}(\rho\Phi)\,d\Omega + \int_{\Gamma_k}(\rho u_i \Phi - \lambda_{ij}\frac{\partial\Phi}{\partial x_j})n_i\,d\Gamma = \int_{\Omega_k} S\,d\Omega \,, \tag{1}$$

where ρ, u_i, λ_{ij} and S are respectively density, components of velocity vector and of diffusivity tensor and rate of volumetric source occuring in Ω_k.

For an infinitesimally small control-volume equation (1) assumes a commonly known local differential form

$$\rho\frac{\partial\Phi}{\partial t} + \frac{\partial}{\partial x_i}(\rho u_i \Phi - \lambda_{ij}\frac{\partial\Phi}{\partial x_j}) = S \qquad \text{for } i,j = 1,2,3 \tag{2}$$

which, through the *continuity condition* for an incompressible fluid, is equivalent to its *non-conservative* form

$$\rho\frac{\partial\Phi}{\partial t} + \rho u_i\frac{\partial\Phi}{\partial x_i} = \frac{\partial}{\partial x_i}(\lambda_{ij}\frac{\partial\Phi}{\partial x_j}) + S \tag{3}$$

The integral formulation (1) forms a basis for setting up equations of both the *Control Volume FDM* (e.g.[4],[9]) and the *Control Volume FEM* (e.g [1],[2],[3]) whereas equation (2) or (3) is used in the weighted residual *Petrov-Galerkin FEM* (e.g. [5]).

A reasonable discrete analogue, correctly modelling a behaviour of the continuum by means of a finite number of nodal unknowns, should ensure a preservation of the physical conservation principle.

In the *FE* discretization procedure the *conservative property* analysis should take into account two features of this numerical model. Namely, the standard *FE* approximation is based on independent interpolations of both the scalar Φ and components of a velocity vector $\bar{u}$ in the local curvilinear coordinates $\bar{\zeta}$ (e.g. [7])

$$\Phi(\bar{\zeta}) = N_m(\bar{\zeta})\Phi_m(t) \qquad \text{for } m = 1,2..N_\Phi \tag{4}$$

$$u_i(\bar{\zeta}) = P_l(\bar{\zeta})u_{i,l}(t) \qquad \text{for } \begin{array}{l} i = 1,2,3; \\ l = 1,2..N_u \end{array} \tag{5}$$

Moreover, the individual nodal equation k in the *PGFEM* cannot in general be considered as the balance equation for any finite control-volume surrounding node k. On the other hand, the *CVFEM* nodal equation is obtained by setting up the local integral balance of Φ within the control-volume

Ω_k associated with node k.

This different meaning of the PGFEM and CVFEM nodal equations is a reason to distinguish two levels of the *conservative property* for the FE analogue, i.e. the *global* and the *local* ones.

The *global conservative property* (*GCP*) is achieved if the FE solution satisfies the integral balance equation (1) within the whole region Ω, irrespective of the grid density element shape, weighting and interpolation functions used.

The FEM analogue is said to have the local *conservative property* (LCP) if (1) is also valid in each Ω_k, confined by imaginary or real boundary surfaces Γ_k.

It can be easily shown that the sufficient condition of the *GCP* is expressed by the requirement that the weighting functions W_k should satisfy

$$\sum_{k=1}^{N_\Phi} W_k = 1 \quad ====> \quad \sum_{k=1}^{N_\Phi} \frac{\partial W_k}{\partial x_i} = 0 \qquad \text{for } i,j = 1,2 \text{ or } 3 \tag{6}$$

The *CVFEM* provides a numerical analogue inherently possessing the *LCP*, as its individual nodal equation is set up by the local balance of the quantity Φ within Ω_k. For diffusive-type problems the *local conservative property* does frequently offer higher accuracy of the *FE* solution in comparison with the one obtained from the only *globally conservative* weighted residual model (e.g. [3],[7]). Furthermore, the *CVFEM* possesses less stringent stability requirements and it assures a preservation of the *discrete maximum principle* for a greater variety of time-space division than the GFEM does [3].

Unfortunately, the superior performance of the CVFEM model is not confirmed for the convection dominated problems, where the substantial mass balance and numerical dispersion errors can deteriorate quality of the solution.

To show this, both techniques have been applied to solve the energy equation for the steady-state forced convection due to the flow of a viscous incompressible fluid inside of the rectangular region $-1 \le x_1 \le 1, 0 \le x_2 \le 1$. The solenoidal velocity field is specified analytically [8] as

$$u_1 = 2x_2(1-x_1^2) \quad \text{and} \quad u_2 = -2x_1(1-x_2^2) \tag{7}$$

The *Dirichlet* condition for temperature Φ is assumed for all boundary surfaces except the outlet one. Along the inlet $\{ -1 \le x_1 \le 0,\ x_2 = 0 \}$ $\Phi = 2(1+x_1)$ is taken, whereas $\Phi = 0$ is maintained at the other boundary surfaces. Along the outlet $\{ 0 \le x_1 \le 1, x_2 = 0 \}$ the condition $\partial\Phi/\partial x_2 = 0$ is used. To verify the accuracy of FEM solutions two different criteria are taken into account. It is a local criterion, based on the maximum value of the absolute relative error (MRE), and the global one, defined in terms of RMS error.

The accuracy of the CVFEM and Galerkin FEM solutions is compared in Figure 1 for various Peclet numbers in the bilinear and biquadratic FE grids.

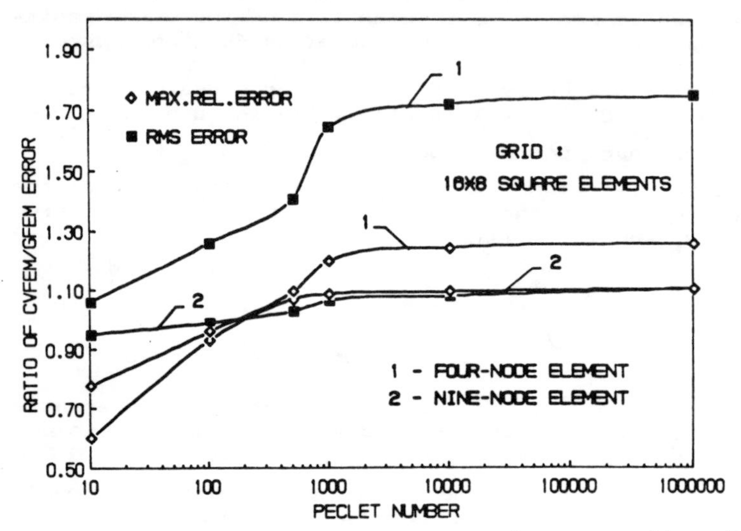

Figure 1 Comparison of FEMs errors for convection-diffusion test problem

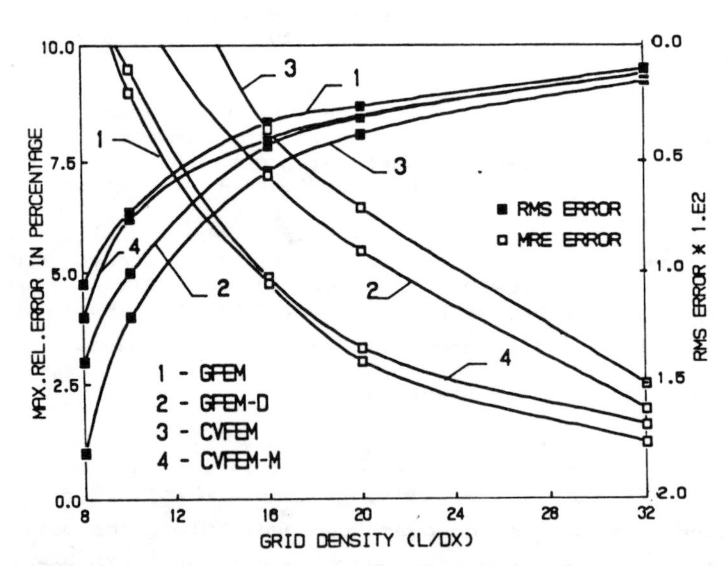

Figure 2 Comparison of FEMs errors for pure convection test problem

For low and medium *Peclet* numbers a reference solution is the one obtained with a grid so fine that further refinements produce only a negligible change in the solutions. For the pure convection transport,i.e. when the *Peclet* number tends to infinity, the reference nodal values are equal to appropriate values of the exact solution. It can be obtained owing to a coincidence of the stream- and isothermal lines. The MRE and RMS errors are compared in Figure 2 for the CVFEM solution and for both weighted residual solutions i.e. the GFEM one which is based on equation (3) and the GFEM-D one which is obtained from the same Galerkin approach but used to equation (2). Thorough analysis of these results leads to the following conclusions.

The control-volume based solution is more accurate only for low Peclet numbers (Figure 1).

Poor performance of the CVFEM and GFEM-D models in the bilinear FE grid is mainly due to a substantial mass balance error. The biquadratic interpolation polynomial P_l ensures the exact piece-wise description of the velocity $\bar{u}$ specified by (7). Unfortunately, the bilinear polynomial P_l gives rather a rough approximation of (7) and thus the continuity condition is not preserved in any point of the domain. In consequence, the mass balance error occurs and in turn it can lead to a significant error in the scalar field calculated. To estimate this inaccuracy, the pure convection test problem has been solved in the isothermal case, i.e. in the case where a unit constant temperature on the inlet and the condition $\partial\Phi/\partial n=0$ on the outlet are taken. The other boundary surfaces are assumed to be adiabatic ($\partial\Phi/\partial n=0$ and $\bar{u}*\bar{n}=0$). The MRE and RMS errors are collected in Table 1 for both the CVFEM and GFEM-D solutions.

TABLE 1 MRE and RMS errors in isothermal flow

GRID / METHOD	TRIANGULAR GRID 32*16 nodes 1024 elements		SQUARE GRID 32*16 nodes 512 elements	
	MRE	RMS	MRE	RMS
CVFEM linear interpol.	17.7%	1.92E-2	8.5%	8.5E-3
CVFEM upwind [1]	11.9%	1.33E-2	—	—
GFEM-D	—	—	6.84 %	6.94E-3

Their substantial values impair the credibility of these discrete models and show a reason of a poor performance of the GFEM-D model in comparison with the GFEM one (Figure 2). The superior accuracy of GFEM can be explained as follows. When the GFEM-D model is used, a sum of convective terms of all nodal equations for element e takes the form

$$\int_{\Omega_e} \frac{\partial}{\partial x_i}\{\rho(P_l u_{i,l})(N_m \Phi_m)\}d\Omega = \int_{\Omega_e} \rho(P_l u_{i,l})(\frac{\partial N_m}{\partial x_i}\Phi_m)d\Omega + \int_{\Omega_e} \rho(\frac{\partial P_l}{\partial x_i} u_{i,l})(N_m \Phi_m)d\Omega \tag{8}$$

providing the condition (6) is satisfied. On the other hand, rough approximation of the solenoidal velocity field generates a redundant amount of mass Δm_e in Ω_e. It in turn results in an additional amount of Φ

$$\Delta m_e \Phi_e = (\int_{\Omega_e} \rho \frac{\partial P_l}{\partial x_i} u_{i,lm} d\Omega)\Phi_e \tag{9}$$

If it is distributed into nodal shares, according to the weighted residual technique used,one obtains the second element integral of the RHS of (8). Next, as this integral comes from the mass balance error,it seems to be reasonable to neglect it. This leads to the GFEM approximation for the convective term in (3). Therefore,the sum of the convective integrals of all GFEM nodal equations in element e can be considered as a net overall convective flux passing through the element boundary but freed from the error caused by improper FE approximation of the velocity $\bar{u}$.

To reduce the CVFEM solution error resulting from a rough velocity interpolation, one should subtract a redundant amount of Φ in Ω_k

$$\Delta m_e \Phi_k = (\int_{\Omega_k} \rho P_l u_{i,l} n_i d\Omega)\Phi_k \tag{10}$$

from the balance (1). This gives the nodal equation for the modified CVFEM,further labelled by the abbrevation CVFEM-M. This equation can be considered as the balance of Φ within Ω_k but calculated for the corrected amount of mass comprised in Ω_k. The improved accuracy of the CVFEM-M solution is confirmed in Figure 2. Moreover, the CVFEM-M and GFEM models ensure nodally exact solution for the isothermal flow test problem.

The above way used to improve the accuracy of the CVFEM solution is similar to the one proposed by Patankar [4] and Raithby & Schneider [9] in the control-volume based finite difference method (CVFDM). Moreover, it also meets a frequently taken a priori assumption that a diagonal term of the convective-diffusive CVFEM matrix should be equal to a sum of absolute values of all its off- diagonal terms.

COMPARATIVE STUDY OF NUMERICAL DISPERSION

The convection dominated transport problem,characterized by large *Peclet* number (*Pe*),is governed by equation (1) or (2) with $\lambda_{ij} \longrightarrow 0$. A common way to compare the accuracy of nu-

merical models utilized in this case is via *Fourier* analysis of the constant velocity,pure convection problem ($Pe=\infty$). When a single *Fourier* mode is placed on a uniform rectangular FE mesh, over which the problem is approximated, the resulting phase speed $\bar{c}$ differs from the phase speed c of the continuous solution due to the numerical dispersion.
On the assumption that only waves of a constant phase angle $\rho=k_j\Delta x_j$ (where Δx_j is the mesh size and k_j is the j-th component of the wave number vector) are chosen, the phase velocity ratio $\bar{c}/c$ of the semi-discrete solution takes the form

$$\frac{\bar{c}}{c} = \frac{\sin(\rho)}{\rho(2+\beta)} * \frac{2\cos(\rho)+\beta}{r_1 + \cos(\rho)*(2r_2+r_3\cos(\rho))} \tag{11}$$

where $\beta=\infty$, 4 or 6 for the CVFDM,GFEM or CVFEM, respectively. The coefficients r_i result from the spatial approximation of the time derivative $\partial\Phi/\partial t$ and they constitute the mass matrix **M** for the bilinear 4-node FE in the manner shown in Figure 4. Their values are given in Figure 4 for both the Lumped Mass Matrix (L) and Consistent Mass Matrix (C) schemes. In CVFEM, the L-scheme is obtained by neglecting spatial changes of $\partial\Phi/\partial t$ within Ω_k whereas the C-scheme results from a use of the element-wise bilinear interpolation for this time derivative.

Comparison of the phase speed ratio $\bar{c}/c$, given in Figure 3, reveals that accuracy of the semi-discrete solution to the pure convection problem is primarily determined by the assumed form of spatial approximation of $\partial\Phi/\partial t$.

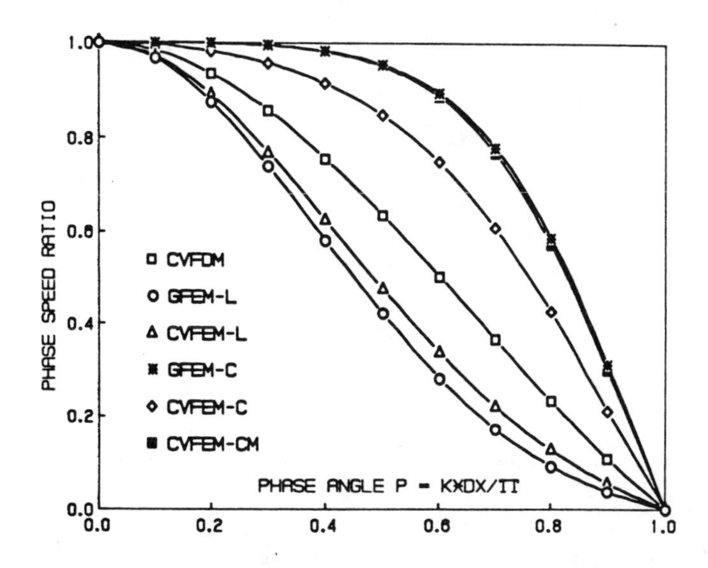

Figure 3 Phase speed ratio for semi-discrete models

The L- mass matrix schemes induce much greater dispersion error than the consistent ones. Moreover, the CVFEM-C is tangibly inferior to the GFEM-C. This is further confirmed in Table 2, where the accuracy inside the asymptotic limit $\Delta x_j \to 0$ for fixed k_j, is compared for all analysed methods.

TABLE 2 Local truncation error ($\bar{c}/c-1$) in semi-discrete approximations of pure convection problem

METHOD	GFEM-L	CVFEM-L	GFEM-C	CVFEM-C	CVFEM-CM
ERROR	$-p^2/3$	$-7p^2/24$	$-p^4/180$	$-p^2/24$	$-p^4/167$

The poor performance of the CVFEM-C in the bilinear FE grid can be considerably improved by a proper choice of the integration point ip within the sub-control volume $\Omega_{k,e}$ of Ω_e (Figure 4).

r_1	r_2	r_3	r_2
r_2	r_1	r_2	r_3
r_3	r_2	r_1	r_2
r_2	r_3	r_2	r_1

MASS MATRIX **M**

L - SCHEMES: $r_1=1$, $r_2=r_3=0$

GFEM-C: $r_1=4/9$, $r_2=2/9$, $r_3=1/9$

CVFEM-C: $r_1=9/16$, $r_2=3/16$, $r_3=1/16$

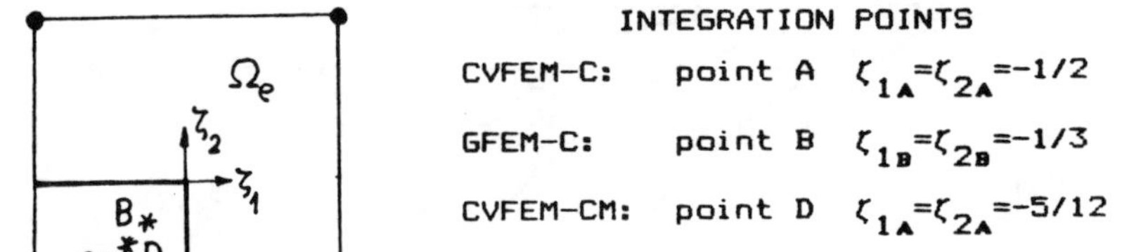

INTEGRATION POINTS

CVFEM-C: point A $\zeta_{1A}=\zeta_{2A}=-1/2$

GFEM-C: point B $\zeta_{1B}=\zeta_{2B}=-1/3$

CVFEM-CM: point D $\zeta_{1A}=\zeta_{2A}=-5/12$

Figure 4 Integration points for consistent mass matrices

This point is used to approximate the temporal integral of equation (1) by means of

$$\int_{\Omega_k} \frac{\partial}{\partial t}(\rho\Phi)\,d\Omega = \sum_e \int_{\Omega_{k,e}} \frac{\partial}{\partial t}(\rho\Phi)\,d\Omega = \sum_e \rho_e \Omega_{k,e} \left(\frac{\partial\Phi}{\partial t}\right)^e_{ip} \tag{12}$$

providing the summation is taken over all elements which share the node k. One can determine the integration point position on requiring that a leading term of the truncation error $(\bar{c}/c-1)$ is to be of the order p^4. The consistent mass matrix model (CVFEM-CM) thus obtained offers the comparable accuracy to the GFEM-C one (Figure 3 and Table 2). With a retention of the nodal symmetry in ζ_i co-ordinates, which implies that $\zeta_{1,ip} = \zeta_{2,ip}$, this gives the integration point *D* (Figure 4) which is placed between the integration points *B* and *A* for the GFEM-C and CVFEM-C, respectively.

The common way to obtain a fully-discrete analogue for convection-diffusion problem consists in a use of the one-step implicit difference scheme

$$\{\phi(t+\Delta t)\} = \{\phi(t)\} + \theta\Delta t\left\{\frac{d\phi}{dt}(t+\Delta t)\right\} + (1-\theta)\Delta t\left\{\frac{d\phi}{dt}(t)\right\} \tag{13}$$

where $0 \le \theta \le 1$, to integrate over time the set of ordinary differential equations issuing from the FEM spatial discretizations. Unfortunately, any temporal approximation generates the additional numerical dissipation and phase error. They can be estimated by means of the *Fourier* wave analysis. On substituting a single Fourier mode into the algebraic nodal equation a value of complex amplification factor G is determined. It depends on the mass matrix model used, the parameter θ and grid Fourier (*Fo*) and Courant (*Cr*) numbers. The value $|G|$ is a measure of dissipation, whereas a divergence in the phase speed ($\bar{c}/c = -\arg(G)/Cr*p$) determines the phase error $\varepsilon = -\arg(G)-Cr*p$.

The exemplifing results, derived from *Fourier* analysis for the pure convection problem are presented in Figure 5.

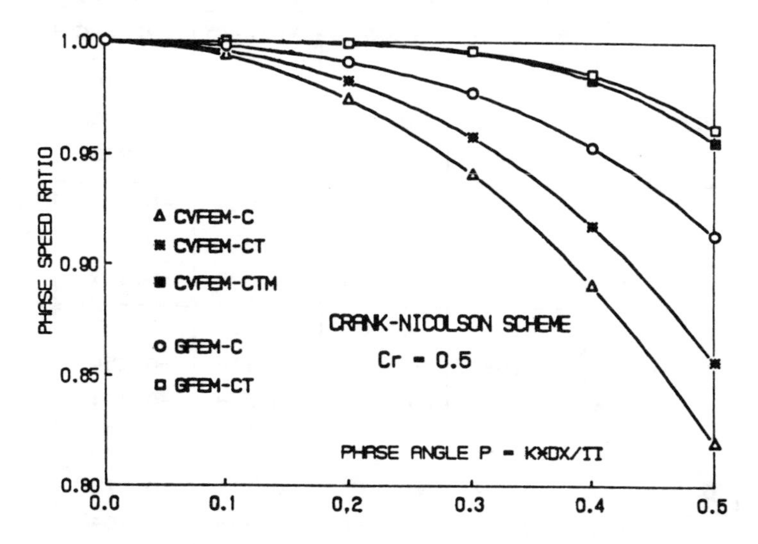

Figure 5 Phase speed ratio for fully-discrete FE models

They show the substantial phase speed error for all so far considered methods, i.e. the GFEM-C, CVFEM-C and CVFEM-CM.

To improve the accuracy of the GFEM-C Donea [10] has proposed the Taylor-Galerkin FE scheme (labelled here by GFEM-CT . It consists in the utilization of the finite difference approximation of $\partial\Phi/\partial t$ in terms of a Taylor series expansion including higher order time derivatives, which are then calculated from equation (2).

This concept can be easily adopted to the CVFEM. For the pure convection problem, the Taylor series expansion, which include time derivatives up to of the third order, is used to approximate $\partial\Phi/\partial t$ in the temporal integral of (1). This leads, through the use of (1), (2) and the Gauss divergence theorem, to the nodal balance equation for the control-volume FEM (CVFEM-CT) of higher accuracy in time. To retain only C^0- continuity of the FE model in the convection-diffusion problem, the Taylor series expansion in time must be limited to the first two terms [10]. In the case of constant coefficients of equation (1), it leads to the Crank -Nicolson time stepping scheme, which is the second order accurate in the time step, in contrast to the only first order accuracy of both the Euler ($\theta=0$) and fully implicit ($\theta=1$) schemes (13).
The considerably improved performance of both the GFEM-CT and CVFEM-CT is visible in Figure 5, where the phase speed errors are compared for the pure convection, and in Table 3, where the phase errors are collected for various *Fo* and *Cr* numbers.

TABLE 3 Phase errors of FEM models for convection-diffusion

Cr	Fo	P	CRANK - NICOLSON SCHEME			$\theta = 1.0$
			CVFEM-CTM	GFEM-CT	CVFEM-CT	CVFEM-C
0.2	0.05	$\pi/4$	-0.001	-0.001	-0.005	-0.010
		$\pi/2$	-0.016	-0.015	-0.048	-0.083
		$3\pi/4$	-0.140	-0.133	-0.220	-0.283
	0.20	$\pi/4$	0.000	0.000	-0.004	-0.022
		$\pi/2$	0.010	0.012	-0.029	-0.142
		$3\pi/4$	0.242	0.288	-0.098	-0.359
0.5	0.05	$\pi/4$	-0.006	-0.006	-0.016	-0.038
		$\pi/2$	-0.067	-0.065	-0.139	-0.254
		$3\pi/4$	-0.391	-0.375	-0.568	-0.734
	0.20	$\pi/4$	-0.004	-0.004	-0.014	-0.066
		$\pi/2$	-0.016	-0.013	-0.101	-0.375
		$3\pi/4$	-0.107	-0.149	-0.336	-0.904

The CVFEM-CT induces the errors which are smaller than the CVFEM-C errors but still much larger than the GFEM-CT ones. However, if **M** matrix is calculated rather by means of (12) at the proper integration points in Ω_k than by the bilinear interpolation of $\partial\Phi/\partial t$ within Ω_e, the model thus obtained (CVFEM-CMT) attains the accuracy which is comparable to the GFEM-CT one (Figure 5, Table 3).

CONCLUDING REMARKS

The control-volume based FEM has two attractive features which justify its increasing popularity. First, its nodal equations are founded on a direct application of the conservation law for the field quantity Φ. This guarantees the local conservative property and simple physical interpretations of the CVFEM model. Second, the commonly appreciated versatility of the FE spatial discretization is retained in the method owing to a use of the piece-wise interpolations for both the element geometry and quantity Φ.

Unfortunately, the local conservative property is not sufficient to ensure the correct and efficient modelling of convection-diffusion problems by means of CVFEM.

It was shown in the paper, that for convection dominated problems the method is inferior to the only globally conservative GFEM. This is due to substantial mass balance errors resulting from rough approximation of the solenoidal velocity field over the bilinear FE grid. The error can be reduced by setting up the balance equation (1) for the corrected amount of mass comprised in control-volume Ω_k.

Moreover,special interpolations are needed in CVFEM to suppress wiggles and to reduce the numerical diffusion and dispersion over the FE grids. This can be achieved by incorporation of more local information, coming from the associated partial differential equation (2), into approximations of each individual term in (1).

Therefore,Baliga and Patankar [1] have proposed exponential interpolation of Φ based on the exact solution to steady-state convection-diffusion equation. Schneider [2] has developed the technique where convected variable Φ at the integration point is determined from the approximate representation of (2) at this point.

Furthermore,the substantial dispersion error of the FE approximations,for transient convection dominated problems, can be reduced by utilizing (2) to calculate higher order time derivatives in a Taylor series expansion for $\partial\Phi/\partial t$. This technique has been successfully used in the GFEM [10]. However, as it is shown in the paper, it gives inconsiderable inprovements in the CVFEM unless the poor bilinear spatial interpolation of $\partial\Phi/\partial t$ is replaced by a special choice of the integration points in (12). Then, the CVFEM solution attains the comparable accuracy to the GFEM one.

The results presented are only the preliminary ones. Hence further study is needed to verify the performance of the control-volume based FEM in discrete analysis of transient convection dominated problems.

REFERENCES

1. Baliga,B.R. and Patankar,S.V. Eliptic Systems: Finite Element Method II, Chapter 11, Handbook of Numerical Heat Transfer, (Ed. Minkowycz, W.J.) John Wiley & Sons, Inc., N.Y., 1988.
2. Schneider,G.E. Eliptic System: Finite Element Method I, Chapter 10, Handbook of Numerical Heat Transfer, (Ed. Minkowycz, W.J.), John Wiley & Sons, Inc., N.Y., 1988.
3. Banaszek,J. Comparison of Control-Volume and Galerkin FEMs for Diffusion-Type Problems, to be published in Numerical Heat Transfer in 1989.
4. Patankar,S.V. Numerical Heat Transfer and Fluid Flow, Mc Graw-Hill Company, N.Y., 1980.
5. Brooks,A.N. and Hughes,T.J.R. Streamline Upwind/Petrov Galerkin Formulations for Convection Dominated Flows with Particular Emphasis on the Incompressible Navier-Stokes Equations, Comp. Meth. Appl. Mech. Eng., vol.32, pp.199-259, 1982.
6. Zienkiewicz,O.C. The Finite Element Method 3rd edition Mc Graw-Hill Company, London, 1976.
7. Schneider,G.E. and Zedan,M. Control-Volume Based Finite Element Formulation of the Heat Conduction Equation, Prog. Astronaut. Aeronaut., vol.86, pp. 305-327, 1983.
8. Smith,R.M. and Hutton,A.G., The Numerical Treatment of Advection: A Performance Comparison of Current Methods Numerical Heat Transfer, vol.5, pp. 439-461, 1982.
9. Raithby,G.D. and Schneider,G.E. Elliptic Systems: Finite Difference Method II, Chapter 8, Handbook of Numerical Heat Transfer (Ed. Minkowycz W.J.), pp. 241-291 John Wiley & Sons, Inc., N.Y., 1988.
10. Donea,J. and Guliani,S. Time Accurate Solution of Advection-Diffusion Problems by Finite Elements,Comp.Meth. Appl.Mech.Eng., vol.45, pp. 123-145, 1984.

Simulation of the Transport Equation by the Network and Transmission-Line Modelling

C.C. Wong, K.K. Cheng

Department of Electronic Engineering, City Polytechnic of Hong Kong, Kowloon Tong, Kowloon, Hong Kong

ABSTRACT

It is shown that an RC network can be used to model the transport equation in two dimensions. The network is then solved by DGEAR of IMSL. Based on the network, a transmission-line model is developed; therefore the problem can also be solved by the method of transmission-line matrix (TLM). The later method does not need a conventional ODE solver and the usual problem of stiffness does not affect the stability of the algorithm. Several numerical examples are described and the results show that the TLM is more efficient.

1 INTRODUCTION

The transport equation or the diffusion-convection equation appear in many branches of engineering or physics, such as diffusion in semiconductor devices and heat transfer problems. Analytical solutions for this type of problems are relatively rare, except for very simple cases, because of the complexity of the problems. Numerical methods are therefore useful and the most commonly used methods are the finite-difference and the finite-element. When the space is discretized by, for example, the method of finite-difference, a set of ordinary differential equations (ODEs) will result. Many numerical schemes are available for solving the set of ODEs and broadly speaking these methods can be classified as implicit and explicit schemes. Implicit schemes are more time consuming to advance one time-step, whereas explicit schemes may suffer from stability problems. Therefore it will be very desirable if an explicit and stable numerical algorithm is available.

In this paper the transport equation is simulated by two methods. First, by finding the response of an appropriate RC network, and secondly, by the method of transmission-line modelling (TLM). The former is based on the work of [1] where the space is discretized by an RC network. From the network a set of simultaneous ODEs is derived. The resulting set of ODEs can be solved by standard solvers such as the DGEAR of IMSL. The second method described is the transmission-line modelling (TLM) which is a numerical routine primarily developed to solve partial differential equations, such as the wave and the diffusion equations [2,3], in the time domain. TLM is explicit and is unconditionally stable. The field region is first discretized by an equivalent transmission- line matrix which is then solved by determining the propagation of pulses along the matrix. The scattering process of these pulses along the transmission-line matrix being monitored continuously and the node voltages determined.

Both methods are used to solve one- and two-dimensional problems. Accuracy and efficiency of the two methods are compared. It is shown that the TLM method is more efficient than the network for the same accuracy requirement.

2 NETWORK AND TRANSMISSION-LINE MODELS

The diffusion of excess carriers into a substrate under the influence of external electric fields can be described by

$$\frac{\partial P_n}{\partial t} = -\frac{P_n}{\tau} - \mu E_x \frac{\partial P_n}{\partial x} - \mu E_y \frac{\partial P_n}{\partial y} + D\left(\frac{\partial^2 P_n}{\partial x^2} + \frac{\partial^2 P_n}{\partial y^2}\right) \tag{1}$$

where

P_n	is the excess carrier concentration (cm^{-3}),
τ	is the carrier life time due to recombination (s),
μ	is the mobility ($cm^2/V/s$),
D	is the diffusion constant (cm^2/s),
E_x and E_y	are the electric fields in the x and y directions respectively (V/cm).

Now, consider a typical RC network model as shown in fig.1; it can be shown, when the mesh size, h, approaches zero, that the equation describing the network is

$$\left(\frac{\partial^2 v}{\partial x^2} + \frac{\partial^2 v}{\partial y^2}\right) - \frac{1}{r}\left(\frac{\partial r}{\partial x}\frac{\partial v}{\partial x} + \frac{\partial r}{\partial y}\frac{\partial v}{\partial y}\right) - v.g.r = r.c\frac{\partial v}{\partial t} \tag{2}$$

where r, g and c are circuit parameters of the appropriate units in per unit length; they are functions of x and/or y. Eq(2) models Eq(1) if

$$\begin{aligned} r.c &= \frac{1}{D} \\ \frac{1}{r}\frac{\partial r}{\partial x} &= \frac{\mu E_x}{D} = A\text{, say} \\ \frac{1}{r}\frac{\partial r}{\partial y} &= \frac{\mu E_y}{D} = B\text{, say} \\ g.r &= \frac{1}{\tau D} \end{aligned}$$

or

$$\begin{aligned} r &= e^{Ax}e^{By} \\ c &= \frac{1}{D}e^{-Ax}e^{-By} \\ g &= \frac{1}{D\tau}e^{-Ax}e^{-By} \end{aligned} \tag{3}$$

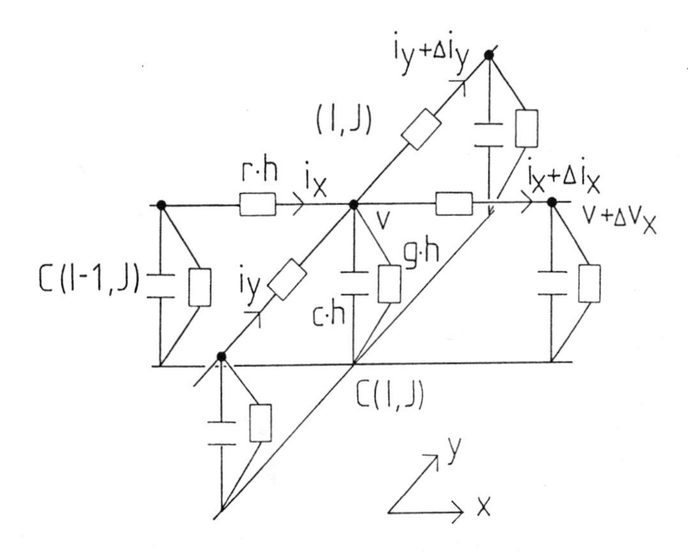

Figure 1: One section of a two-dimensional RC network.

and the node voltage 'v' represents the excess carrier concentration. A and B of eqs(3) are assumed constants or are known priori, however, this is not a limitation and nonlinearities can also be represented as in [1]. It is obvious, from the network, that a set of ODEs can be derived and they can be integrated numerically by the standard Runge-Kutta or other packages. However, it has been shown [4,5] previously that the problem may be stiff or non-stiff depending on the convective term and the mesh size (i.e. the term $E\mu h/D$ determines the stiffness of the set of ODEs when $2D/h^2 >> 1/\tau$). This will affect the choice of the solution method since an efficient scheme for non-stiff problems is not necessarily suitable for stiff problems.

An alternative way to model the above problem is by the TLM. In order to produce a transmission-line model the capacitor lumped at node (I,J) is assumed to be divided into 4 capacitors as in fig.2 where $CX(I,J)$ is made up of one-quarter of $C(I,J)$ (of fig.1) and one-quarter of $C(I-1,J)$ (of fig.1). Similarly $CX(I+1,J)$ is made up of one-quarter of C(I,J) and one- quarter of $C(I+1,J)$. Therefore, for a two-dimensional transmission-line model

$$c = \frac{e^{-Ax}e^{-By}}{2D}$$

A more complicated model (this will be termed model 2 in the following sections) where $RG(I,J)$ is also split is shown in fig.3. Therefore, the transmission-line model can be developed and one section is shown in fig.4 where the former (termed model 1) is used. $ZX(I,J)$ is the characteristic impedance of the transmission-line modelling $CX(I,J)$ and equals $\Delta t/CX(I,J)$ where Δt is the time-step (i.e. the time required for the pulse to travel along the transmission-line which represents the capacitor). The transmission-line model is solved by monitoring the transmission and scattering of node voltages. More detail discussions about the

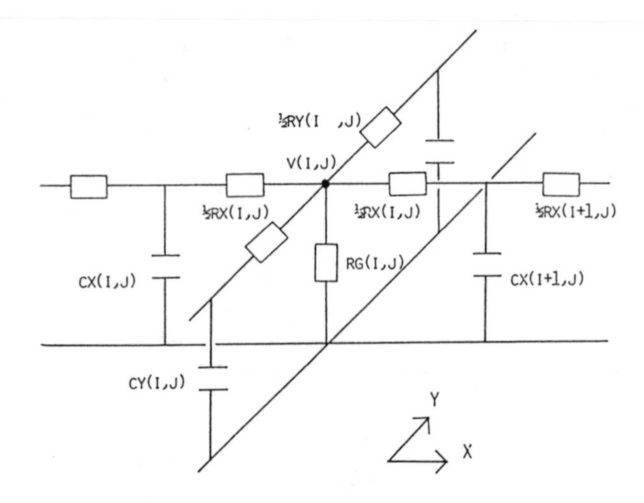

Figure 2: One section of a modified two-dimensional RC network.

TLM can be found in references [3,6-7]. A brief description on applying the TLM is given in the following section using a one-dimensional problem as an example.

3 RESULTS AND DISCUSSIONS

The course of diffusion (assumed one-dimensional) of the excess carriers (created by external injection at $t = 0$) from an infinitesimally thin layer in the presence of an external electric field and with combination is given by

$$P_n(x,t) = \frac{Q}{2\sqrt{\pi Dt}} e^{-\frac{(x-E\mu t)^2}{4Dt}} e^{-\frac{t}{\tau}}$$

where Q is the excess carrier concentration in per unit area and

$$Q = N_o h \tag{4}$$

where N_o and h are the concentration in per unit volume and the mesh size respectively. It is assumed that there are no initial excess carriers at t=0. Eq(4) is the initial condition required by the network or the TLM.

<u>Network and TLM Solutions</u>

When modelled by the network the set of ODEs is solved by the DGEAR of IMSL using Adam's code which is designed for non-stiff equations. Data used are

$$\begin{aligned} D &= 10 - 200\ cm^2/s \\ \mu &= 2000\ cm^2/V/s \\ \tau &= 5 * 10^{-7} - 5 * 10^{-6}\ s \\ E &= 10 - 40\ V/cm \\ Q &= 200\ cm^{-2} \\ h &= \text{various} \\ \text{specimen length} &= 0.2\ cm \end{aligned}$$

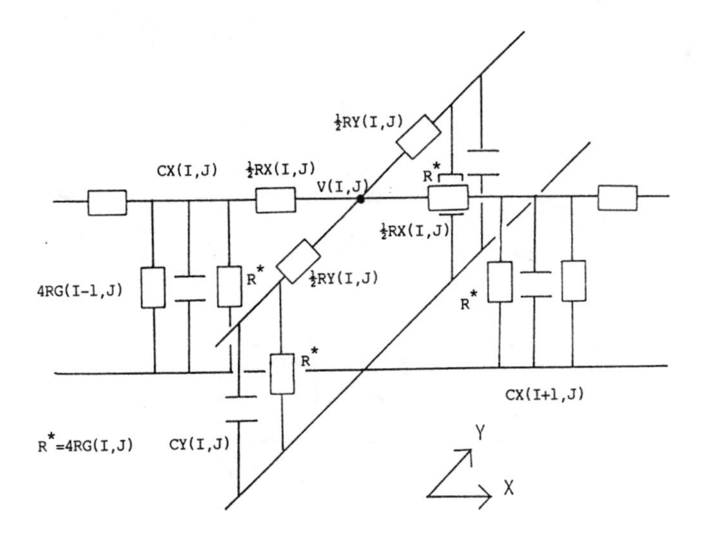

Figure 3: Another alternative for fig.1 with $C(I, J)$ and $RG(I, J)$ split.

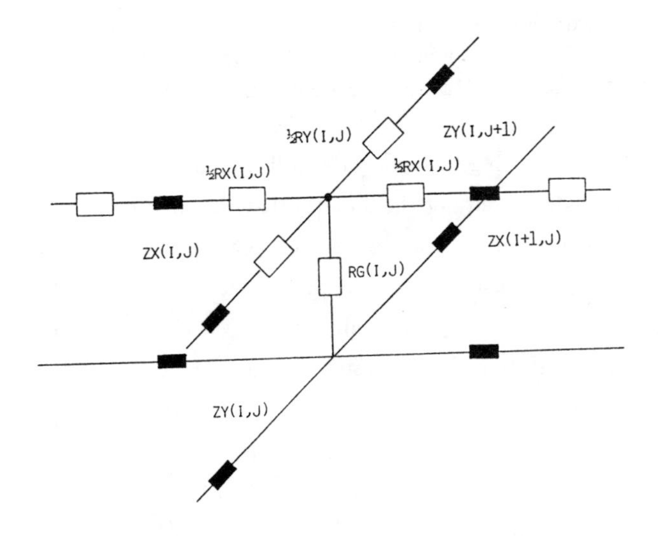

Figure 4: A transmission-line model (model 1) of fig.2.

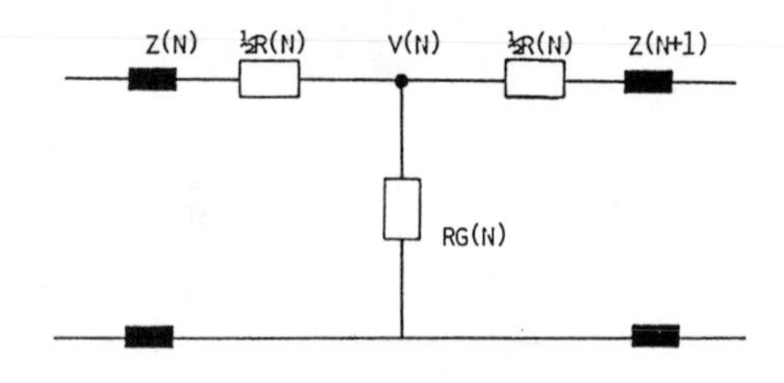

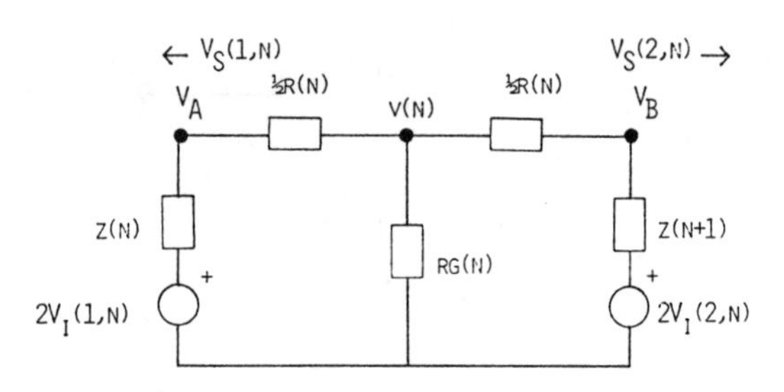

Figure 5: (a) A one-dimensional transmission-line model (model 1) (b) Thevenin's equivalent circuit of (a)

and the excess carriers are initialized at the middle of the specimen. Note that the problem may be stiff or non-stiff depending on the parameters and the mesh size. However, only one ODE solver is used here because the aim of the present paper is not to compare various ODE solvers.

On the other hand, solving the transmission-line model cannot utilize a standard package. Indeed the novelty of the TLM is that the solution of the matrix does not require conventional integrating schemes. The node voltages (=carrier concentration) are determined through the propagation of voltage pulses which can be conveniently monitored by Thevenin's equivalent circuit. A one-dimensional equivalent circuit using model 1 is shown in fig.5 where the subscripts i and S indicate the incident and scattered voltages respectively. The node voltage $V(N)$ is given by

$$V(N) = \frac{\dfrac{2V_i(1,N)}{ZEF1(N)} + \dfrac{2V_i(2,N)}{ZEF2(N)}}{\dfrac{1}{ZEF1(N)} + \dfrac{1}{ZEF2(N)} + \dfrac{1}{RG(N)}}$$

where

$$ZEF1(N) = \frac{1}{2}R(N) + Z(N)$$

$$ZEF2(N) = \frac{1}{2}R(N) + Z(N+1)$$

For the scattered voltages,

$$\begin{aligned} V_S(1,N) &= V_A - V_i(1,N) \\ V_S(2,N) &= V_B - V_i(2,N) \end{aligned} \tag{5}$$

where

$$V_A = \frac{\frac{V(N)}{\frac{1}{2}R(N)} + \frac{2V_i(1,N)}{Z(N)}}{\frac{1}{\frac{1}{2}R(N)} + \frac{1}{Z(N)}}$$

$$V_B = \frac{\frac{V(N)}{\frac{1}{2}R(N)} + \frac{2V_i(2,N)}{Z(N+1)}}{\frac{1}{\frac{1}{2}R(N)} + \frac{1}{Z(N+1)}}$$

At the next time-step, the scattered voltages become the incident voltages at the neighbouring nodes

$$V_i(1,N+1) = V_S(2,N)$$
$$V_i(2,N-1) = V_S(1,N)$$

A similar equivalent circuit can be produced for model 2 and similarly for two-dimensional models. It is clearly seen that the TLM process is explicit. Furthermore, the process is unconditionally stable because it models the propagation of pulses along a transmission-line matrix which is completely passive [8]. However, the time-step which affects the accuracy cannot be arbitrary chosen. In order to control the error due to time discretization a time-step-doubling (see [9], for example) method for error estimation is incorporated in the TLM routine. The local error is estimated by comparing the rms norm of the solution at $time + 2 * \Delta t$ with the solution using a time-step of $2\Delta t$. A relatively small time-step is used during starting and it can change adaptively when the estimated error is smaller than a certain tolerance ($10^{-4} - 5 * 10^{-4}$). A typical TLM result as compare with the exact solution is shown in fig.6.

Comparison of Results

The one-dimensional transport equation is studied in detail for both methods with different values of D, E, τ and h. Accuracy and efficiency of the network (solved by DGEAR) and TLM are shown in Table 1 and fig.7 respectively. All results are obtained from a VAX8800.

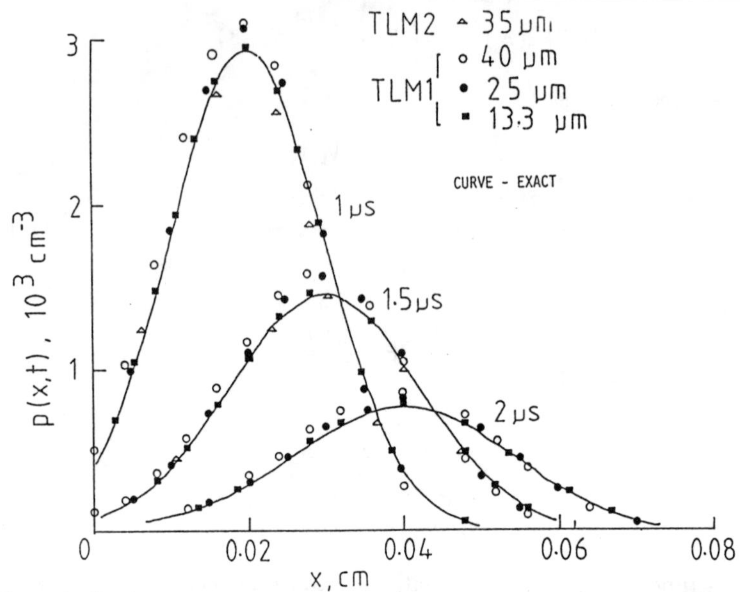

Figure 6: Carrier concentration against distance. Excess carriers are injected at $t = 0, x = 0$. Other data as in Table 1.

Table:1 Accuracy of the TLM and the network at different observation times for the one-dimensional problem.

	Observation time (μs)			
	1.0	1.5	2.0	
Method		rms error (%)		h (cm)
TLM1	3.72	5.14	6.45	
TLM2	3.84	3.93	3.87	$4*10^{-3}$
Network	2.81	2.98	3.06	
TLM1	1.17	1.32	1.85	
TLM2	0.99	0.99	0.93	$2*10^{-3}$
Network	0.77	0.81	0.87	
TLM1	<0.2*	<0.2*	0.33	
TLM2	<0.2*	<0.2*	0.29	$1*10^{-3}$
Network#	<0.2*	0.21	0.22	

TLM1: model 1 TLM2: model 2 Network: solved by DGEAR * Only the 4th significant figure of the peak value is different from the exact value

Double precision needed for the DGEAR

Other data are:
$D=50\ cm^2/s$, $E=10\ V/cm$, $\tau = 10^{-6}\ s$, $\mu = 2000\ cm^2/V/s$.

The independent variable in fig.7 is the rms error (in %), therefore h used in the network and the TLM are not identical. The rms error is defined as

$$\text{RMS error} = \sqrt{\frac{\sum_{i=1}^{N}\left(\frac{X_i - X}{X_{max}}\right)^2}{N}}$$

where

X_i and X	are the numerical and exact results respectively,
X_{max}	is the maximum value within the samples at that observation time (=2 μs, for this example).

The results suggest that the TLM is more efficient for the same accuracy requirement (see also Table 2).

The improvement is more significant when the accuracy requirement is tightened or when the system becomes large. The gain will increase further when the problem is diffusion dominated. TLM model 2, though more complicated than model 1, is slightly more accurate and therefore the overall efficiency of the two models are similar. A comparison of the finite-difference with the network and the TLM has also been studied. Results show that the behaviour of the finite-difference and the network are similar to those reported in [1] and therefore they are not included here. It should be noted the exact solution requires an infinitesimally thin source and an infinite medium; both cannot be exactly represented in the numerical process.

Several two-dimensional problems have also been examined. The specimen used for simulation is a square plate of $0.14cmX0.14cm$. A line of impurities is injected in the middle of the plate (fig.8). Table 2 shows the results of the network and the TLM (model 1) when the line of injected impurities is very long such that the variation of the carrier concentration along the line 1-1' is practically similar to the one- dimensional case. The corresponding CPU times (on a VAX8800) are included. Various lengths of impurity injection have been used and it can be shown that the results provided by the two methods are very close. In general, the difference between the peak values is less than 2%. As in the one-dimensional case, the TLM is more efficient. Different values of E_y have been used while keeping E_x constant at 10V/cm. An example is shown in fig.9 where two values of E_y are used. At the same observation time, the peak values are almost identical, only the relative positions of the peak values have changed. The velocity is governed by the magnitude of the resultant electric field.

4 CONCLUSIONS

Numerical results show that both the network and the transmission-line model can be used to form a space discretized model of the transport equation. The network can be solved by standard ODE solvers. In this paper the DGEAR of IMSL is used. For consistence only the code for non-stiff problems is used.

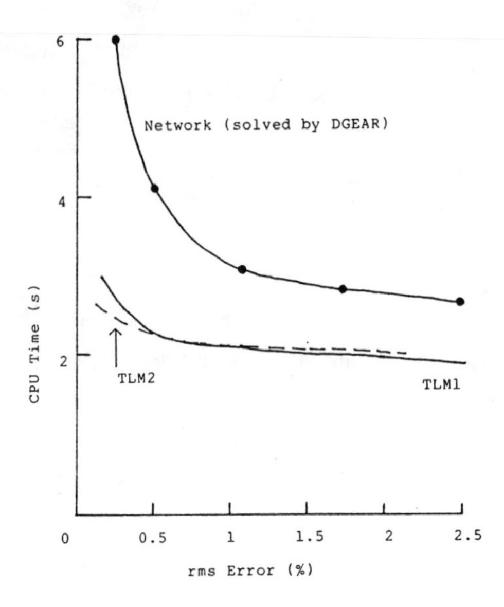

Figure 7: CPU time (on a VAX 8800) required to solve the one-dimensional problem by the network and TLM. Other data as in Table 1.

Table 2: Comparison between two-dimensional numerical solutions with the exact equation (for one dimension).

Method	Observation time (μ s) 1.0	1.5	$h(*10^{-3}cm)$
	rms error (%) [CPU time (s)]		
TLM1	1.70 [4.8]	2.24 [5.1]	2.333
Network	1.24 [41.4]	1.31 [55.6]	2.333
TLM1	0.98 [11.6]	1.21 [14.8]	1.75
Network	0.73 [129.2]	0.86 [173.2]	1.75
TLM1	0.57 [47.2]	0.76 [60.8]	1.333
Network	0.48 [329.2]	0.64 [546.7]	1.333

Other data as in Table 1.

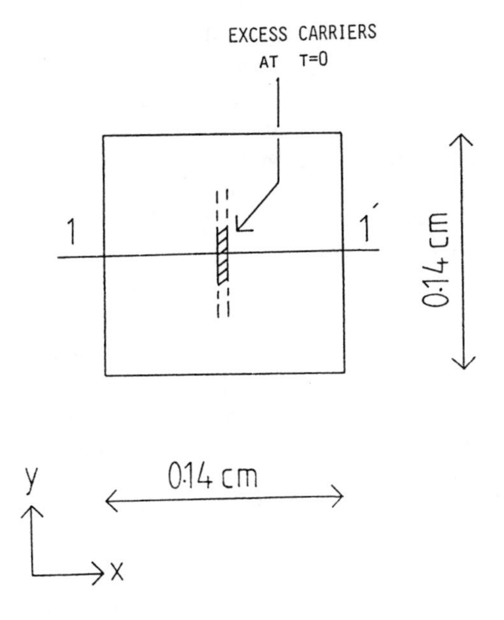

Figure 8: A two-dimensional sample and the injected excess carriers.

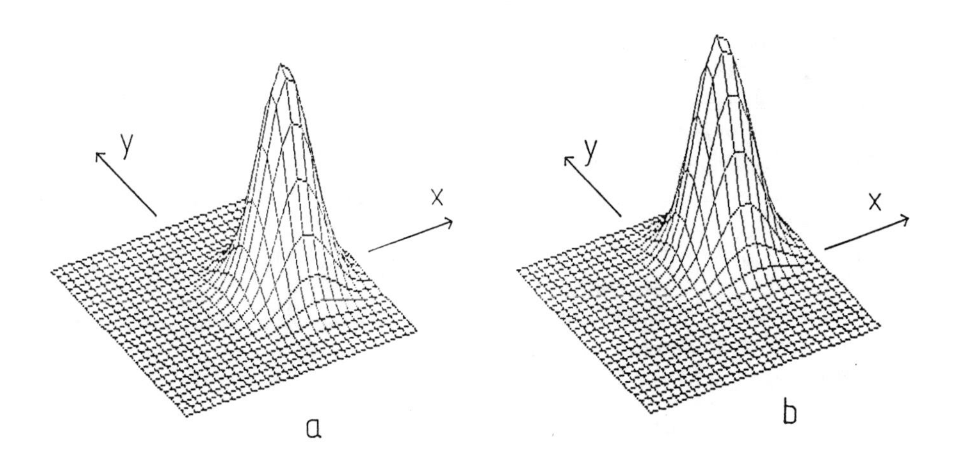

Figure 9: Plots of the concentration distribution of the carriers when the observation time = $1\mu s$ and impurity length = $0.003cm$; (a) $E_y = 0.01\ V/cm$, (b) $E_y = 10\ V/cm$. Other data as in Table 1.

The problems may, however, be stiff, and thus the efficiency will be affected. Therefore, to maximize efficiency, the stiffness of the problem must be known. In practice, this may not be possible. On the other hand, the TLM method is based on the propagation of voltage pulses along a transmission-line matrix and the efficiency is not affected by the stiffness of the problem. The time-step can be changed adaptively and controlled by the accuracy requirement. A comparison between the two methods suggests that the TLM is more efficient and merits further investigation in this application area.

REFERENCES

[1] C. C. Wong, *"A network model for the diffusion-convection equation,"* Int. J. Cir. Theor. Appl. Vol. 16, pp.147-156, 1988.

[2] P. B. Johns, *"The solution of inhomogeneous waveguide problems using a transmission-line matrix,"* IEEE Trans. Microwave Theoy Tech. MTT-22, 209-215, 1975.

[3] P. B. Johns, *"A simple explicit and unconditionally stable numerical routine for the solution of the diffusion equation,"* Int. J. Num. Meth. Eng. Vol. 11, pp.1307-1328 1977.

[4] G. F. Carey and K. Sepehrnoori, *" Gershgorin theory for stiffness and stability of evolution systems and convection-diffusion, "* Comp. Meths. Appl. Mech. Eng. Vol. 22, pp.23-48 1980.

[5] K. Sepehrnoori and G. F. Carey, *"Numerical integration of semidiscrete evolution systems,"* Comp. Meths. Appl. Mech. Eng. Vol. 27, pp.45-61 1981.

[6] D. de Cogan and M. Henini, *"Transmission-line matrix (TLM): a novel technique for modelling reaction kinetics"*, J. Chem. Soc., Faraday Trans. Vol. 2, pp.843-855 1987.

[7] J. W. Bandler, P. B. Johns and M. R. Rizk, *" Transmission-line modelling and sensitivity evaluation for lumped network simulation and design in the time domain,"* J. Franklin Institute Vol.304, pp.15-23 1977.

[8] C. A. Desoer and E. S. Kuh, Basic circuit theory McGraw-Hill, New York, 1969, p.816.

[9] R. Alexander, *"Implicit Runge-Kutta methods for stiff ODEs,"* SIAM J. Numer. Anal. Vol.14, pp.1006-1021 1977.

[10] S. Pulko, A. Mallik and P. B. Johns, *"Application of transmission-line modelling (TLM) to thermal diffusion in bodies of complex gemetry,"* Int. J. Num. Meth. Eng. Vol.23, pp.2303-2312 1986.

On the Solution of the Advection-Diffusion Equation on a Non-Uniform Mesh by a RMP-Accelerated Multi-Grid Method

A.S-L. Shieh

Idaho National Engineering Laboratory, PO Box 1625, Idaho Falls, ID 83415-2508, USA

ABSTRACT

Ordinary multigrid methods with linear interpolation quite often exhibit a slow rate of convergence for problems the solutions of which have steep gradients. In this work an accelerated technique based on residual minimizing prolongation methods is used to improve dramatically the rate of convergence of the underlying multigrid method for the solution of advection-diffusion equation on nonuniform meshes. The method is particularly suitable for use on supercomputers because of the use of accelerated point Jacobi schemes and multigrid V-cycles.

INTRODUCTION

In this work we are primarily concerned with the solution of advection-diffusion equation that arises from convective or turbulent heat transfer problems on supercomputers. Although multigrid methods (see e.g. Philips and Schmidt[1,2], Brandt[3], Barkai and Brandt[4]) in general are highly regarded as fast elliptic solvers, their performance on supercomputers deteriorate if too much computation is done on coarse grids or if the underlying relaxation method is not fully vectorizable or parallelizable. Hence in this work we only consider simplified V-cycles with accelerated Jacobi schemes. Our numerical experiments, however, indicate that within this context, conventional multi-grid methods using linear interpolations have a rather slow convergence rate if the advection-diffusion equations with rapid varying solutions are solved on nonuniform meshes. This is in part because linear interpolation is not suitable to be used for solutions that have large gradients. Another problem is that the eigenstructure of the underlying matrix may not permit efficient use of ordinary interpolation techniques. For a pure diffusion problem, the eigenvalues are real and the eigenvectors have certain smoothness properties that permit conventional interpolational techniques. When the Peclet number is not small for an advection-diffusion problem, however, both the eigenvalues and the eigenvectors of the resulting matrices are complex with, depending on how the advective terms are discretized, relatively small real parts. The eigenvectors are not necessarily smooth and the classical interpolation methods may not work very well. Furthermore, it is typical of these matrices that they are derived from numerical procedures in which large variations in mesh size have been used to improve local field resolution. This clustering has the effect of greatly increasing the condition number of the basic matrix which in turn greatly influences the convergence rate of a relaxation scheme. The conventional approach is therefore to use a combination of scaling and preconditioning before applying the multigridding procedure. In this work, we propose an alternative based on a residual minimizing prolongation method that eliminates the need of either solving $A^TAx = A^Tb$ or using a special relaxation method such as Kaczmark method (equivalent to applying Gauss-Seidel on A^TA)). As a result, the rate of convergence is much faster than that of the conventional approach. The residual minimizing prolongation method is discussed in section 2 while some aspects of accelerated Jacobi schemes are discussed in section 3. Numerical comparisons with multigrid methods that use linear interpolations for various choices of Peclet number and mesh spacings are presented in section 4.

RESIDUAL MINIMIZING PROLONGATION (RMP)

Suppose we are solving the Dirichlet problem of the advection-diffusion equation

$$-\frac{1}{Pe} Lu + D_x(Pu) + D_y(Qu) = S(x,y) \qquad (1)$$

Here L denotes the Laplacian and Pe is the Peclet number. D_x and D_y denote partial differentiation with respect to x and y respectively and P and Q are functions of x and y. Let U be the discrete variable for u in Equation (1). Let

$$-\frac{1}{Pe}L_{h,k}U + D_h(PU) + E_k(QU) = S$$

be the discrete analog of Equation (1). Here $L_{h,k}$ denotes the five-point discrete Laplacian and D_h and E_k are discret operators corresponding to a finite difference scheme for the advective terms. The basic scheme considered in this work is the first order upwind difference scheme described below. See also the end of section 3 for a description of how the algorithm discussed here can be adapted to solving problems discretized with higher order upwind differencing schemes or central differencing scheme. Let h and k be respectively the mesh size in the x and y coordinate directions. Define $D_h(PU)$ at a point (x,y) to be

$$D_h(PU) = [P(x+h/2,y)U_e - P(x-h/2,y)U_w]/h \tag{2}$$

where U_e is given by

$$U_e = \begin{cases} U(x,y) & P \geq 0 \\ U(x+h,y) & P \leq 0 \end{cases} \tag{3}$$

Similar expressions are used for U_w and for $E_k(QU)$.

Suppose there are only two nonuniform grids with meshes uniform in each coordinate axis direction where the coarse grid is obtained from the fine grid by deleting every other grid line of the fine grid. Assume that the solution is already known on the coarse grid. Below we describe how the solution is prolongated to the fine grid.

Let h(2h) and k(2k) be the fine (coarse) grid mesh size. Let A_h be the set of mesh points whose mesh neighbors in the y-coordinate directions are coarse grid mesh points. Let B_h be the set of remaining mesh points in the complement of the coarse grid. Let C_h be the set of mesh points whose mesh neighbors in the x-axis directions are coarse grid mesh points. Assume that the aspect ratio a = h/k is greater than 1. At each point on A_h, the following equation is formed

$$[-U(x,y-k)+2U(x,y)-U(x,y+k)]/(Pe.k^2) + [-U(x-2h,y)+2U(x,y)-U(x+2h,y)]/(Pe.4h^2) + D_{2h}(PU) + E_k(QU) = S(x,y)/2 + [S(x-h,y)+S(x+h,y)]/4$$

Here $D_{2h}(PU)$ is defined in the same way as $D_h(PU)$ with h in Equations (2) and (3) replaced by 2h. Assume that U is known on the coarse grid. Then the values of U are obtained on A_h by solving linear tridiagonal systems of the form

$$[-1-2hPeP \quad 8a^2+2+4ahPeQ+2hPeh \quad -1]\ U = G.$$

Once U is determined on A_h, its values on B_h are determined by operator based prolongation by solving tridiagonal equations of the form

$$[-1-kPeQ \quad 2a^2+2+hPeP+ahPeQ \quad -1]\ U = G.$$

Because the tridiagonal systems are strongly diagonally dominant, they can be solved efficiently by iterative methods. If a < 1, we may solve the problem on C_h first or we may obtain U simultaneously on both A_h and C_h

first before we use operator based prolongation to obtain U at the remaining mesh points. The latter approach is, in general, the most efficient approach to RMP prolongation on supercomputers. The numerical results reported in this work, however, are obtained by the first approach.

RMP-ACCELERATED MULTIGRID METHOD

The multigrid method consists of solving approximately the problem on the fine grid levels before correcting on the coarse grid levels. Because of difficulties associated with representing high frequencies on the coarse grid, it is desirable to damp out the rapidly varying parts of the errors on the fine grids.

The basic relaxation scheme used on each grid level (except the finest grid level) is an accelerated point-Jacobi scheme with acceleration parameters chosen as follows:

$$w_j = 2.0/(a+\cos[(2j-1)\pi/2n]), \quad j=1,2,.,n/2, \tag{4}$$

$$w_j = w_{j-n/2}, \quad j=n/2+1,..,n. \tag{5}$$

where $a = 1.0 + 2.5/n^2$, and n is the number of consecutive relaxation sweeps used on that grid level. On the finest grid level of each cycle, the parameters are chosen differently. They are set to be

$$w = 2.0/(a+\cos(\pi/6), \tag{6}$$

where $a = 1.0 + 2.5/9$. The parameter sequence (4) with j=1 to n/2 replaced by j=1 to n has also been used (Gentzch[5], Kim[6]). The major difference here is that we do not attempt to dampen the low frequency components of the error in the relaxation step in our algorithm. Hence no high frequency component of the error is magnified (Lomax and Maksymiuk[7]). In our numerical experiments, we've found that this approach is more compatible with the algorithm used in this work.

Below we briefly describe the simplified V-cycle used in this work. Suppose there are M grid levels with level M being the finest grid level. Assume that we are solving the following discrete problem on level M.

$$L^M U^M = F^M.$$

Let u be an approximate solution on level M. Form the residual

$$R^M = F^M - L^M u.$$

Obtain the weighted residual R^{M-1} on level M-1 from R^M

by defining

$$R^{M-1}(i,j) = 0.5R^M(i,j) + 0.125(R^M(i,j-1) + R^M(i,j+1) + R^M(i-1,j) + R^M(i+1,j)).$$

Define $R^1, R^2, \ldots, R^{M-2}$ in a similar manner. We first solve the residual correction equation

$$L^1V^1 = R^1$$

approximately on level 1. We then prolongate the solution to level 2. Repeat the process until we reach level M. The corrected solution u+v where v is an approximation to the solution of

$$L^MV^M = R^M,$$

is our corrected approximation to U^M after one simplified V-cycle. The above can be represented in a flow diagram as follows.

3*M - compute weighted residuals - 2M*1 -> 2M*2 -> (2M-2)*3 -> (2M-2)*4 -> ......-> 8*(M-1) -> 3*M - compute u+v

Here M is the number of grid levels, -> denotes RMP prolongation, and p*q denotes that p relaxation sweeps are used on level q. For example for M=5, 10 accelerated Jacobi sweeps are used on levels 1 and 2 while 8 sweeps are used on levels 3 and 4. In the above flow diagram, the first 3*M corresponds to using three sweeps on level M for the original discrete problem while all the other p*q's correspond to solving approximately the residual correction equation on level q using p sweeps. Before the first V-cycle, it is advantageous to use the following I-cycle to obtain an initial guess on level M if RMP is used.

2M*1 -> 2M*2 -> ......-> 8*(M-1) -> 3*M.

Here p*q corresponds to solving the original discrete problem approximately using p sweeps on level q.

The above algorithm is intended only for first order upwind difference schemes. If instead a high order difference scheme is used for the advective term, we recommend the following procedure. Define the companion problem to be the discrete problem that used the first order upwind differencing scheme. First solve the companion problem approximately using one V-cycle. Then compute the residuals using the high order difference scheme and solve the residual correction equation corresponding to the companion problem using one V-cycle. Repeat the process until convergence. While variants of this approach have been suggested by other authors (e. g. Philips and

Schmidt[2]), it has not been tested within the context of the algorithm used here. We are currently investigating this approach and the results will be reported in a later work.

NUMERICAL RESULTS

Numerical results for the advection-diffusion problem with first order upwind differencing are obtained on the CRAY-XMP at the Idaho National Engineering Laboratory. The discrete problem is solved both on a rectangle $[0,1]x[0,0.5]$, with $h=1/N$ and $k=1/2N$ and on a square $[0,1]x[0,1]$ with both h and k equal to $1/N$. Here N is the number of mesh points in the x-coordinate direction. The exact solution of the discrete problem for the test case is given by

$$U(x,y) = 500\exp(-50((1-x)^2+y^2) + 100(1-y)x$$

The right hand side forcing terms of the discrete problem are computed from the exact solution using the discrete operator with different values for the Peclet number (Pe). The functions P and Q in Equation (1) are taken to be the constant function one. The following results are obtained.

Region	N	Pe	No.V	Residual-RMP	Residual-Interp
Rectangle	65	1	3	1.34672E-03	54.76325
Rectangle	65	1	4	3.46850E-05	9.42319
Rectangle	65	10	1	1.30835E-02	10.16720
Rectangle	65	10	4	2.60488E-07	5.64724E-2
Square	65	10	1	1.53541E-03	0.28442
Square	65	10	4	2.44879E-08	1.49692E-4
Rectangle	129	10	1	0.13517	11.57251
Rectangle	129	10	4	1.17256E-05	0.18229
Rectangle	129	100	1	1.22620E-02	0.15783
Rectangle	129	100	4	1.52372E-06	3.70297E-4

Table 1
Numerical Comparison of RMP-accelerated Multigrid Method with Interpolation Based Multigrid Method

Here no.V represents the number of V-cycles used; residual-RMP and residual-interp are the sum of the squares of the residuals scaled by $1/N^4$ computed after the indicated no. of V-cycles when the V-cycle is used with the RMP prolongation method and when the same algorithm is repeated with RMP replaced by linear interpolation respectively. In both cases an I-cycle is used to obtain the initial guess just before the first V-cycle is used. The nonuniform mesh with aspect ratio equal to 2 for the rectangular region seems to pose more of a problem for the interpolated multigrid method than for the RMP-accelerated multigrid method. The rate of convergence for both methods actually improve for increased Peclet numbers because of the use of the first order upwind differencing scheme. Although the nonuniform mesh is actually uniform in each coordinate direction separately, the computer code used here can handle highly nonuniform mesh. The rate of convergence does slow down even for the RMP-accelerated multigrid method when the aspect ratio is large. We are currently investigating the use of alternating line Jacobi method as our basic relaxation method on each grid level for such situations. The above comparison to interpolated multigrid method is not totally fair to conventional multigrid methods since these methods work best in either a W-cycle or a full multigrid cycle setting (e. g. Barkai and Brandt[4]). It is our view, however, that the V-cycle is more suitable for applications on supercomputers because of the relatively less time spent on the coarse grids. On a massively parallel machine, for example, most of the processors will be occupied at least half of the time which the V-cycle is used assuming that the number of the processors is not more than the number of mesh points.

CONCLUSIONS

The RMP-accelerated multigrid method is an efficient solver for the advection-diffusion equation on nonuniform mesh at least for moderate values of the aspect ratio. The method is especially suitable for supercomputers because of the use of point Jacobi method and V-cycles.

ACKNOWLEDGMENTS

This work is prepared under U. S. Department of Energy Contract No. DE-AC07-76-ID01570, supported by the INEL Long-Term Research Initiatives Program. The author would like to thank Professor Dr. George S. Dulikravich and Dr. George Mesina for valuable discussions.

REFERENCES

1. Philips, R. E., and Schmidt, F. W., "Multi-Grid Techniques for the Numerical Solution of the Diffusion Equation", Numer. Heat Transfer, 7, 251-268 (1984).

2. Philips, R. E., and Schmidt, F. W., "Multi-Grid Techniques for the Solution of the Passive Scalar Advection-Diffusion Equation", Numer. Heat Transfer, 8, 25-43 (1985).

3. Brandt, A., "Multi-Level Adaptive Solutions to Boundary-Value Problems", Mathematics of Computation, 31, 333-390 (1977).

4. Barkai, D., and Brandt, A., "Vectorized Multigrid Poisson Solver for the CDC CYBER 205", Applied Mathematics and Computation, 13, Nos. 3 and 4, 215-228 (1983)

5. Gentzch, W., "Uber ein verbessertes explizites Einschrittverfahren zur Losung parabolisher Differentialgleichungen", DFVLR, Institut fur Theoretische Stromungsmechanik, Gottingen, Deutschland, 1980.

6. Kim, K., "Acceleration of Elliptic Partial Differential Equations by Using Super-Step Method", Project No. 2 of AERSP 597C, Dept. of Aerospace Engineering, Penn State University, November, 1987.

7. Lomax, H., and Maksymiuk, C., "Eigensystem Analysis of Classical Relaxation Techniques with Applications to Multigrid Analysis", NASA TM-88377, March, 1987.

SECTION 4: INVERSE PROBLEMS

Boundary Element Method for Inverse Heat Conduction Problems

C. Le Niliot(*), F. Papini(*), R. Pasquetti(**)
(*) Laboratoire Systèmes Energétiques et Transferts Thermiques, URA, CNRS 1168, Centre de St Jérôme, Avenue Escadrille Normandie Nièmen, 13397 Marseille, France
(**) Laboratoire de Mathématiques, URA, CNRS 168, Université de Nice, Parc Valrose, 06034 Nice, France

ABSTRACT

This paper deals with the resolution of inverse heat conduction problems by using a boundary element formulation. In stationary and transient cases, the basic discretized equations are expressed and associated, in order to obtain satisfying results, to regularization procedures over space and time (transient case). An example of application, involving the exploitation of infrared thermography data, is presented.

INTRODUCTION

This paper deals with the resolution of stationary and transient Inverse Heat Conduction Problems (IHCP) by using the Boundary Element Method (BEM). As shown in fig. 1, for a domain Ω of boundary Γ, the IHCP we are interested in consist in the determination of temperatures and fluxes over some parts of Γ, when knowing temperature and fluxes over some other parts and/or temperatures at some internal points; elsewhere one can have the classical boundary conditions of Dirichlet, Neuman or Fourier. The BEM appears well adapted to this type of problem because of the direct connections given by the Boundary Integral Equations (BIE) between boundary and domain variables.

Nevertheless IHCP are ill posed problems /1/ in the sense that minor deficiences in the inputs (temperature measurements) can have a drastical incidence over the outputs, and so numerical regularization procedures, over space and time in the transient case, are required. In the first part of this paper the stationary IHCP is revisited /2,3/, in order to show simply how to set the regularization over space. After remembering the transient BIE, when using a time and space dependent fundamental solution, we set the regularization over time, by using some "future time steps", and so we show how to express transient IHCP by BEM.

All along this paper it is supposed that the conductivity λ can be temperature dependent; that is the reason why our basic variables are the Kirchoff transform Ψ of the temperature T and the flux density p, instead of T and p/λ. Let us remind that the Kirchoff transform permits to linearize the heat conduction equation, by using the identity :

$$\nabla (\lambda \nabla T) = \Delta \psi \qquad \text{with} \qquad \psi = \int_0^T \lambda (T')\, dT'$$

Our investigations are related to a software developped in our laboratory /3,4,5/; this software is applicable to different types of direct and inverse heat conduction problems :
- 2D and 3D
- stationary and transient
- domain with subregions
- temperature dependent conductivity
- non linear boundary conditions.

For the boundary discretization, linear over time and constant over space elements are used.

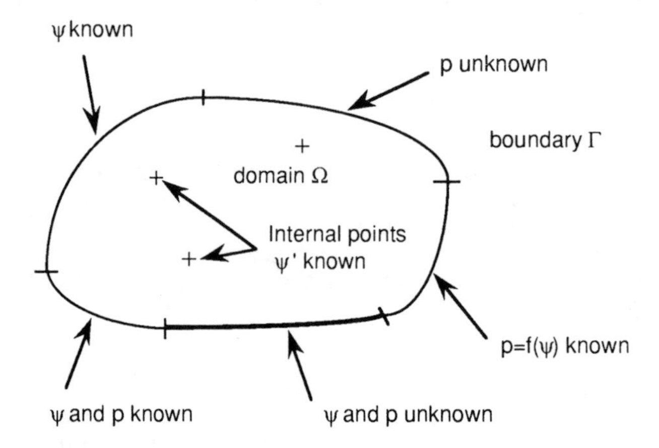

STATIONARY INVERSE HEAT CONDUCTION PROBLEMS

Let us remember again the BIE for stationary non linear heat conduction; for a domain Ω of boundary Γ :

$$c\, \psi_M + \int_\Gamma \psi\, q^* d\Gamma = \int_\Gamma p T^* d\Gamma + \int_\Omega g T^* d\Omega$$

with M : point of Γ or Ω, Ψ : Kirchoff transform of the temperature, p surfacic heat flux and c=1, if M is in Ω and c< 1, if M is on Γ (c=0.5 if Γ is smooth at M). T^* is the stationary fundamental solution and q^* is its normal derivative; with r the distance from M to a point P of Γ :

$$T^* = \frac{1}{2\pi} \ln \left(\frac{1}{r} \right) \quad \text{in 2D,} \qquad T^* = \frac{1}{4\pi r} \quad \text{in 3D}$$

When considering direct problems, if M is on Γ, one has with this BIE the starting point of BEM for potential problems. For M in Ω, the BIE, which then connect internal temperatures and boundary fields, are usually used for an a posteriori computation.

Discretized equation

When using constant over space elements, i.e. the temperature and the surfacic flux are constant over an element, one demonstrates that the discretized BIE can be written :

$$\mathbf{C}\Psi + \mathbf{H}\Psi = \mathbf{G}P + S$$

with Ψ: vector of temperature Kirchoff transforms over n boundary elements
P : vector of the surfacic heat fluxes
H and **G** : matrices of dimension (n, n)
C : diagonal matrix of dimension (n, n)
S : vector associated to heat sources

If there is n' internal points one obtains :

$$\Psi' + \mathbf{H}'\Psi = \mathbf{G}'P + S'$$

with : Ψ' : vector of temperature Kirchoff transforms at the n' internal points
H' and **G'** : matrices of dimension (n', n) .

To solve the IHCP it is necessary to connect these two equations, in order to get the basic dicretized system :

$$\begin{bmatrix} \mathbf{H}+\mathbf{C} \\ \mathbf{H}' \end{bmatrix} \psi = \begin{bmatrix} \mathbf{G} \\ \mathbf{G}' \end{bmatrix} P + \begin{bmatrix} S \\ -\psi' + S' \end{bmatrix}$$

Inverse problem resolution

rearranging the unknowns in a vector X yields the system of linear equations :

$$\mathbf{A}\,X = B$$

This system has (n+n') equations and m unknowns (the number m is problem dependent). A sinequanon condition to find out a solution is of course n+n' greater than or equal to m. For m=n+n', if **A** is regular we have simply for X : $X = \mathbf{A}^{-1} B$

Nevertheless, in the general case, n+n'>m.

The classical way to get a solution in this general case is to solve in the sense of the minimization of the distance (**A**X, B) ; with min for minimum and arg for argument :

$$X = \arg \{ \min_X \| \mathbf{A}X - B \|^2 \}$$

If one uses an Euclidian norm, elsewhere justified by statistical considerations/6/, one has to solve a quadratic optimization problem which solution is (^{T}: transposition):

$$X = (\mathbf{A}^T \mathbf{A})^{-1} \mathbf{A}^T B$$

Now, to avoid a too large sensitivity to experimental errors, it is necessary to use a regularization procedure /7/.With **R** and μ a regularization matrix and coefficient respectively, we write :

$$X = \arg \{ \min_X \{ \| \mathbf{A}X - B \|^2 + \mu \| \mathbf{R}X \|^2 \} \}$$

The matrix **R** is used to set the order of the regularization process and the coefficient μ to set the magnitude of the regularization; in this case we have to solve:

$$X = (\mathbf{A}^T\mathbf{A} + \mu\,\mathbf{R}^T\mathbf{R})^{-1}\,\mathbf{A}^T B$$

TRANSIENT INVERSE HEAT CONDUCTION PROBLEMS

For transient and non linear thermal diffusion (temperature dependent conductivity), when the derivatives with respect to space and time of the diffusivity can be neglected, the time and space dependent BIE is /8/:

$$c\psi_{M,t_F} + \int_{t_0}^{t_F}\int_{\Gamma} a\psi q^* d\Gamma\, dt = \int_{t_0}^{t_F}\int_{\Gamma} apT^* d\Gamma\, dt + \int_{t_0}^{t_F}\int_{\Omega} agT^* d\Omega\, dt + \int_{\Omega} \psi_0 T^* d\Omega$$

with the additive notations a : thermal diffusivity, t / t_0 / t_F time, initial time, final time, Ψ_0 initial condition . T^* is a Green function which depends on space and time and q^* is its normal derivative:

$$T^* = \frac{1}{(4\pi a\tau)^{s/2}} \exp\left(\frac{-r^2}{4a\tau}\right) H(\tau)$$

with H : the Heaviside function, $\tau = t_F - t$ and s : space dimension. In T^*, the thermal diffusivity a, which is a function of the temperature, is space and time dependent.

Discretized equation

Let us assume that the domain integral associated to initial conditions vanishes (e.g. the initial temperature field is constant, stationary or a study temporal domain is used /3/) and use constant over space and linear over time elements, i.e. temperatures and fluxes are constant over an element and linearly variable between two successives times. With these assumptions one gets the discretized BIE /9/ :

$$\mathbf{C}\,\psi_K + \sum_{k=1}^{K} (\mathbf{H1}_{k,K}\psi_{k-1} + \mathbf{H2}_{k,K}\psi_k) = \sum_{k=1}^{K} (\mathbf{G1}_{k,K}P_{k-1} + \mathbf{H2}_{k,K}P_k) + S_K$$

where k refers to the current time and K to the resolution time and with :

Ψ_k: vector of temperature Kirchoff transforms over n boundary elements at time k

P_k : vector of the surfacic heat fluxes at time k

$\mathbf{H1}_{k,K}$, $\mathbf{H2}_{k,K}$, $\mathbf{G1}_{k,K}$ and $\mathbf{G2}_{k,K}$: matrices of dimension (n, n)

C : diagonal matrix of dimension (n, n)
S_K : vector associated to heat sources.

By extracting from this form the boundary variables at time K, one obtains :

$$(\mathbf{C} + \mathbf{H}2_{K,K})\,\psi_K = \mathbf{G}2_{K,K}\,P_K + V_{K,K} + S_K$$

with :

$$V_{K,K} = \mathbf{G}1_{K,K}p_{K-1} - \mathbf{H}1_{K,K}\psi_{K-1} + \sum_{k=1}^{K-1} (\mathbf{G}1_{k,K}P_{k-1} + \mathbf{G}2_{k,K}P_k - \mathbf{H}1_{k,K}\psi_{k-1} - \mathbf{H}2_{k,K}\psi_k)$$

When the BIE is expressed for n' internal points, one obtains in a same way :

$$\psi'_K + \mathbf{H}2'_{K,K}\,\psi_K = \mathbf{G}2'_{K,K}\,P_K + V'_{K,K} + S'_K$$

with Ψ'_k: vector of temperature KIrchoff transforms over n' internal points at time k
$\mathbf{H1}'_{k,K}$, $\mathbf{H2}'_{k,K}$, $\mathbf{G1}'_{k,K}$ and $\mathbf{G2}'_{k,K}$: matrices of dimension (n', n)
S'_K : vector associated to heat sources.

By connecting the BIE associated to boundary elements and to internal points, one gets a discretized system similar to the stationary basic one :

$$\begin{bmatrix} \mathbf{C}+\mathbf{H}2_{K,K} \\ \mathbf{H}2'_{K,K} \end{bmatrix} \psi_K = \begin{bmatrix} \mathbf{G}2_{K,K} \\ \mathbf{G}2'_{K,K} \end{bmatrix} P_K + \begin{bmatrix} V_{K,K}+S_K \\ -\psi'_K+V'_{K,K}+S'_K \end{bmatrix}$$

An approach similar to the stationary one is now possible but not efficient because of the damping and lagging effects of the heat conduction operator. In fact, to determine what happens at time K, it is necessary to take into account some 'future time steps". Several methods are then possible as reviewed in /1/. If R future time steps are considered, as suggested by Beck one can anticipate the variations of the unknowns; for example, it can be simply supposed that $\Psi_{K+r} = \Psi_K$,$P_{K+r} = P_K$, $1 \leq r \leq R$, or assumed linear, quadratic etc.. variations of Ψ and P during the R future time steps.

A different approach to avoid such anticipations is to solve for all the Ψ_{K+r} ,P_{K+r} , $0 \leq r \leq R$ and to use a regularization procedure. Such an approach is more powerful than the Beck's one but involves heavier computation.

At time K+r, the discretized BIE can be written :

$$\mathbf{C}\psi_{K+r} + \sum_{k=K}^{K+r} (\mathbf{H}1_{k,K+r}\psi_{k-1} + \mathbf{H}2_{k,K+r}\psi_k) = \sum_{k=K}^{K+r} (\mathbf{G}1_{k,K+r}P_{k-1} + \mathbf{H}2_{k,K+r}P_k) + V_{K,K+r} + S_{K+r}$$

with

$$V_{K,K+r} = \mathbf{G}1_{K,K+r}P_{K-1} - \mathbf{H}1_{K,K+r}\psi_{K-1} +$$

$$\sum_{k=1}^{K-1} (\mathbf{G}1_{k,K+r}P_{k-1} + \mathbf{G}2_{k,K+r}P_k - \mathbf{H}1_{k,K+r}\psi_{k-1} - \mathbf{H}2_{k,K+r}\psi_k)$$

A similar form can obviously be obtained for the internal points. Finally, one can gather in a matrical form from the BIE expressed at K+r, $0 \leq r \leq R$:

$$(\mathbf{C}+\mathbf{H})\begin{bmatrix} \psi_K \\ : \\ \psi_{K+R} \end{bmatrix} = \mathbf{G}\begin{bmatrix} P_K \\ : \\ P_{K+R} \end{bmatrix} + S$$

with :

$$[\mathbf{H}]_{i+1,i+1} = \begin{bmatrix} \mathbf{H}2_{K+i,K+i} \\ \mathbf{H}2'_{K+i,K+i} \end{bmatrix} \quad , 0 \leq i \leq R$$

$$[\mathbf{H}]_{i+1,j+1} = \begin{bmatrix} \mathbf{H}1_{K+j,K+i} + \mathbf{H}2_{K+j-1,K+i} \\ \mathbf{H}1'_{K+j,K+i} + \mathbf{H}2'_{K+j-1,K+i} \end{bmatrix} \quad , 0 \leq j < i \leq R$$

$$[\mathbf{H}]_{i+1,j+1} = \begin{bmatrix} 0 \\ 0 \end{bmatrix} \quad , 0 \leq i < j \leq R$$

The elements of **G** are similar with **G1** and **G2** instead of **H1** and **H2.**

$$S_{i+1} = \begin{bmatrix} V_{K,K+i} + S_{K,K+i} \\ -\psi'_{K+i} + V'_{K,K+i} + S'_{K,K+i} \end{bmatrix}, 0 \leq i \leq R$$

$$[\mathbf{C}]_{i+1} = \begin{bmatrix} \mathbf{C} \\ 0 \end{bmatrix}, 0 \leq i \leq R$$

Now, one can gather the unknows at times K+r in vectors X_{K+r} and obtain the linear system :

$$\mathbf{A}\,X = B$$

where **A** is a block triangular matrix and X a vector such as

$$X = (X_K , X_{K+1} , ... , X_{K+R})^T$$

This system has generally more unknows than equations, as in the stationary case. Finally, by minimizing an Euclidian norm we can write :

$$X = (\mathbf{A}^T \mathbf{A})^{-1} \mathbf{A}^T B$$

Nevertheless, to get satisfying results regularizations over space and time have to be used.

<u>Regularization over space</u>

In the transient case regularization over space has to occured at each time K+r, $0 \leq r \leq R$, and so the regularization term must be written :

$$\mu \sum_{k=K}^{K+R} || \mathbf{R}_k X_k ||^2$$

If one introduces a block diagonal regularization matrix such as :

$$\mathbf{R} = \mathrm{diag}\,(\mathbf{R}_{K+r}) \qquad , \quad 0 \le r \le R$$

then we have :

$$\mu \sum_{k=K}^{K+R} \| \mathbf{R}_k X_k \|^2 = \mu \| \mathbf{R}\, X \|^2$$

When taking into account this regularization term, as in the stationary case one gets :

$$X = (\mathbf{A}^T\mathbf{A} + \mu\, \mathbf{R}^T\mathbf{R})^{-1}\, \mathbf{A}^T B$$

<u>Regularization over time</u>

Let us note x_K one of the componant of X_K which has to be regularized; the regularization term associated to x_K can be written :

$$\eta \, \| \mathbf{Q}\, (x_K , x_{K+1} , \ldots\ldots , x_{K+R})^T \|^2$$

with $\mathbf{Q}$ a (R+1,R+1) regularization matrix and η a regularization coefficient. As over space the matrix $\mathbf{Q}$ is used to set the order of the regularization and the coefficient η to set its magnitude.

Let us introduce the projection operator $\mathbf{P}$ such as :

$$(x_K , x_{K+1} , \ldots\ldots , x_{K+R})^T = \mathbf{P}\, X$$

and consider all the L componants of X which have to be regularized; then one gets for the additive regularization term :

$$\eta \sum_{l=1}^{L} \| \mathbf{Q}\, \mathbf{P}_l\, X \|^2$$

Taking into account this new regularization term one gets :

$$X = (\mathbf{A}^T \mathbf{A} + \mu\, \mathbf{R}^T R + \eta \sum_{l} \mathbf{P}_l^T\, \mathbf{Q}^T \mathbf{Q}\, \mathbf{P}_l)^{-1}\, \mathbf{A}^T B$$

Finally, X_K must be extracted from X and all the X_{K+r}, $r \neq 0$, rejected; then, the procedure can start again for the calculation of X_{K+1}.

APPLICATION

At time of submission of this paper, the transient part of our software is not yet available; that is the reason why only results obtained in the stationary case are here produced again /3/.

To qualify our approach of IHCP an experiment has been set up in our laboratory :

it is a plexiglass bar (50*50*400mm) which center is drilled to get a hole (d 30mm) inside which are placed two electrical heaters and some insulation. Each heater is constitued by a quarter circle heating film (fig.2) . The conductive transfers are assumed bidimensional and the symetry of the system is such that all the informations, necessary to solve the IHCP, are contained on one side of the bar .

The experiment consists in infrared thermographic measurements by using an infrared scanner (Agema 880) connected with a micro-computer (80386-20MHz). With such a system, the real time measurment (infrared pictures 128*64 pixels with a 25Hz frequency) is directly transfered to the RAM of the computer /10/.

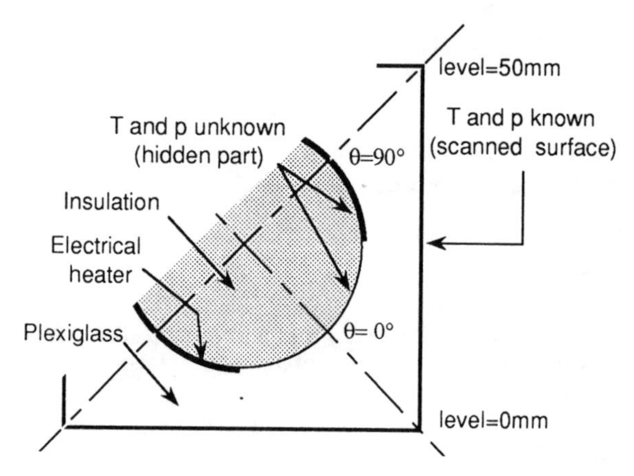

Figure.2: Experimental dispositive design.
A temperature distribution on the scanned side of the bar is presented in fig.3.

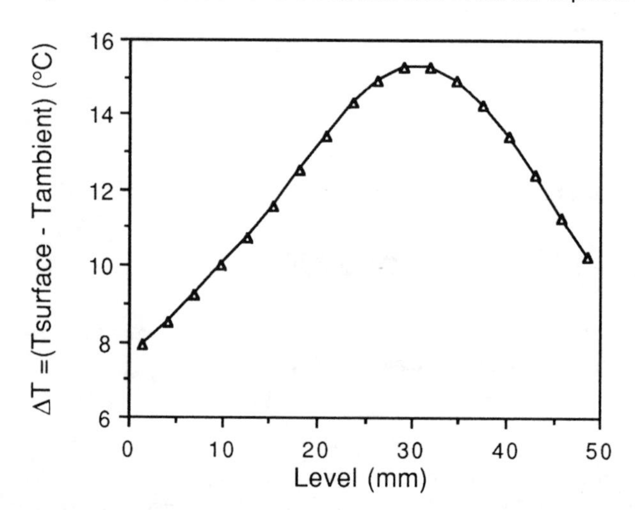

Figure.3 : thermal profile along the bar side .

The computations were proceeded by BEM and by a a Superposition method. In the stationnary case the Superposition method supposes a preliminary computation, by the means of a direct model, of the temperature θ_{ij} on the element j when a unit heat density flux is applied on the element i. Then, with the assumption of linearity we get :

$$\Theta_j = \sum_i \theta_{ij} \varphi_i$$

with : φ_i flux density on the element i

Θ_j temperature on the element j

Using as previously a space regularization procedure, this relation permits the computation, in the sense of least squares, of the φ_i knowing the Θ_j. Let us note that on the contrary of our Boundary Element approach, the Superposition method doesn't allow non linear boundary conditions and so a global heat transfer coefficient (considered in the preliminary computation) must be used to take into account the radiative and convective heat losses.

The results of both methods, when using a second order regularization over flux density are presented in fig. 4, which plots the heat flux density applied in the cylinder vs the polar angle. In fig. 5 we point out the influence of the regularization coefficient μ. These results settle down the good agreement of the two methods and thus can be considered as encouraging for the future. Nevertheless, the heater between 45 and 90° is not clearly defined; essentialy that demonstrates the necessity of very accurate measurments for the computation of IHCP .

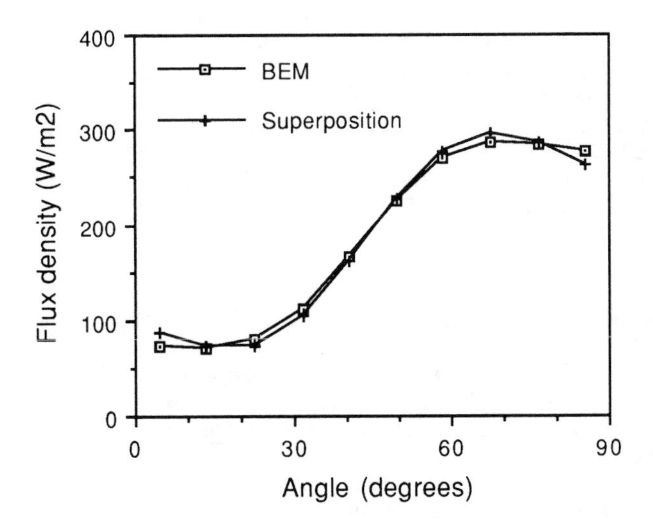

Figure.4:Boundary Elements and Superposition results

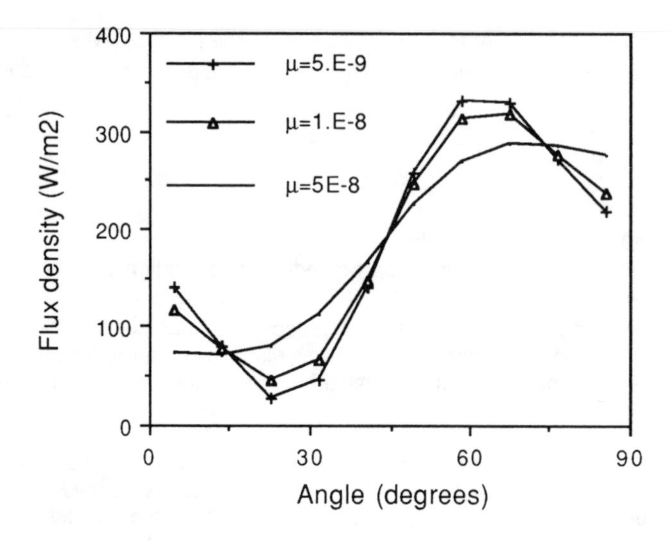

Figure.5:Boundary Elements results for some values of the regularization coefficient μ.

CONCLUSION

In this paper a Boundary Element Inversion Method has been proposed to solve stationary and transient non linear IHCP. In the transient case, this inversion method uses a time and space dependent fundamental solution; moreover, the BIE are associated to regularization procedures over space and time and the notion of "future time steps" is utilized, in order to take into account the specific properties of the diffusion operator. This inversion method, which does not use the superposition principle and preliminary results of a direct model, is still applicable when considering non linear boundary conditions.

REFERENCES

1. J.V. Beck B. Blackwell, C.R.St.Clair - Inverse heat conduction, ill posed problems. Wiley Interscience - 1985

2. R.Pasquetti, C. Le Niliot - Conduction inverse par éléments de frontiere cas stationnaire - Rev. Phy. App. 25 (1990) 99-107.

3. R.Pasquetti, C. Le Niliot - Direct and Inverse potential problems by BEM. EuroBEM - Nice - May 90.

4. R. Pasquetti, A. Caruso, C.A. Bebbia - Methode des elements de Frontiere : résolution numériquedes problèmes de diffusion thermique - Rev . Gen. Th. n° 328, pp. 228-234, april1989.

5. R. Pasquetti, A. Caruso - A boundary element approach for transient and non linear thermal diffusion - to be published in Numerical Heat Transfer.

6. P. Eykhoff - System identification, parameter and state estimation. Wiley Interscience - 1974

7. A. Thikonov, V. Arsenine - Méthode de résolution des problèmes mal posés -

Editions de Moscou, 1976

8 C.A. Brebbia - Application of the boundary element method for heat transfer problems - conference "Modélisation et simulation en thermique", september1984, ENSMA Poitiers.

9. C.A. Brebbia, J.C.F. Telles and L.C. Wrobel - Boundary Element Techniques, Springer-Verlag, 1984.

10. P. Guillemant, F. Papini et al. - Une évolution expérimentale dans la restitution des champs de température - Congress "visualisation et traitement d'images", Belfort -may 1988

An Iterative Solution Scheme to Estimate Local Heat Transfer Coefficient Distribution Along the Surface of an Axisymmetric Body

H.J. Kang, W.Q. Tao

Department of Power Machinery Engineering, Xi'an Jiaotong University, Xi'an, Shaanxi 710049, China

ABSTRACT

In this paper a new type of inverse heat conduction problem is presented and an approximate solution scheme is proposed. The solution procedure is iterative in nature. The feasibility of the proposed solution scheme is demonstrated by three examples.

INTRODUCTION

To increase the mechanical strength of glassware, tempering is an efficient technique widely used in glass technology. In this process temperature of the glassware is raised to a value in the neighborhood of the softening point. It is then removed from the heating furnace, and the surfaces are quickly cooled. Several cooling media can be used. Impingement air jet is utilized very often. A well-organized cooling system may result in a internal stress pattern which can greatly enhance the mechanical strength. The satisfactory shape of the stress distribution curve depends upon the temperature range over which the rapid cooling takes palce, the local cooling rate and the geometric shape of the glassware. A well-organized cooling system is the key to achieve high productivity of tempered glassware. A poor cooling system causes an unsatisfactorary pattern of the internal stress which may give rise to cracking glass workpiece. It is consisdered that a satisfactorary internal stress pattern for a specific workpiece may be formed if the temperatures of different locations in the body midplane can approximately reach a certain value at the same time. This specific temperature (strictly speaking, a range of temperature) is called transition temperature [1] . It is usually in the order of 450°C - 500°C. In this paper the local heat transfer distribution which may result in a satisfactorary stress pattern will be called favorable. A favorable local heat transfer distribution can serve as a guideline when one wants to design the cooling system for the specific glassware.

From heat transfer point of view, the above-mentioned engineering problem raises a new type of inverse heat concudtion problem. It is of inverse type in that some conditions of the interior points are known while the boundary conditions are to be found. However, this problem differs from a commonly-referred inverse problem. In the commonly-referred inverse heat conduction problem, the temperature history of some interior point is known and the surface temperature or heat flux are to be determined. While in the present problem the known condition is the temperature variation range and the corresponding time duration, rather than the temperature history.

Over the past two decades, various solution methods have been proposed to the inverse heat conduction problems. Most of these methods are presented in [2] . However, all these methods are designed to solve the problem with known temperature history at least of one interior point. As this is indicated by the definition given in [2] : " The IHCP is the estimation of the surface heat flux history given one or more measured temperature histories inside a heat-conducting body." To the authors' knowledge, the inverse problem described in this paper has not been solved in the literature.

The complexity of the present problem is in the following two aspects. (1) The temperature distribution is two-dimensional and the distribution of the surface heat flux is multiple. A schematic view of a workpiece cross-section is shown in Fig. 1, where the local heat transfer coefficient may be a function of r and z. (2) The workpiece physical properties are functions of temperature, thus it is a nonlinear problem. With these points in mind, we decided to develop an iterative solution procedure which could provide a favorable local heat transfer distribution. The basic idea of the solution procedure is as following. First a uniform distribution of the local heat transfer coefficient is assumed, and a computation of direct heat conduction problem is performed for one time step. The resultant temperatures of the selected points in the midplane will not, in general, be the same. Then taking the temperature of a specific point (say, point a in Fig. 1) as a reference, the local heat transfer coefficients corresponding to the other midplane points are revised according to the temperatures of the reference point and the point studied. The computation is re-performed for the first time step until the computed temperatures of selected points in the midplane are approximately the same. In this solution procedure, the key is to design a way by which the local heat transfer coefficients are revised. In order that this method can be used for complex geometry, the finite element approach is used.

PROBLEM FORMULATION

The finite element formulation of the direct problem is first briefly described, followed by formulation of the inverse problem investigated in this paper. Consider transient heat conduction in a homogeneous isotropic axisymmetric solid body shown in Fig. 1. In the abscence of heat source inside the domain, the governing equation is

$$\rho r c \frac{\partial T}{\partial \tau} = k\left(r \frac{\partial^2 T}{\partial z^2} + r \frac{\partial^2 T}{\partial r^2} + \frac{\partial T}{\partial r}\right) \tag{1}$$

with initial condition of the type

$$T = T_o \quad \text{in } \Omega \tag{2}$$

and boundary condition of third kind

$$-k \frac{\partial T}{\partial n} = h(T - T_f) \quad \text{on } \Gamma \tag{3}$$

By using the Galerkin's method the above governing equation may be recast in a set of first order differential equations in time. This discretized procedure is well documented in the literature, such as [3] , and will not be enumerated here. In the matrix form, these equations are

$$[K]\ \{T\} + [C]\left\{\frac{\partial T}{\partial \tau}\right\} = \{R\} \tag{4}$$

where $[K]$ is the conductance matrix, $[C]$ is the capacitance matrix and $\{R\}$ is the load vector which arises from the surface convective heat transfer coefficients.

Introducing the assumptions

$$\frac{\partial T}{\partial \tau} = (T^{n+1} - T^n)/\Delta \tau \tag{5a}$$

$$T = (1 - f)T^n + fT^{n+1} \tag{5b}$$

and substituting these expressions in equation (4) gives

$$\left[f\ [K] + \frac{C}{\Delta \tau}\right]\{T\}^{n+1} = \left[-(1 - f)\ [K] + \frac{C}{\Delta \tau}\right]\{T\}^n + \{R\} \tag{6}$$

where $\{T\}^{n+1}$ is unknown and $\{T\}^n$ is known from the previous time step. If f=0 the resulting algebraic equations are uncoupled and can be solved from step to step with ease. The full implicit choice(f=1) requires the simultaneous solution of the algebraic equations. To save the computer time and storage, the explicit scheme(f=0) was used in the present work.

For the inverse problem studied in this work, the heat transfer coefficient in equation (3) is unknown. It is a function of r and z. The problem is to estimate a favorable distribution of h(r,z) such that the temperatures of the selected points in the body midplane (say, points a-f in Fig. 1) can approximately reach the same value over the same time duration.

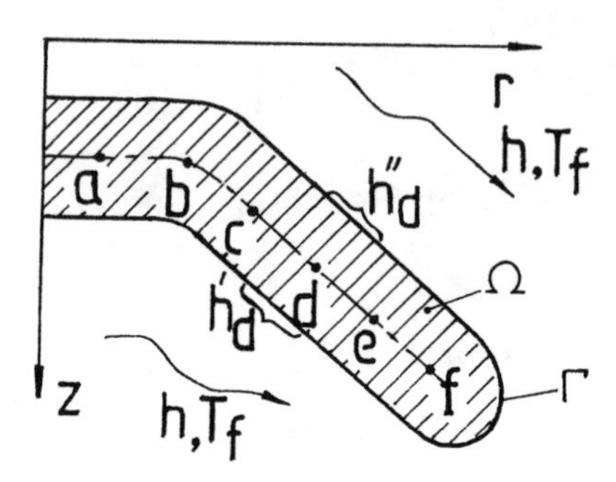

Fig. 1 Geometry of heat-conducting body

SOLUTION PROCEDURE

For most application cases, the geometric shape of the glassware to be tempered are of cone-type, which has a narrow-and-long cross-section. To assume a reasonable distribution of the local heat transfer coefficient, experimental studies of the local heat transfer for circular jet impingement on surfaces of cone-type cavities were performed [4] . It is found by the experiments that within the parameter range used in engineering, the local heat transfer coefficient on the impinged surface are in the order of 300 - 600 $W/(m^2 \cdot ^\circ C)$. The local heat transfer coefficient on the surfaces corresponding to the reference point is taken as 300 $W/(m^2 \cdot ^\circ C)$.

From physical intuition it was considered that for a fixed point in a narrow-and-long region(say, point d in Fig. 1), the temperature variation with time might be dominated by the local heat transfer coefficients on its two surfaces(h'_d and h''_d in Fig. 1), while the effects from the neighbouring points were weak. To demonstrate reasonableness of the intuition, a preliminary computation was made for an axisymmetric body shown in Fig. 2 The temperature variation with time for the midplane points were computed by using 2-D and 1-D transient models. It is found that for a large Biot number, say, Bi=1, the temperature variations of most of the midplane points can be well predicted by a one-dimensional model. A typical result is shown in Fig. 3 for point D of Fig. 2. The results of the preliminary computation makes it clear that the improvement of the local heat transfer coefficient may be conducted with a one-dimensional model.

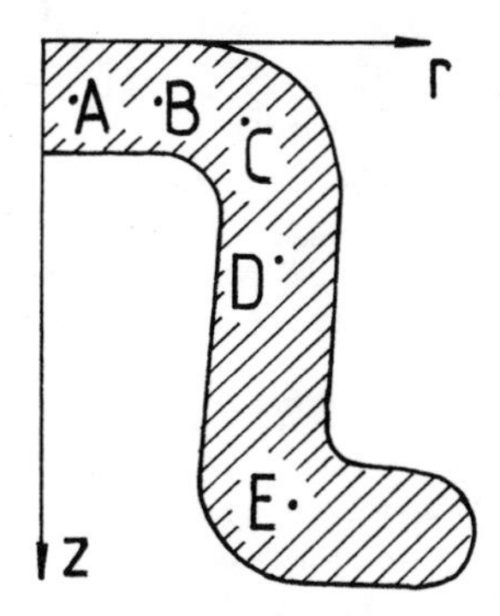

Fig. 2 Geometry of preliminary computation

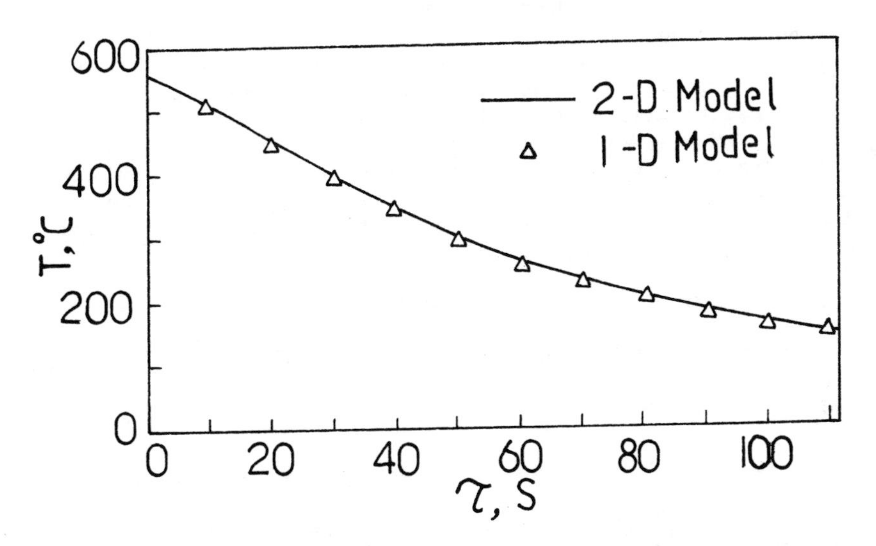

Fig. 3 Results of preliminary computation

Attention is now focused on the correction equation for the local heat transfer coefficient. As indicated above, the temperature history of a selected point in the midplane is mainly controlled by the local heat transfer coefficients on the two parts of the surfaces corresponding to the point. Taking point d as an example. Its temperature history is dominated by the heat transfer coefficients h_d' and h_d''. Our computation practice shows that the following correction equation can work satisfactorily:

$$\overline{h_d'} = h_d' + \left[(T_a - T_a^{\circ})/(T_d - T_d^{\circ}) - 1\right] h_d' \cdot \omega \qquad (7)$$

$$\overline{h_d''} = h_d'' + \left[(T_a - T_a^o)/(T_d - T_d^o) - 1\right] h_d'' \cdot \omega \qquad (8)$$

where T_a^o and T_d^o are the initial temperatures of points a and d, T_a and T_d are the temperatures at the end of the first time step, the symbol "—" indicates the improved value and ω is a weighting factor. Experience shows that the value of 0.618 for ω works effectively.

After the first time step computation was performed under the assumption of uniform distribution of the local heat transfer coefficients, the correction of the heat transfer coefficients was conducted for each selected midplane point. The corrected distribution was then used to re-performed the computation of the first time step and the local heat transfer coefficients were re-corrected according to the computed temperatures of the midplane points via equations (7) and (8). This computation was performed again and again until a favorable distribution was found, by which the temperatures at the selected points at the end of the first time step were approximately the same. This favorable distribution was then used in the computations of the successive time steps. The variable physical properties were treated in a explicit manner. That is, for the computation of (n+1)th time step, the physical properties were determined by the temperatures obtained at the nth time step. Using the data given in [5,6] , two equations were fitted for determining thermal conductivity and specific heat :

$$k = 2.95\times10^{-8}\, T^2 + 0.00223T + 0.927, \ W/(m\cdot{}^{\circ}C) \qquad (9)$$

$$c = 4186.8\times \left[(0.000504T + 0.1746)/(0.00146T + 1)\right], \ J/(kg\cdot{}^{\circ}C) \qquad (10)$$

$$(30\,^{\circ}C \leqslant T \leqslant 600\,^{\circ}C)$$

The glass density was taken as a constant (2500 kg/m^3).

Because of the variable physical properties the finite element matrix coefficients were not constant, rather, they were re-calculated for each time step during the marching procedure

RESULTS AND DISCUSSIONS

The above-presented solution procedure was used to estimate the surface heat transfer coefficient distribution for three axisymmetric geometries: two cone-type bodies and one cap-type article. The presentation begin with the results of the cylindrical cavity, followed by those of the conical cavity and the cap-type body.

Cylindrical cavity

The cross-section of the body and the FEM discretization are shown in Fig. 4. The reference point was point a and the reference heat transfer coefficient was 300 W/(m^2.°C). The ambient temperature and the initial temperature were 30 °C and 600°C, respectively. After four iterations, a favorable distribution of the heat transfer coefficient was obtained. The results are h_2=370, h_3=350 and h_4=300 W/(m^2.°C). The computed temperature variations of the midplane points a to e with these heat transfer coefficients as boundary conditions are shown in Fig. 5. It can be seen that even with a rather coarse discretization of the surface element, the maximum difference between the temperatures of a to e is about 6% of the average value.

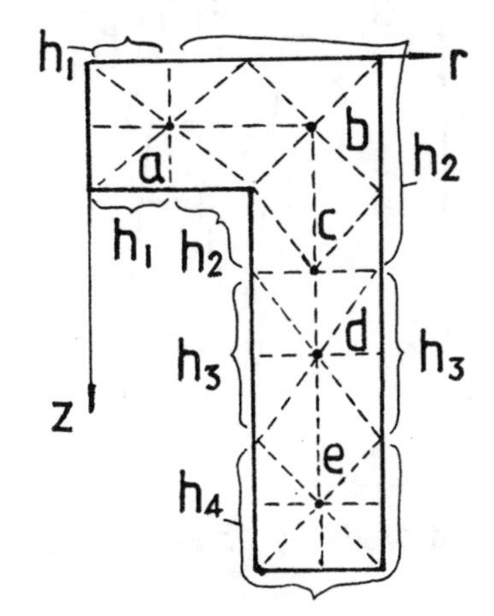

Fig. 4 Geometry of example 1

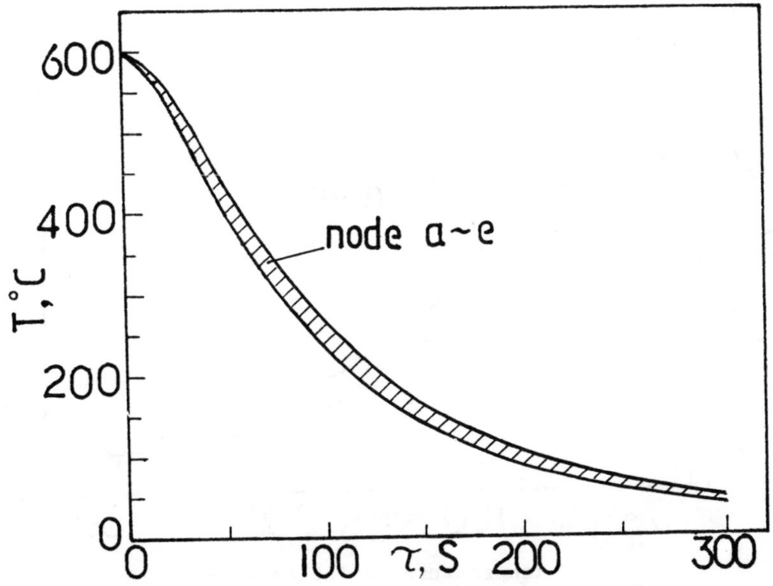

Fig. 5 Temperature variation of example 1

Conical cavity

The configuration and the FEM discretization are presented in Fig. 6. Except the geometry, all other conditions were the same as those of example 1. For this problem the local heat transfer coefficients were only devided into three levels. The

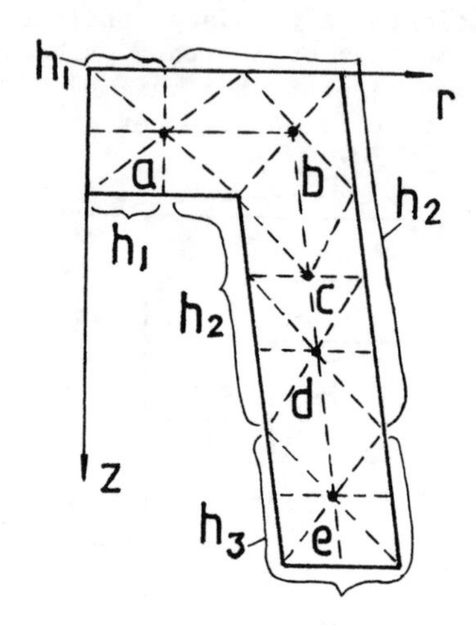

Fig. 6 Geometry of example 2

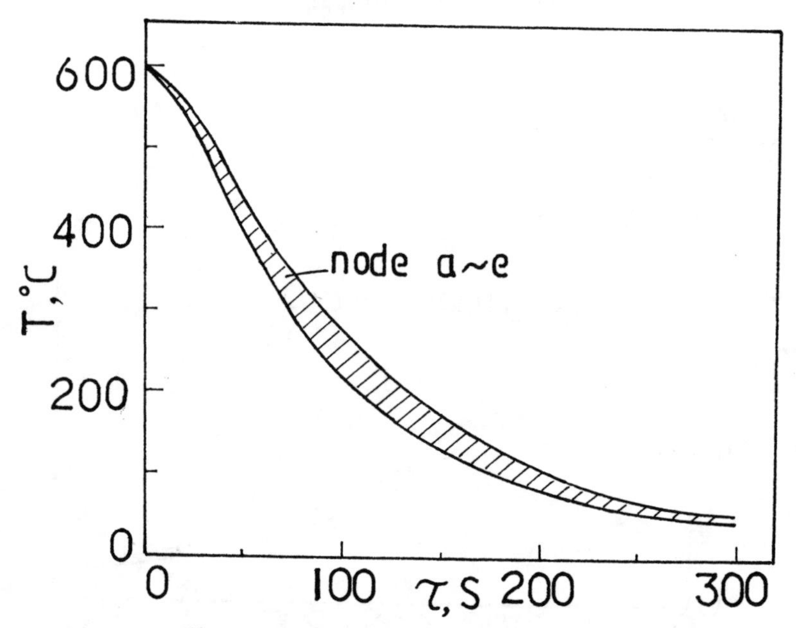

Fig. 7 Temperature variation of example 2

estimated values of h_2 and h_3 are 330 and 280 W/(m^2·°C), respectively. The variation of temperature with time for points a to e are shown in Fig. 7. The maximum difference is about 7% of the average value.

Cap-type body

Finnaly the code developed was used to estimate a favorable distribution for a cap-type body shown in Fig. 8. The selected midplane points were a to l. All the computation conditions were the same as those of example 1 except the body shape. The estimated favorable local heat transfer distribution is shown in the same figure by numbers(in SI units). The temperature variations for the selected points are presented in Fig. 9 by solid lines. In the range of the transition temperature of glass(470 - 480°C), the realtive difference is about 3%.

For the third example a computation was made with the assumption of uniform heat transfer coefficient along the whole surface. It was found that for the midplane points the maximum temperature occurred at node f while the minimum temperature occurred at node l. The temperatures of other midplane points were in between. The temperature variations of these two nodes are presented in Fig. 9 by dashed lines. From this figure effectiveness of the estimated heat transfer coefficient distribution can be recognized very clearly. It is worth noting further that the temperature differences between the midplane points may be reduced further by the above-described solution procedure if more surface elements are used, over each of which the local heat transfer coefficient is taken as a constant.

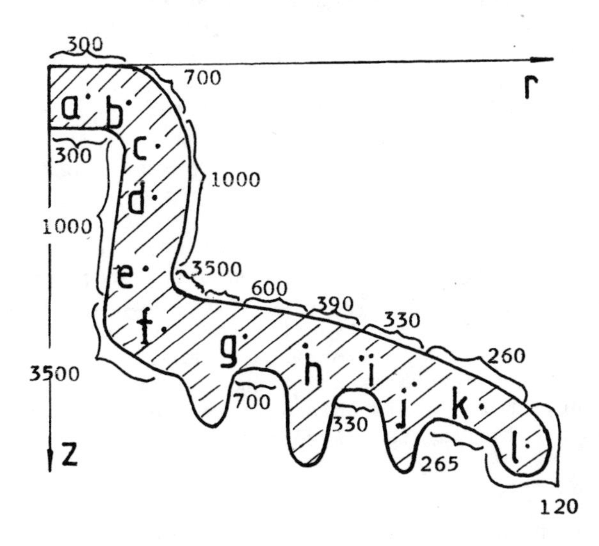

Fig. 8 Geometry of example 3

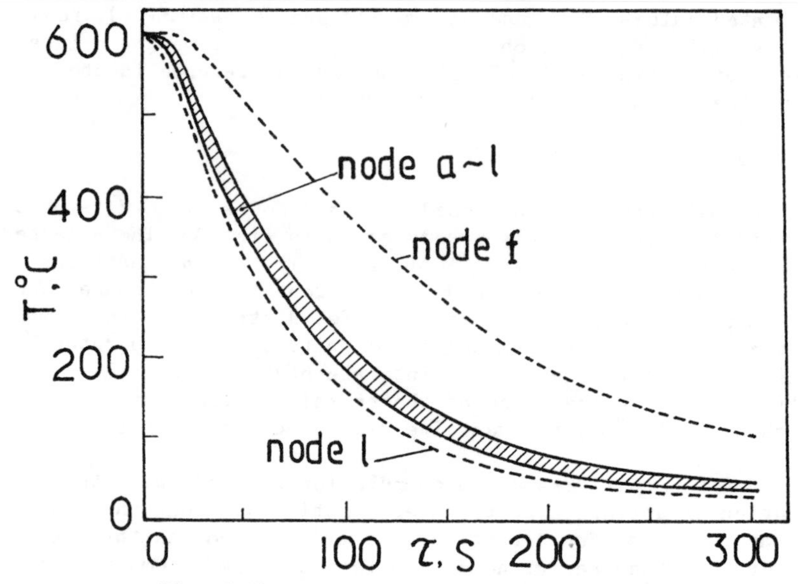

Fig. 9 Temperature variation of example 3

The time step used in the present computation was 20 seconds. It was found that to decrease the time step the sizes of the elements should be refined, otherwise numerical oscillation would occur. The critical grid Fourier number was about 0.1.

Attention is now turned to the characteristics of the estimated local heat transfer coefficient distribution common to the three examples. It can be seen that the local heat transfer coefficients on the surfaces of corner part of the bodies are higher than those on the flat surfaces. An auxiliary computation was done by the present authors for an axisymmetric boay with finite difference method. The same characteristcs was found from the estimated local heat transfer coefficient distribution.

It is obvious that the local heat transfer coefficient distribution obtained by the present solution procedure is, by no means, unique. However, it is believed that the distribution pattern of the local heat transfer coefficient is qualitatively the same for different estimations(with different number of elements, different reference point and corresponding heat transfer coefficient, etc.) It is this distribution pattern that can be served as a guideline in designing a cooling system.

CONCLUSIONS

In the present work a new type of inverse heat conduction

probelem was formulated and an iterative solution procedure was proposed for axisymmetric body with narrow-and-long solution domain. The desired local heat transfer coefficient distribution should meet the requirement that the temperatures of the midplane points could approximately reach the same value at the same time. The key point of the iterative solution procedure was the way of revising the assumed local heat transfer coefficients. The proposed method was used to solve three examples and the results were presented, indicating the feasibility of the proposed solution procedure. The estimated distribution pattern of the local heat transfer coefficient may be served as a guideline in designing a cooling system.

NOMENCLATURES

c	= specific heat
$[C]$	= capacitance matrix
f	= weighting factor
h	= heat transfer coefficient
k	= thermal conductivity
$[K]$	= conductance matrix
r	= radius
$\{R\}$	= load vector
T	= temperature
$\{T\}$	= temperature vector
T_f	= temperature of ambient fluid
z	= axial coordinate

Subscripts

$-$	= corrected value
o	= initial value
$n, n+1$	= number of time step

Greek symbols

Γ	= boundary of solution domain
$\Delta\tau$	= time step
Ω	= solution domain
ω	= weighting factor
ρ	= density
τ	= time

ACKNOWLEDGMENT

This work was supported by the National Natural Science Foundation of China.

REFERENCES

1. McLellan, G. W. and Shand, E. B., Glass Engineering Handbook, Third Edition, McGraw-Hill, New York, 1984

2. Beck, J. V., Blackwell, B. and St. Clair, C. R., Jr., Inverse Heat Conduction, Ill-Posed Problems, John Wile & Sons, New York, 1985

3. Huebner, K. H. and Thornton, E. A., The Finite Element Method for Engineers, Second Edition, John Wiley & Sons, New York, 1982

4. Kang, H. J. and Tao, W. Q., Experimental Study of Heat/Mass Transfer for Circular Jet Impingement on Surface of Cone-Type Cavities, be presented at the International Conference on Energy Conversion and Energy Source Engineering, October 5 - 8, 1990, Huhan, China

5. Touloukian, Y. S. et al., Thermal Conductivity - Nonmetalic Solids, IFI/Plenum, New York, 1970

6. Matviev, M. A. et al., Glass Chemistry and Technology Calculation (in Russian), Chemical Engineering Press, Moscow, 1972

Applying BEM and the Sensitivity Coefficient Concept to Inverse Heat Conduction Problems

K. Kurpisz, A.J. Nowak

Institute of Thermal Technology, Silesian Technical University, 44-101 Gliwice, Konarskiego 22, Poland

ABSTRACT

In the paper inverse heat conduction problems are considered. Proposed algorithm of estimation temperature and heat flux consists in two stages. First stage is to determine sensitivity coefficients which are used by the procedure called function specification method. This procedure makes the final results stable.
The Boundary Elements Method was applied to obtain sensitivity coefficients and also within the function specification method.
Numerical example is included.

INTRODUCTION

The problem considered herein is the estimation of the temperature and heat flux on the surface as a function of time from transient temperature measurement at interior locations of a heat conducting body. Such a problem belongs to inverse heat conduction problems that are known as ill-posed. It means that they are very sensitive to measurement errors and to obtain stable and accurate results special numerical techniques should be employed.

Generally, solution procedure consists of two stages: solving the direct boundary value problem and stabilizing the results. Thus, one has to select both above mentioned techniques. Since all unknowns are associated with the boundary only it seems to be natural to employ the boundary element method (BEM) to solve the direct boundary problem. Another advantage of applying BEM is that contrary to other numerical methods it does not need any internal cells. Thus, the location of internal points where temperature is measured can be chosen in a quite arbitrary way. When obtained integral equations are discretized constant elements, for the sake of simplicity, are assumed. The numerical technique used in this paper to make the result stable is based on the sensitivity coefficient concept and function specification method. Approach, originally proposed by Beck [1] for 1-D problems is extended for 2-D or 3-D problems. The proposed procedure can also be adjusted to non-linear problems.

PROBLEM FORMULATION

The temperature distribution inside a heat conducting body can be described by the following boundary problem:

- differential equation

$$\frac{\partial T(t,r)}{\partial t} = a\,\nabla^2 T(t,r) \tag{1}$$

where:

T - temperature as a function of time,
t - time,
r - space variable vector,
a - diffusivity coefficient,

- boundary conditions

$$-\lambda\,\frac{\partial T}{\partial n}\bigg|_{S_A} = q(t) \tag{2}$$

$$-\lambda\,\frac{\partial T}{\partial n}\bigg|_{S_B} = q_B(t) \tag{3}$$

where:

λ - thermal conductivity coefficient,
n - outward normal to the boundary,
$q(t), q_B(t)$ - heat flux on the surface S_A and S_B, respectively, and $S_A \cap S_B = S$ -external surface of the body,

- initial condition

$$T(t,r) = T_0(r), \quad t \to 0. \tag{4}$$

The thermal diffusivity and conductivity are postulated to be known and constant.

For direct heat conduction problems heat fluxes as well as the initial condition, λ and a are known and the objective is to determine the temperature distribution $T(t,r)$. For inverse problems the boundary condition is not specified for a part or, in particular, on the whole external surface of the body. Instead, some measured temperature histories are given at interior locations at discrete times as follows

$$T(t_k, r_i) = U_{k,i}, \quad k=1,2,\ldots,M, \ i=1,2,\ldots,I \tag{5}$$

where number I is the total number of interior locations and M is the total number of discrete times. The values $U_{k,i}$ are called internal temperature responses. The objective is to estimate the unknown heat flux at discrete times t_k

$$q_k = q(t_k), \ k=1,2,\ldots,M. \tag{6}$$

Hence, the inverse problem consists of the following equations:

- differential equation (1),
- boundary condition e.g. (3),
- initial condition (4),
- internal responses (5).

The unknown boundary condition is chosen in the form of the second kind because the heat flux is more difficult to calculate accurately than the surface temperature. In addition, when knowing the heat flux it is easy to determine the temperature distribution and finally a convective heat transfer coefficient, if needed (provided the ambient temperature is known). The known boundary condition can be chosen in any form and this does not introduce any new difficulty.

As a consequence of equation (6) the boundary condition (2) can be replaced by

$$-\lambda \left.\frac{\partial T}{\partial n}\right|_{S_A} = \begin{cases} q_k = \text{const}, \ t_{k-1} < t \le t_k, \\ q(t) = f(t), \ t > t_k. \end{cases} \tag{7}$$

When determining q_k the heat flux $q(t)$ for $t > t_k$ can be an arbitrary function of time because the temperature distribution at $t=t_k$ as well as the component q_k are independent of the function $f(t)$.

SENSITIVITY COEFFICIENTS

For the inverse heat conduction problem considered herein the sensitivity coefficient is the first derivative of a dependent variable that is temperature T

L

with respect to an unknown parameter that is a heat flux component q_k and is defined by

$$Z^k(t,r) = \partial T(t,r)/\partial q_k. \tag{8}$$

For $t<t_{k-1}$ the coefficient $Z^k(t,r)=0$, that is the body has not yet been exposed to q_k. For times greater than t_{k-1} equations (1) - (7) are differentiated with respect to q_k to obtain

$$\frac{\partial Z^k(t,r)}{\partial t} = a\ \nabla^2 Z^k(t,r), \tag{9}$$

$$-\lambda \frac{\partial Z^k}{\partial n}\bigg|_{S_A} = \begin{cases} 1, & \text{for } t_{k-1}<t\le t_k, \\ \gamma & \text{for } t>t_k \end{cases} \tag{10}$$

$$-\lambda \frac{\partial Z^k}{\partial n}\bigg|_{S_B} = 0, \tag{11}$$

$$Z^k(t_{k-1},r) = 0. \tag{12}$$

The variable γ depends on the form of function $f(t)$ in equation (7).

The most important conclusion that can be drawn from equations (9)-(12) is that the same differential equation holds for $Z^k(t,r)$ as for $T(t,r)$. Also boundary conditions are of the same type. However the known boundary condition as well as the initial condition become homogeneous. The right hand side of the unknown boundary condition becomes equal to one except for times $t>t_k$ when it is equal to γ. It means that the same computer procedure can be used for the solution of $Z^k(t,r)$ as for $T(t,r)$ solutions and the boundary and initial conditions are the only changes have to be made.

The boundary problem (9)-(12) can be solved by any numerical method but in this paper BEM is applied [2,3]. The main reason for that choice is the method does not require any domain discretization. This means that unknowns are calculated only at boundary nodes and selected internal points. These internal points are identical with locations of temperature sensors and their number is limited.

As a result of discretization of time and space

one obtains a set of discret values of sensitivity coefficients $Z^{k,j}_{m,i}$. Sensitivity coefficients $Z^{k,j}_{m,i}$ are the first derivatives of temperature $T_{m,i}$ at time t_m and location r with respect to the heat flux compo- $q_{k,j}$ prescribed at the boundary node $r_j \in S_A$ at time t_k.

Since boundary condition (10) causes a thermal shock to achieve appropriate accuracy time steps selected in calculations should be fairly small. Therefore formulation with time-dependent fundamental solution [2] is used in this stage of analysis. Hence, discretized boundary integral equation has the following form

$$\mathbf{H}\mathbf{Z}^k_m = \mathbf{G}\mathbf{Q}_k \tag{13}$$

where vectors $\mathbf{Z}^k_m$ and $\mathbf{Q}_k$ contain sensitivity coefficients and heat flux analog respectively. Influence matrices **H** and **G** for constant elements are defined as usual [2]

$$(H)_{ij} = \frac{1}{2}\,\delta_{ij} - \frac{1}{2\pi}\int_{\Gamma_j} \frac{1}{r}\exp\left(-\frac{r^2}{4at}\right)\frac{\partial r}{\partial n}\,d\Gamma$$

$$(G)_{ij} = -\frac{1}{4\pi\lambda}\int_{\Gamma_j} Ei\left(\frac{r^2}{4at}\right)d\Gamma$$

Notice that according to the homogeneous initial condition (12) there is no domain integral in equation (13).

The values of $Z^{k,j}_{m,i}$ are obtained by solving equation (13) in double looped process. Internal loop is associated with time steps (k=1,2,...,M; m=1,2,...,M) whereas external loop refers to the boundary elements where heat flux $q_{k,j}$ is prescribed (j=1,2,...,N; i=1,2,...,I). It is worth pointing out that number N of heat flux components $q_{k,j}$ at time t_k should be less or equal to the number of sensors I.

FUNCTION SPECIFICATION APPROACH FOR SOLVING THE INVERSE PROBLEM

Because of the linearity of the considered inverse problem it is possible to utilize the principle of superposition, i.e. the influence of the heat flux components $q_{k,j}$ on the temperature distribution can be taken into account successively.

The problem is solved in a sequential manner. It is assumed that the temperature distribution and heat flux components are known at times t_{k-1}, t_{k-2}, ..., and it is desired to determine the heat flux components at time t_k. The temperature field $T(t,r)$ depends in a continuous manner on the unknown heat flux components. Therefore, the temperature $T_{m,i}$ can be expanded in a Taylor series about an arbitrary but known values of heat flux $q^*_{m,i}$ (cf. [1])

$$T_{m,i} = T^*_{m,i} + \sum_{j=1}^{N} \left. \frac{\partial T_{m,i}}{\partial q_{m,j}} \right|_{q_{m,j}=q^*_{m,j}} (q_{m,j} - q^*_{m,j}), \tag{14}$$

where $T^*_{m,i}$ is the temperature at time t_m and at location r_i with $q_{m,j} = q^*_{m,j}$ over $t_{m-1} < t \leq t_m$. The higher derivatives in equation (14) are equal to zero due to the linearity of the considered problem. Introducing sensitivity coefficients to equation (14) we obtain

$$T_{m,i} = T^*_{m,i} + \sum_{j=1}^{N} Z^{m,j}_{m,i} (q_{m,j} - q^*_{m,j}). \tag{15}$$

The values of $q^*_{m,j}$ can be chosen in an arbitrary way but a common choice [1] is

$$q^*_{m,j} = q_{m-1,j}, \quad \text{and} \quad q^*_{1,j} = 0. \tag{16}$$

There are N unknown values of $q_{m,j}$ in equation (15) at time t_m. If the N values of temperature $T_{m,i}$ were known (equation (5)) the inverse problem could be solved because the values of $q_{m,j}$ might be derived from the system of N algebraic equations of (15) type. In fact, because of the ill-posed nature of inverse problems, the numerical results would be very unstable and inaccurate.

One possible approach to overcome stability problems is to record temperature histories at more than N locations (i.e. I>N). However, as for multidimensional problems the number N can be fairly large there may be no possibility to introduce more measurement sensors.

The more effective procedure called function specification method was introduce by Beck [1]. The heat flux $q_{k,j}$ at time t_k will influence the temperature field at times t_{k+1}, t_{k+2}, ..., called future time steps. Therefore, temperature T at those steps carries information about the unknown heat flux components and can be employed to estimate $q_{k,j}$. The basic concept in the function specification procedure is that the heat flux is assumed to be constant over r future steps and equal to the heat flux at time t_k. This is a temporary assumption made only to determine the heat flux components at time t_k. This assumption can be expressed by the equation

$$q_{k,j} = q_{k+1,j} = \dots = q_{k+r,j}, \tag{17}$$

which is equivalent to assuming $\gamma=1$ in equation (10). The unknown values of $q_{k,j}$ (j=1,2,...,N) can be determined with the least squares method that minimizes the error between the computed and measured sensor temperatures. The objective function is

$$\Phi = \sum_{m=1}^{r} \sum_{j=1}^{N} (U_{k+m-1,j} - T_{k+m-1,j})^2 \rightarrow \min. \tag{18}$$

Solving this problem one obtains the following system of algebraic equations

$$\mathbf{A}_k \mathbf{Q}_k = \mathbf{B}_k, \quad k=1,2,\dots,M, \tag{19}$$

where the components of $\mathbf{A}_k$ matrix are

$$(A_k)_{i,j} = \sum_{m=1}^{r} \sum_{l=1}^{N} Z^{m,i}_{m,l} \, Z^{m,j}_{m,l}, \quad i,j=1,2,\dots,N, \tag{20}$$

the components of $\mathbf{B}_k$ vector are

$$(B_k)_i = \sum_{l=1}^{N} \left[\sum_{m=1}^{r} \sum_{l=1}^{N} Z^{m,i}_{m,l} \, Z^{m,j}_{m,l} \right] q^*_{k,l} +$$

$$+\sum_{m=1}^{r}\sum_{l=1}^{N} Z_{m,l}^{m,i}(U_{k+m-1,l} - T^{*}_{k+m-1,l}) \;, \; i=1,2,\ldots,N \qquad (21)$$

and

$$\{Q_k\}_j = q_{k,j}$$

The computed heat flux components are retained only over the time interval $t_{k-1}<t\le t_k$. For each subsequent time interval new components are calculated.

The computer procedure of determining the temperature distribution at time t_k consists of the following steps:

- first: the sensitivity coefficients are calculated; they remain unchanged at times t_{k+1}, t_{k+2},... for linear problems as the boundary and initial conditions do not depend on time. It should be stressed that this step is carried out only once,
- second: for an arbitrarily assumed heat flux $q^{*}_{k,j}$ temperatures $T^{*}_{k,j}$, $T^{*}_{k+1,j}$,...,$T^{*}_{k+r-1,j}$,are determined; this step of analysis can be performed again using BEM. This time, however, formulation with time-independent fundamental solution is applied. Multiple Reciprocity approach [4,5] permits to obtain boundary only formulation of the problem. This technique is recognized to be quite efficient when thermal shock does not occur or is fairly weak.
- third: the heat flux components $q_{k,j}$ are calculated from equation (19),
- fourth: the temperature distribution is calculated from equation (15).

TEST CASES AND CONCLUSIONS

Numerous examples were carried out in order to examine the accuracy and stability of the present method. The temperature histories which were necessary to solve the test cases were obtained from the direct boundary problems of heat condition. To make the cases more realistic errors were added to the exact temperatures. The simulated temperatures were given by

$$U_{k,j} = \bar{U}_{k,j} + (1 - 2\rho)\delta$$

where:

$\bar{U}_{k,j}$ - the exact temperature at node r_j and at time t_k,

ρ - the random value, produced by a random value generator of a uniform distribution within the range [-1,1],

δ - the maximum absolute error.

The test cases presented herein refer to 2-D transient heat conduction within the region shown in Figure 1. The problem is formulated in a dimensionless form. Surfaces x=0 and y=0 are assumed to be insulated. The boundary conditions of the third kind are prescribed on the other surfaces with the Biot number defined as follows

$$Bi_x = \frac{h_x d_x}{\lambda} , \quad Bi_y = \frac{h_y d_y}{\lambda} ,$$

where:

h_x, h_y - convective heat transfer coefficients along x and y axis respectively,

d_x, d_y - the size of the plate in x and y direction respectively.

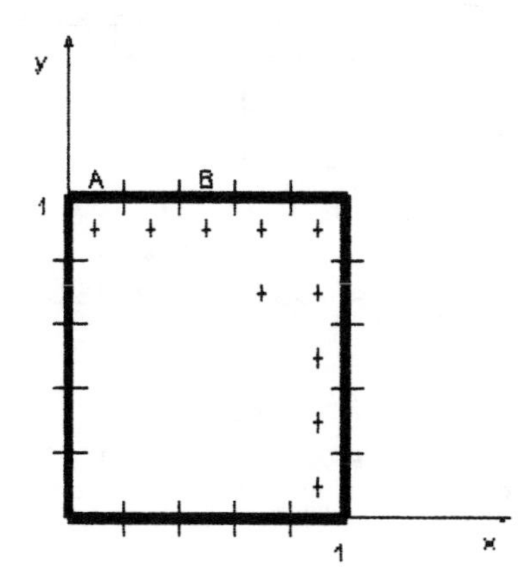

Figure 1. *Model for 2-D heat conducting body showing the location of temperature sensors*

The test cases were calculated for different Biot numbers, time steps, maximum absolute errors etc.

Figure 2 displays results of the heat flux estimation at point A for both Biot numbers equal to 0.5. The solid line corresponds to the exact solution derived from the direct boundary problem. Symbols in Figure 2 are associated with results obtained with the described method for different values δ of absolute errors of measurements. Calculations were carried out with the time step $\Delta t=0.75$, assuming the uniform initial condition. Temperature sensors were located at internal points marked by + in Figure 1.

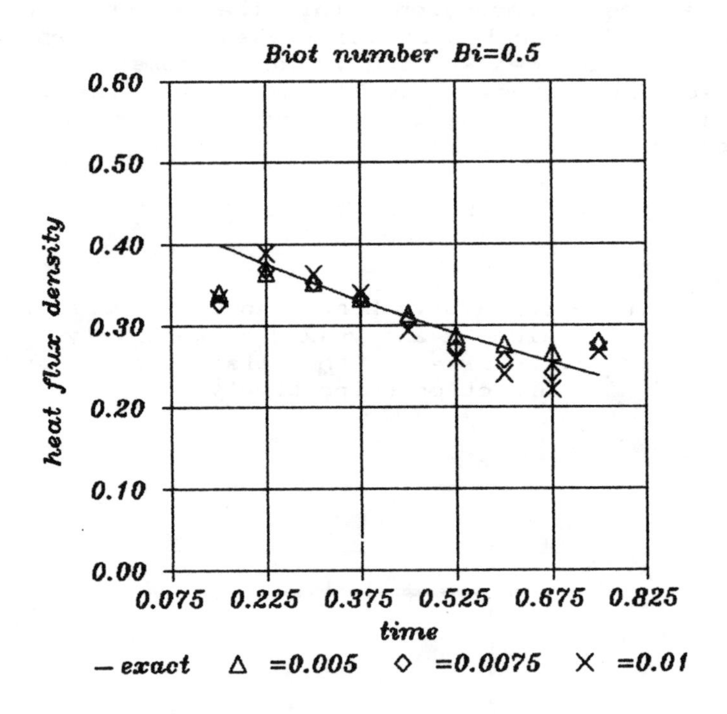

Figure 2. *Dimensionless heat flux at point A as a function of dimensionless time for* Bi=0.5.

Figure 3 presents the same quantity but for different Biot numbers both equal to 1.

In Figure 4 estimation of temperature at point B for Biot numbers equal to 0.5 is shown.

Calculations carried out so far proved that the proposed approach can produce results which are stable and accurate enough for a wide range of problems. It is worth pointing out that they are weakly sensitive to measurements errors. Temperature is estimated much more accurately than the heat flux. The largest differences between an exact solution and results produced by the present method occurred at the vicinity of corners. The greater the Biot number is the less accurate results are.

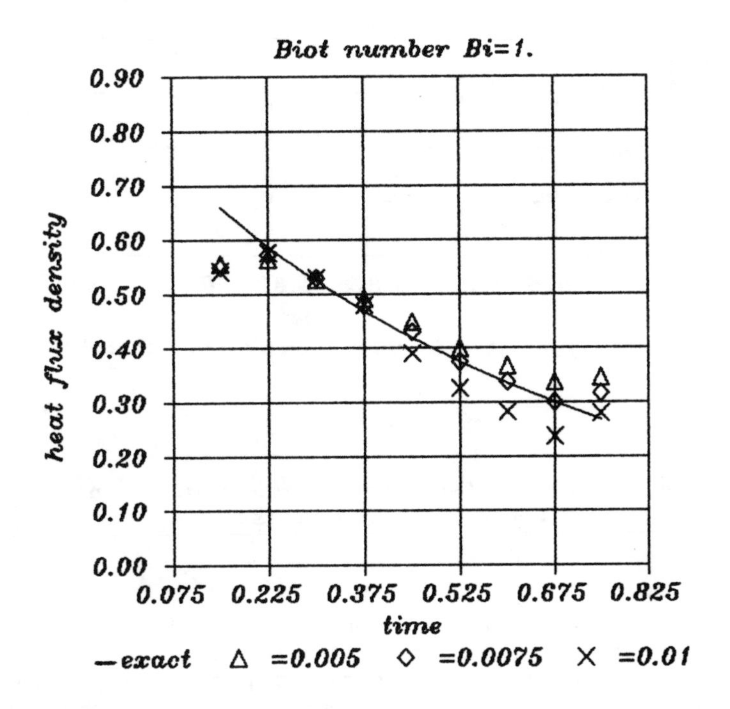

Figure 3. *Dimensionless heat flux at point A as a function of dimensionless time for* Bi=1.

It should be added that we found some cases where solutions were not satisfactory. This probably results from the fact that when thermal shock occurs, in the solution obtained by any numerical method some perturbations can be observed at the beginning of the transient process. These perturbations affect the accuracy of sensitivity coefficients. Therefore, it is very important to apply special techniques like smoothing algorithm [6] in order to cope with thermal shocks. Although presented results can not be treated

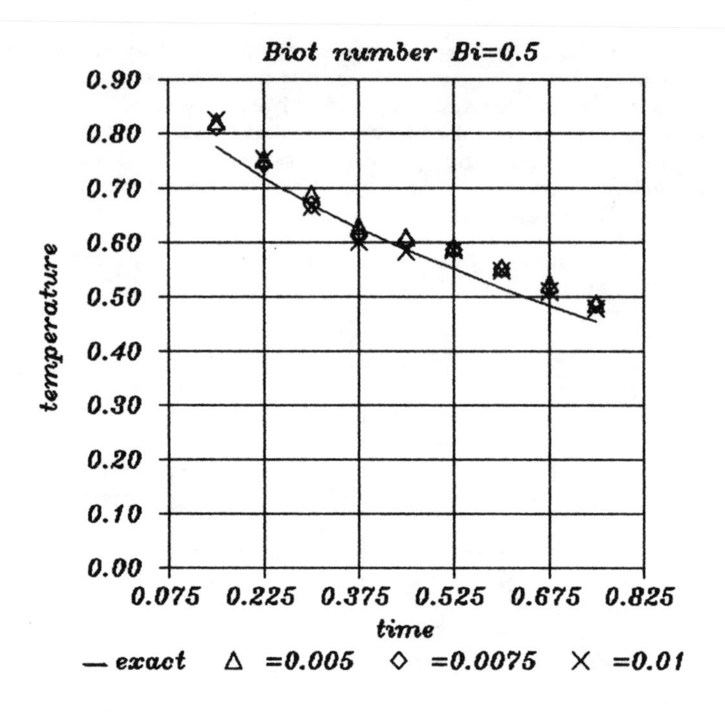

Figure 4. *Dimensionless temperature at point B as a function of dimensionless time.*

as a convincing argument for applying BEM and sensitivity coefficient concept to solving inverse problems in comparison with results published by Tanaka [7] they seem to be very promising. However, much more research has to be carried out in this important topic.

ACKNOWLEDGEMENTS

Financial support obtained from Ministry of National Education within the Central Plan 02.18 coordinated by the Technical University of Silesia is gratefully acknowledged herein.

REFERENCES

1. J.V. Beck at all: Inverse Heat Conduction, J. Wiley Intersc. Publ., N.Y. 1985.

2. C.A. Brebbia, J.C.F. Telles and L.C. Wrobel, Boundary Element Techniques: Theory and Applications in Engineering, Springer - Verlag, Berlin, 1984.

3. C.A. Brebbia and J. Dominguez, Boundary Elements - An Introductory Course, Comp. Mech. Publications, Mc Graw-Hill Book Co., 1988.

4. A.J. Nowak, C.A. Brebbia: The Multiple Reciprocity Method - A new approach for transforming BEM domain integrals to the boundary. Eng. Analysis with Boundary Elements, vol. 6, No. 3, 1989

5. A.J. Nowak : The Multiple Reciprocity Method of solving heat conduction problems. Proc. 11th BEM Conference, Cambridge, Massachusetts, USA, (ed. C.A. Brebbia & J.J. Connor) Springer-Verlag, vol. 2, 1989, pp 81-95

6. O.C. Zienkiewicz: The Finite Element Method, Mc Graw-Hill, New York, 1977.

7. M. Tanaka, Some recent advances in boundary element research for inverse problems, Proc. 10th BEM Conference, Southampton, UK, (ed. C.A. Brebbia) Springer - Verlag, vol. 1, 1988, pp 567-582

Determination of Characteristics in Thermal Interaction of Materials with High-Enthalpy Gas Flows through Methods of Inverse Problems

O.M. Alifanov, E.A. Artyukhin, A.V. Nenarokomov
Moscow Aviation Institute, named after Sergo Ordzhonikidze, Volokolamskoe Highway 4, Moscow, 125871, USSR

ABSTRACT

The algorithm suggested for specific unknown characteristics of heat-and-mass exchange on surfaces of bodies are based on the instrument of inverse heat conduction problems, which, at present, are widely used in studying the processes of heat-transfer inside the materials. Methods of inverse problems allow to elaborate mathematically grounded algorithms of search for unknown heat exchange characteristics by the results of indirect measurements.

INTRODUCTION

The given papers considers methods and algorithms for determining unknown heat transfer characteristics, based on the instrument of inverse heat transfer problems. Methods of inverse problems help us to develop mathematically grounded algorithms of search for unknown heat transfer characteristics. Maximum correspondence is achived hereby for heat values calculated using the assumed mathematical model with the values measured.

The aim of this work is to construct an experiment-design procedure for determining a set of unknown characterics, heat transfer on the surface through additional temperature measurements inside the material. Under analysis is the question of uniqueness of solution of problem formulated.

The unknown vector of characteristics is selected by means of minimization through the conjugate gradient method for the functional of mean-square deviation of temperature experimental values from the designed values, corresponding to a given value of the vector sought for. Due to a mathematical ill-posedness of the problem on determination of unknown characteristics it becomes important to choose a criterion for halting the process of functional minimization. In solving the problem of minimization, an agreement between the value of residual functional and the measurement error was taken as such a criterion according to the principle of iterative regularization. An expression has been obtained for the gradient of the functional minimized by means of a boundary-value problem, conjugated with a linearized initial problem. To solve the inverse problem one may apply a finite difference method. The results of mathematical modelling have been carried out with respect to the problem on determination of heat transfer characteristics on the surface of carbon materials.

THE ALGORITHM DEVELOPMENT

Algorithm presented below have been developed following an assumption that a heat transfer inside the structure can be described by a one-dimensional generalized heat conduction equation. It should be noted, however, that a transition to two- and three-dimensional statements does not meet any principal difficulties.

Under the assumption made, the following mathematical model of heat transfer inside the multilayered system is considered:

$$C_\ell(T)\frac{\partial T_\ell}{\partial \tau} = \frac{\partial}{\partial x}\left(\lambda_\ell(T)\frac{\partial T_\ell}{\partial x}\right) + K_\ell(T)\frac{\partial T_\ell}{\partial x} + S_\ell(T), \qquad (1)$$

$$\tau \in (\tau_{min}, \tau_{max}],\ x \in (X_{\ell-1}, X_\ell),\ \ell = \overline{1,L},\ X_0 = 0,\ X_L = b(\tau)$$

$$T_\ell(x,0) = T_{0\ell}(x),\ x \in [X_{\ell-1}, X_\ell], \qquad (2)$$

$$-\lambda_1(T)\frac{\partial T_1}{\partial x}(0,\tau) = q_1(\tau), \qquad (3)$$

$$-\lambda_L(T)\frac{\partial T_L}{\partial x}(\beta(\tau),\tau) = q_\lambda(\tau), \quad (4)$$

$$\lambda_\ell(T)\frac{\partial T_\ell}{\partial x}(X_\ell,\tau) = \lambda_{\ell+1}(T)\frac{\partial T_{\ell+1}}{\partial x}(X_\ell,\tau),\ \ell=\overline{1,L-1}, \quad (5)$$

$$-\lambda_\ell(T)R_\ell\frac{\partial T_\ell}{\partial x}(X_\ell,\tau) = T_\ell(X_\ell,\tau)-T_{\ell+1}(X_\ell,\tau), \quad (6)$$

$$\beta(\tau)=\beta(\tau_{min})-\int_{\tau_{min}}^{\tau} V\,d\tau, \quad (7)$$

where l is the number of layers in the system. As usually a boundary condition on the external surface is specified from the heat balance equation on the surface:

$$q_\lambda = q_0 + q_R + q_\varepsilon + q_i + q_t, \quad (8)$$

where q_0 is density of the convective heat flux, supplied to the surface of the body; q_R is density of the radiation flux, supplied to the body; q_ε is density of the heat flux, radiated by the surface; q_i is density of heat flux, blocked in the result of injection of gaseous (liquid) products into the boundary layer; q_t is density of heat flux, blocked as result of physical and chemical transformations on the surface.

We can propose a formalized and generalized form of the equation (8) representation, the structure of which allows to consider a sufficiently wide range of pfenomena on the surface:

$$q_\lambda(\tau)=H(\overline{G},\overline{u}), \quad (9)$$

where $\overline{G}$ is the function vector, the components of which characterize the external conditions, they are usually time-dependent and determine a particular heat transfer process; $\overline{u}$ is the function vector, the component of which are the material charactiristics (integral emissivity factor, etc.). It is assumed that all components $\overline{u}$ is unknown.

As an additional information, needed in solving inverse probllems, we use temperature values taken at several points inside the specimen:

$$T_{exp}(Y_{m\ell},\tau) = f_{m\ell}(\tau), \quad m=\overline{1,M_\ell}, \ \ell=\overline{1,L}, \tag{10}$$

where M_ℓ, $l=\overline{1,L}$ is the number of thermosensors in every layer of the system.

Basing on the experience of solving boundary and coefficient inverse heat conuction problems, it seems of reason to solve the above formulated inverse problem by extreme methods, by minimizing the root-mean-square deviation of design temperatures from these experimentally measured:

$$J(\bar{u}) = \sum_{n=1}^{N} \sum_{\ell=1}^{L_n} \sum_{m=1}^{M_{n\ell}} \int_{\tau_{min}}^{\tau_{max}} (T_\ell^n(Y_{m\ell}^n,\tau) - f_{m\ell}^n(\tau))^2 d\tau, \tag{11}$$

where n is the number of experiments (for uniquness of solution), l is the number of layers in every specimen under study, M_ℓ is the number of thermosensors in every layer.

Due to a mathematical ill-posedness of the problem on determining the heat transfer characteristics on the surface of the body the presence of solution $\bar{u}$, which can have nothing in common whith a solution sought for. To overcome this difficulty we can make use of the controlling properties of optimization gradient methods, which effectively allow to start an iterative process from a distant estimate of characteristics under stydy and abruptly slow down when nearing the functional minimum.

As the number of iterations increases, the solution of an inverse problem may start oscillating, gradually loosing its smoothing nature, necessity arises to halt the iterative process, without reaching the solution's oscillations.

An important item here is the choice of criterion for the iterative process halt. On the principle of generalized residual a limitation to a value of minimized functional can be taken as such criterion:

$$J(\bar{u}) \le \delta^2, \tag{12}$$

where δ^2 is the total measurement error, metrically

matched with the purpose functional.

Instead of unknown continuous functions u_i, $i=\overline{1,N_i}$ we consider functions $\tilde{u}_i(p_i, g^*)$ with a given structure and being dependent on the argument g^* and vector of unknown parameters $\overline{p}_i$. This is possible when the unknown functions are approximated by a system of base functions:

$$u_i \simeq \sum_{k=1}^{N_i} p_{ki}\, \varphi_{ki}(g^*) \tag{13}$$

particucularly, in the present work as φ_{ki}, $k=\overline{1,N_i}$ we used cubic B-splines. Thus, the condition (8) is rewritten as:

$$q_\lambda(\tau) = H(\overline{G}, \overline{p}), \tag{14}$$

where $\overline{p}=\{p_k\}_1^{N_p}$, $N_p=\sum_{i=1}^{N_u} N_i$.

Having the gradient of the minimizing functional by components we can solve the inverse problem (1)-(7),(14),(10),(11) by a numerical gradient method of absolute minimization. The iteration process is constructed as follows:

1) an initial approximation of unknown parameters $\overline{p}_0$ is given;

2) a value of vector $\overline{p}$ is determined in the iteration followed:

$$\overline{p}^{\,s+1} = \overline{p}^{\,s} + \alpha^s \overline{g}^{\,s}, \quad s = 0, 1, \ldots \tag{15}$$

$$\overline{g}^{\,s} = -(\overline{J}'_p)^s + \beta^s \overline{g}^{\,s-1},$$

where $(\overline{J}'_p)^s$ is the gradient of functional in the running iteration, β^s is the parameter, depending on the minimization method, as for example: $\beta = 0$ for the steepest descent method; the descent step α^s is sampled from the condition:

$$\min_{\alpha^s \in R^+} J\,(\overline{p}^{\,s} + \alpha^s \overline{g}^{\,s}). \tag{16}$$

The utmost difficulties in realizing gradient methods are connected with calculations of a gradient of the functional minimized (11).

We can show that for the problem in question it is equal to:

$$J'_p = -\sum_{n=1}^{N} \int_{\tau_{min}}^{\tau_{max}} \psi^n_{m\ell}(\beta^n(\tau),\tau)\frac{\partial H}{\partial p}(G^n,\bar{p})\,d\tau, \qquad (17)$$

where $\psi^n_{m\ell}(x,\tau)$ is a solution of the boundary-value problem, adjoined with a linearized form of the initial problem.

COMPUTATIONAL EXPERIMENT

A successive application of methods for heat exchange studies, based on the solution of inverse problems, needs thorough development of computational algorithms, as well as selection on the number of specimens treated and the number of thermosensors, etc.

The most universal approach hereby is a computational experiment, which is made in the following way: first we solve a direct heat exchange problems in the specimen, on the assumption that all coefficients of a mathematical model are known, using the obtained domain of temperatures in the supposed points of sensor installation, then we form additional information necessary for solving an inverse problem, and after that we solve an inverse problem on determination of heat transfer charactfristics on the surface.

Random errors in the additional information are formed as follows:

$$\tilde{f}(\tau) = f(\tau)(1 + \omega\delta(\tau)), \qquad (19)$$

where $f(\tau)$ is the exact dependence, ω is a randum value of distribution by normal law with a dispersion equalling 1 and with mean value equalling 0, δ is a relative error.

Simulation for experimental data treatment was performed with respect to a problem of determining a semi-spherical integral emissivity factor ε and total thermal effect of sublimation ΔQ_W of a carbon-type material.

The heat balance equation on the external surface was written as:

$$q_L^n(\tau) = q_w^n(\tau) + \sigma \varepsilon T^4(b(\tau), \tau) + \Delta Q_w \cdot G_w . \quad (20)$$

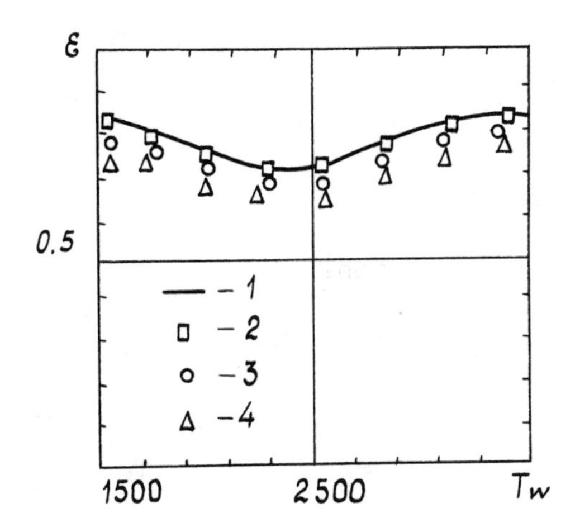

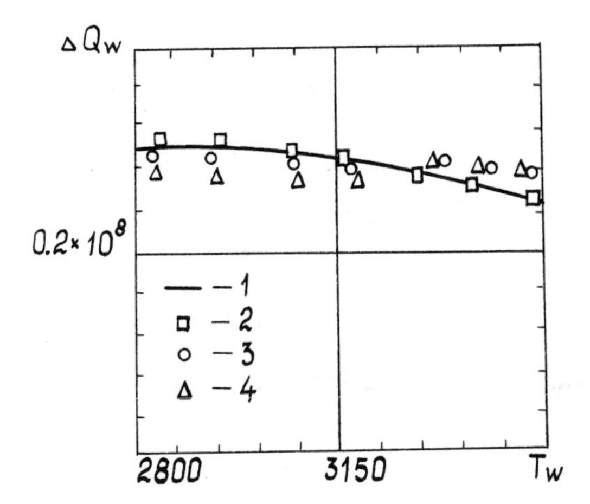

Figure 1. 1 - the given values; 2 - $Y_1^n = b^n(\tau)$; 3 - $i_n=4$; 4 - $i_n=0$; n=1,2.

The results of the analysis are presented in Fig. 1 illustrating, as example, the influence of the thermosensor attachment depth (considered was one sensor in every experiment), the co-ordinate of their fixation varied as follows:

$$Y_1^n = i_1 \Delta Y^n, \ n=\overline{1,2}\ , \ i_1=\overline{0,4}\ , \ \Delta Y^n = \beta^n(\tau_{max}^n)/4 \quad (21)$$

The relative error of measurements δ was assumed equalling 0.05 .

CONCLUSIONS

We can make the following conclucions:

1) In the case under consideration the fixation of thermosensors to a depth greater than the thickness of the material ablation layer will result in considerable errors of the solution being obtained.

2) Information about the surface temperature, obtained with an error (δ=0.05), appears to be more usefull then exact measurements in the depth of the material ablation layer.

3) The adding of the second thermosensors into each experiment will improve the solution accuracy when they are fixed in the close proximity of the final boundary of material ablation (with regard for inadmissibility of mutual action of thermosensors).

4) The adding of the third thermosensors will not improve the quallity of a solution irrespective of co-ordinates for their fixation.

5) The error of the solution of the inverse problem in neither case essentially (more than 3%) do not exceed the errors in the assignment of initial data.

The suggested algorithm as a whole seems to be effective for experimental data processing when studiyng the interaction of materials with the environment in the unsteady-state conditions.

REFERENCES

1. Artyukhin, E.A. and Nenarokomov, A.V. Identification of teristics for materials surface interaction with gas flow, Journal of engineering physics, Vol.46, No.4, 1984.

2. Alifanov, O.M. and Nenarokomov, A.V. The effect of different parameters on the accuracy of solution of the inverse heat conduction problem in parameterized form, Journal of engineering physics, Vol.56, No.3, 1989.

SECTION 5: NUMERICAL AND COMPUTATIONAL TECHNIQUES

ACFD: An Asymptotic/Computational Approach to Complex Problems

H. Herwig, K. Klemp, M. Voigt
Institut für Thermo- und Fluiddynamik, Ruhr-Universität, 4630 Bochum, West Germany

ABSTRACT

Solving complex flow and heat transfer problems numerically created a new branch of fluid dynamics: CFD. In this paper we infer that it might not always be the best strategy to simply delegate the problem to the computer. Instead we suggest to combine CFD and asymptotic methods to a mutually profitable alliance, which we call ACFD: Asymptotic computational fluid dynamics.

INTRODUCTION

In a time of rapid computer development one might expect that more and more – and finally all – problems will be solved by simply delegating them to the computer. Undoubtedly flow and heat transfer problems of realistic complexity can only be solved with the help of CFD (computational fluid dynamics). But, simply delegating the problem to the computer might not always be the best strategy.

Basically there are three objections to be raised against a "blind computer approach":

(1) loss of physical insight; the mere numbers of a specific solution can hardly be interpreted in physical terms like "dominating effect", "negligible influence" and so on.

(2) very limited results; a specific numerical solution will hold for one set of initial and boundary conditions and specific values for the parameters of the problem. Results for neighbouring values of these quantities can only be provided by running through the whole procedure again.

(3) extensive time of computation; often unnecessary complex equations are solved. This, together with the large number of runs to account for parameter variations, leads to a considerable waste of computational time.

Alternatively we suggest another approach to complex problems which we call the ACFD approach (asymptotic computational fluid dynamics). It is characterized by thoroughly combining asymptotic and numerical aspects in solving a problem.

In this paper we will first introduce the general procedure then give a detailed example and finally return to the procedure itself for a more detailed description.

Altogether three advantages characterize this asymtotic/computational approach:

(A1) The influence of solution parameters is revealed quite clearly which often gives good physical insight.

(A2) The final results hold for a whole range of parameters or coordinates and thus are more general than specific numerical solutions.

(A3) The equations that have to be solved finally are often much simpler from a mathematical point of view due to asymptotic expansions of the quantities under consideration. This can save a whole amount of computational time.

THE ACFD–APPROACH (GENERAL PROCEDURE)

A theoretical approach to a physical problem always starts with the choice of an appropriate mathematical model. In most cases this will be a set of differential equations. Here the first asymptotic element is involved since neglecting certain physical effects that might be accounted for in a more rigorous approach always means that the associated dimensionless parameters are set equal to zero or infinity. For example, neglecting compressibility effects in a flow problem is equivalent to setting the Mach–number Ma equal to zero. Actually this first step usually is done in a somewhat heuristic manner though one should keep in mind that dropping parts of the complete equations always means that higher order effects are neglected from an asymptotic point of view.

If now these effects nevertheless should be accounted for in a more rigorous approach these higher order effects often can be taken into account as far as necessary instead of reformulating the whole problem to include the effects neglected in the first approach.

From these considerations the procedure in the ACFD approach is:

[I] Start from the "complete" set of equations for a problem under consideration. Here "complete" means: including all effects that might be of importance.

Nondimensionalize the equations "properly". Here "properly" means: nondimensionalize with characteristic quantities in the sense of dimensional analysis. Nondimensional parameter combinations P_i like the Reynolds or Mach number now appear in the equations.

[II] Decide which effects can be neglected completely by setting P_i or P_i^{-1} equal to zero.

[III] Decide which effects can be accounted for approximately by asymptotic expansions for $P_i \to 0$ or $P_i^{-1} \to 0$.

AN EXAMPLE

As an example the laminar entrance flow of water into a heated channel (T_W = const > T_∞) is considered. Figure 1 shows the flow situation as well as the viscosity law of water. The oncoming uniform velocity profile is changed to the fully developed parabolic profil in the entrance region of the channel. It is a mutually coupled momentum and heat transfer process when variable fluid properties are considered.

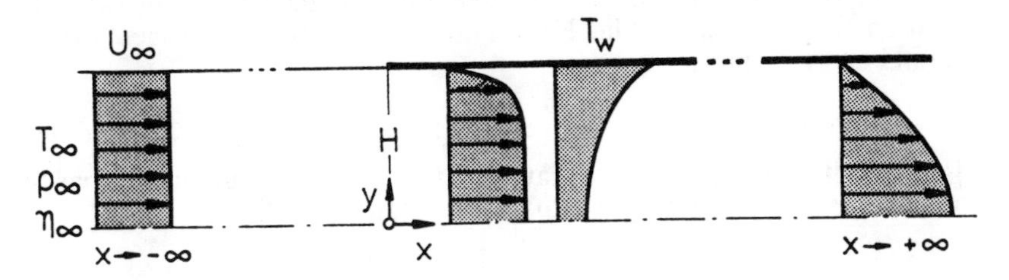

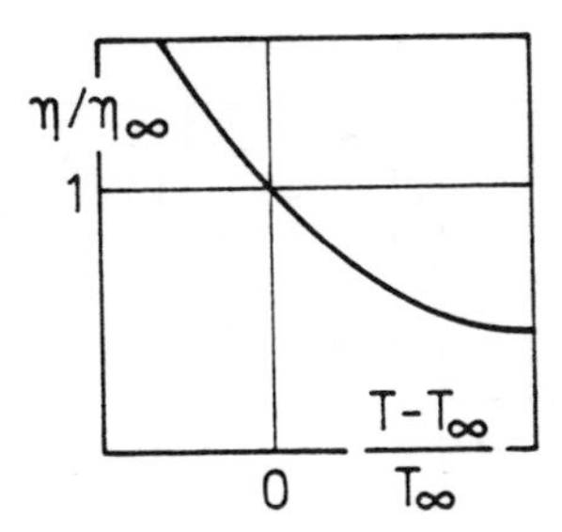

Fig. 1a: Flow situation

Fig. 1b: Viscosity law

To simplify matters we a priori assume some effects to be completely negligible in the sense of point [II] of the previous chapter:

(a) Of all property dependencies $\rho(T,p)$, $\eta(T,p)$ ... only the temperature dependence of viscosity is considered, i.e. asymptotically:

$$K_\alpha := \left[\frac{T}{\alpha}\frac{\partial \alpha}{\partial T}\right]_\infty = 0 \quad ; \quad \alpha = \rho,\lambda,c_p \tag{1}$$

$$\hat{K}_{\alpha} := \left[\frac{p}{\alpha}\frac{\partial \alpha}{\partial p}\right]_{\infty} = 0 \quad ; \quad \alpha = \rho, \lambda, c_p \text{ and } \eta \tag{2}$$

(b) Viscous dissipation is neglected, i.e. asymptotically:

$$Ec := \frac{U_{\infty}^2}{c_p\ T_{\infty}} = 0 \quad (Ec \hat{=} \text{Eckert number}) \tag{3}$$

(c) Buoyancy effects are neglected, i.e. asymptotically:

$$Fr^{-1} := \frac{\sqrt{g\ L}}{U_{\infty}} = 0 \quad (Fr \hat{=} \text{Froude number}) \tag{4}$$

Based on these assumptions a mathematical model for the entrance flow situation is provided by the Navier–Stokes equations together with the thermal energy equation.

Nondimensionalized with U_{∞}, H, $T_W - T_{\infty}$, ρ_{∞}, η_{∞} according to Figure 1a there are five equations for the two unknown velocity components u,v, pressure p, temperature Θ and viscosity η:

$$\frac{\partial u}{\partial x} + \frac{\partial v}{\partial y} = 0 \tag{5}$$

$$u\frac{\partial u}{\partial x} + v\frac{\partial u}{\partial y} = -\frac{\partial p}{\partial x} + \frac{2}{Re}\left\{2\frac{\partial}{\partial x}\left[\eta\frac{\partial u}{\partial x}\right] + \frac{\partial}{\partial y}\left[\eta\left[\frac{\partial u}{\partial y} + \frac{\partial v}{\partial x}\right]\right]\right\} \tag{6}$$

$$u\frac{\partial v}{\partial x} + v\frac{\partial v}{\partial y} = -\frac{\partial p}{\partial y} + \frac{2}{Re}\left\{2\frac{\partial}{\partial y}\left[\eta\frac{\partial v}{\partial y}\right] + \frac{\partial}{\partial x}\left[\eta\left[\frac{\partial u}{\partial y} + \frac{\partial v}{\partial x}\right]\right]\right\} \tag{7}$$

$$u\frac{\partial \Theta}{\partial x} + v\frac{\partial \Theta}{\partial y} = \frac{2}{Re\ Pr}\left\{\frac{\partial^2 \Theta}{\partial x^2} + \frac{\partial^2 \Theta}{\partial y^2}\right\} \tag{8}$$

$$\eta = \exp\left[\frac{\Theta}{(\Theta + A)\ B}\right] \tag{9}$$

with the definitions: $Re := \frac{\rho\ U_{\infty}\ 2H}{\eta_{\infty}}$; $Pr := \frac{\eta_{\infty}\ c_p}{\lambda}$

This system of equations is nonlinear. In addition there is a strong mutual coupling between the momentum equations and the energy equation through $\eta(\Theta)$.

Conventional CFD solution
Since here $\eta(\Theta)$ is the viscosity law of water (with special constants A and B for specific numbers of T_W and T_∞) the solution of equations (5) – (9) subject to the appropriate boundary conditions will only hold for a special heat transfer situation (T_∞, T_W) of a special fluid (water).

Three solutions of this kind are provided in Herwig et al. [1] for the following three flow and heat transfer situations:

$$\begin{aligned}
&(1) \quad Re = 10 \,;\, Pr = 10 \,;\, T_\infty = T_W \\
&\qquad \rightarrow A^{-1} = 0 \qquad \text{(isothermal case: } \eta = 1\text{)} \\
&(2) \quad Re = 10 \,;\, Pr = 10 \,;\, T_\infty = 281\ K \,,\, T_W = 301\ K \\
&\qquad \rightarrow A = 7 \,;\, B = -0{,}25 \qquad (10) \\
&(3) \quad Re = 10 \,;\, Pr = 10 \,;\, T_\infty = 281\ K, T_W = 321\ K \\
&\qquad \rightarrow A = 3{,}5 \,;\, B = -0{,}25
\end{aligned}$$

In Figure 2 the velocity gradient at the wall is shown at two x–positions, one immediately after the inlet ($x = 0{,}1$) and one far downstream ($x = 10$).

These solutions only hold for the special situations described by (10).

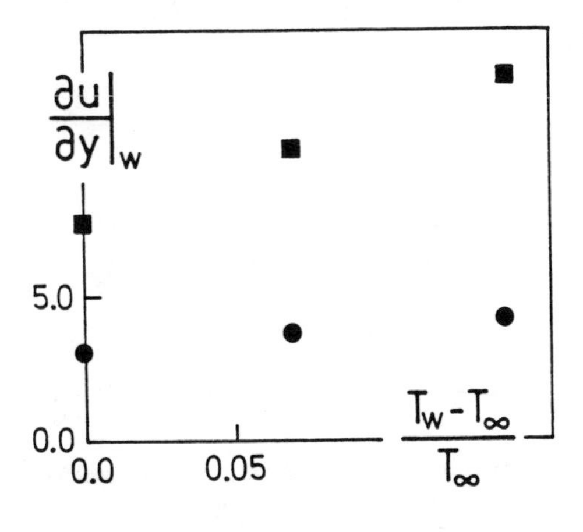

Fig. 2:

Velocity gradient at the wall

■■ at x = 0,1

●● at x = 10

Alternative ACFD solutions

General viscosity law Solutions of the basic equations (5) – (9) would be far more general if the specific viscosity law could be replaced by a "general viscosity law".

Such a general law emerges when η is expanded into a Taylor series with respect to temperature. After such an expansion equation (9) will be replaced by

$$\eta = 1 + \epsilon K_\eta \Theta + O(\epsilon^2) \; ; \; \epsilon := \frac{T_W - T_\infty}{T_\infty} \; , \; K_\eta := \left[\frac{T}{\eta}\frac{\partial\eta}{\partial T}\right]_\infty \tag{11}$$

For $\epsilon \to 0$ this is an asymptotic approximation of any viscosity law. Different fluids are characterized by different numbers of K_η. For water under the conditions (10) K_η for example is –8,05.

If now equation (9) is replaced by (11) solutions are no longer restricted to certain fluids. A solution can be found for all values of K_η by a classical regular perturbation procedure, s. for example Van Dyke [2]. Analogous to the expansions of η all dependent variables are assumed to be of the general form

$$a = a_0 + \epsilon K_\eta a_1 + O(\epsilon^2) \quad ; \quad a \mathrel{\hat{=}} u,v,p,\Theta \tag{12}$$

After inserting the expansions (11) and (12) into the basic equations two sets of equations emerge, one for u_0,v_0,p_0,Θ_0 and one for u_1,v_1,p_1,Θ_1. For details s. Herwig et al. [1].

These equations can be solved as soon as the parameters Re and Pr are specified. According to the expansions (11) and (12) these solutions are free of ϵ and K_η, i.e. free of T_∞,T_W and a specific viscosity law.

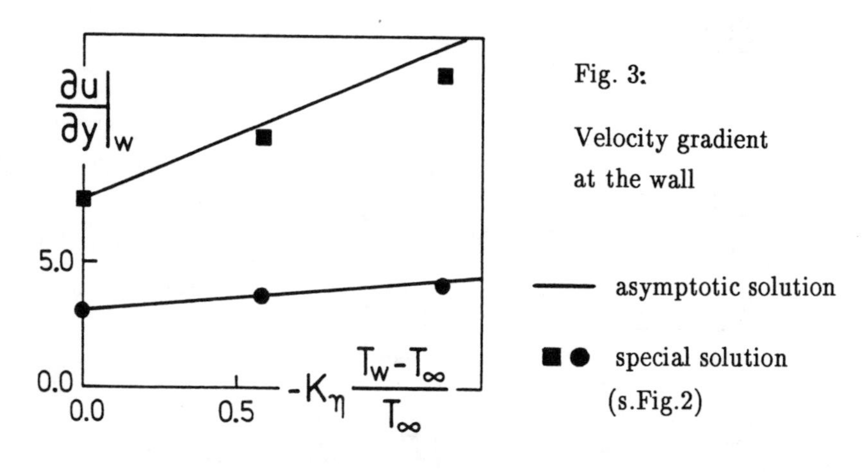

Fig. 3: Velocity gradient at the wall

These general results can be applied to the specific situation of the previous chapter. In Figure 3 the asymptotic results are compared to those of Figure 2 (special solutions for water). Deviations for increasing heat transfer rates (increasing $T_w - T_\infty$) are due to higher order effects which are neglected in our approach.

These solutions are far more general than those for water alone. But there are still two parameters that have to be specified for a numerical solution:

(1) the Reynolds number Re ,

(2) the Prandtl number Pr or alternatively the Peclet number Pe = Pr Re.

To get an even more general result according to point [III] of the ACFD approach we can try to find approximate asymptotic solutions for $P_i \to 0$ or $P_i^{-1} \to 0$, i.e. for limiting values of Re for example.

Entrance flow for high Reynolds numbers Asymptotically this is the limit $Re \to \infty$. In an extensive study of this problem Van Dyke [3] has shown that two asymptotic expansions are needed in the limit of infinite Reynolds number, one in the inlet region (streamwise coordinate: x) and one far downstream (streamwise coordinate: $\tilde{x} = x/Re$). Both expansions are matched for $x \to \infty$ and $\tilde{x} \to 0$, respectively. Here the downstream expansion is the so–called secondary expansion, the first term of which gets its boundary condition at $\tilde{x} = 0$ from matching with the primary expansion. Physically this corresponds to a plug flow situation at $\tilde{x} = 0$, i.e. $u = 1$.

This one term downstream expansion will tentatively be taken as an asymptotic approximation for the whole entrance flow problem. The equations now are of boundary layer type and are sometimes called slender channel equations, see Williams [4]. Instead of equations (5) – (8) we now have to solve ($\tilde{x} = x/Re$, $\tilde{v} = v\,Re$):

$$\frac{\partial u}{\partial \tilde{x}} + \frac{\partial \tilde{v}}{\partial y} = 0 \tag{13}$$

$$u\frac{\partial u}{\partial \tilde{x}} + \tilde{v}\frac{\partial u}{\partial y} = -\frac{\partial p}{\partial \tilde{x}} + 2\frac{\partial}{\partial y}\left[\eta\frac{\partial u}{\partial y}\right] \tag{14}$$

$$0 = -\frac{\partial p}{\partial y} \tag{15}$$

$$u\frac{\partial\Theta}{\partial\tilde{x}} + \tilde{v}\frac{\partial\Theta}{\partial y} = \frac{2}{Pr}\frac{\partial^2\Theta}{\partial y^2} \qquad (16)$$

with the appropriate boundary conditions. If now this approach is combined with the general viscosity law (11) equations (13) – (16) again can be expanded according to equation (12). Instead of equations (13) – (16) we then have one set of equations for $u_0, \tilde{v}_0, p_0, \Theta_0$ (which is identical to (13) – (16) with $\eta = 1$ in equation (14)) and one for $u_1, \tilde{v}_1, p_1, \Theta_1$.

In Figure 4 the asymptotic result for constant properties (i.e. $u_0, \tilde{v}_0, p_0, \Theta_0$) is compared to two results for the complete equations (5) – (9), also for constant properties. For Reynolds numbers above ~ 50 there is an excellent agreement between both theories.

In this theory the Prandtl number is the only parameter of the problem. Even with variable viscosity no additional parameter is added to the equations since according to the expansions (11) and (12) the final result for the Nusselt number for example is:

$$Nu = Nu_{cp}[1 + \epsilon K_\eta N + O(\epsilon^2)] \quad \text{with} \quad N = N(\tilde{x}, Pr) \qquad (17)$$

In Figure 5 the "viscosity influence function" N for Pr = 1 is compared to the equivalent function of the complete solution based on equations (5) – (8) and (11). Again the approximation is satisfactory for high Reynolds numbers.

THE ACFD – APPROACH (DETAILED)

In the general procedure of our ACFD approach three major steps [I] – [III] have been identified which now will be explained in detail.

[I] Basis equations

The choice of basic equations which actually is a choice of a mathematical model for a certain physical problem under investigation is always an implicit application of asymptotic considerations (or at least can be interpreted in this way). In this sense for example

— "twodimensional equations" apply in situations for which $W/L \to \infty$ ($W \hat{=}$ width ; $L \hat{=}$ length)
— "steady equations" apply in situations for which $t/t_0 \to \infty$ ($t \hat{=}$ time ; $t_0 \hat{=}$ characteristic time constant of a nonperiodic process)

One should always keep in mind that whenever a theoretical approach

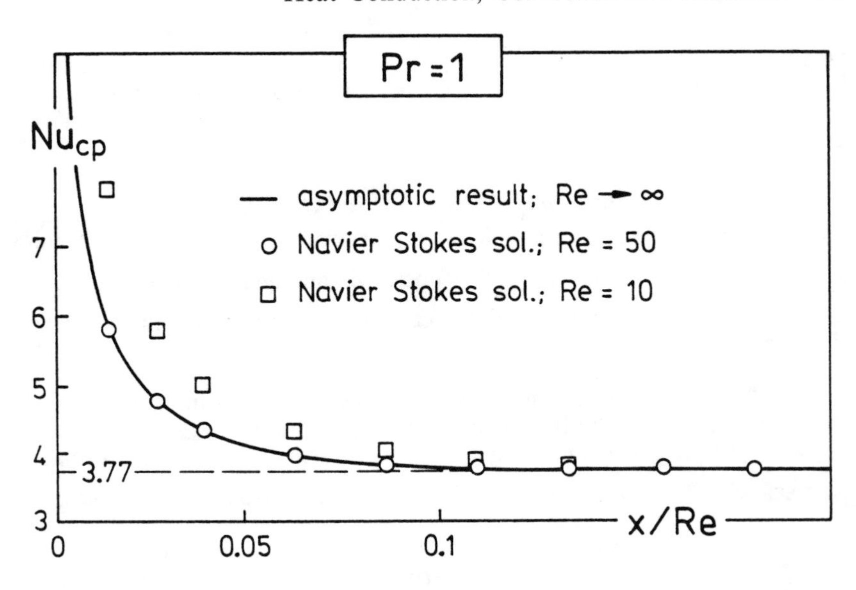

Fig. 4: Nusselt number for constant properties

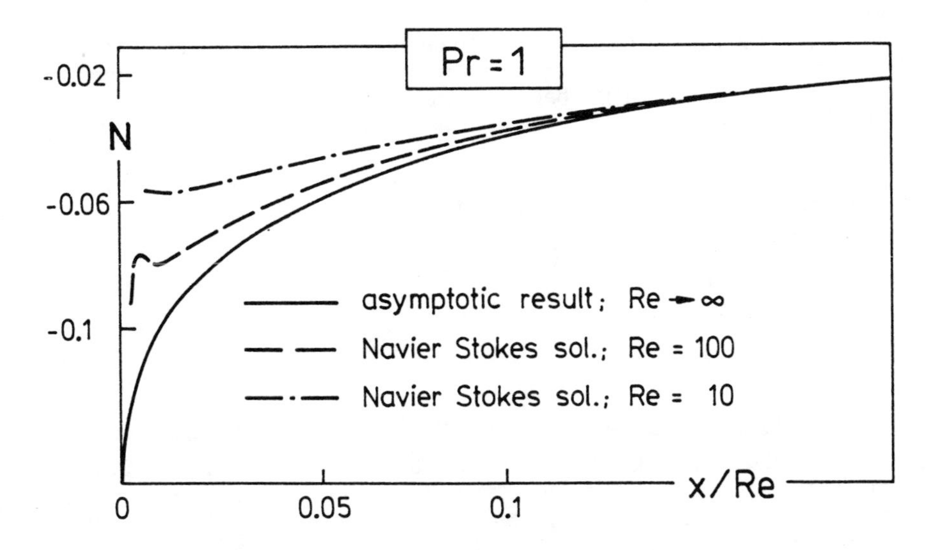

Fig. 5: Viscosity influence function N

fails not the equations are "wrong" but that the mathematical model may be inadequate to describe a certain physical problem. Asymptotic considerations will always be a good guideline in choosing the adequate mathematical model.

[II] Neglecting certain effects completely ($P_i = 0$ or $P_i^{-1} = 0$)
After a proper nondimensionalisation parameters P_i like the Reynolds number Re or the K–number K_η appear in the equations. Setting one or more of these parameters equal to zero or infinity corresponds to neglecting certain effects completely.

This will be a reasonable approximation in certain situations, but it also may fail completely in others. In the entrance flow example setting $K_\eta = 0$ in equation (11) is a good approximation to the whole problem, whereas with $Re^{-1} = 0$ in equations (6) and (7), we would fail completely.

From an asymptotic point of view both approximations are different in character. The influence of variable viscosity is a regular perturbation problem whereas the Reynolds number influence is of singular perturbation type.

Simply setting $P_i = 0$ or $P_i^{-1} = 0$ in the basic equations of the problem under consideration will lead to a reasonable approximation in the whole solution domain only when it is a regular perturbation problem with respect to $P_i \to 0$ or $P_i \to \infty$.

A careful inspection of the complete entrance flow problem reveals:

▶ singular perturbation behaviour for
— $Re \to \infty$
— $Pr \to 0$; $Pr \to \infty$ (18)

▶ regular perturbation behaviour for
— $Re \to 0$
— $Fr \to \infty$
— $Ec \to 0$
— $Ma \to 0$ (19)
— $K_\alpha, \hat{K}_\alpha \to 0$

[III] Accounting for certain effects asymptotically ($P_i \to 0$ or $P_i^{-1} \to 0$)
For a regular perturbation problem the approximation can be uniquely improved in the whole solution domain by adding higher order approximations to the analysis. The procedure is straightforward as was demonstrated in the entrance flow example ($\eta = 1 + \epsilon K_\eta \Theta + \ldots$).

For singular perturbation problems things are much more complicated. Due to the singular nature of the limiting process there are at least two different expansions in two different parts of the solution domain.

If now an approximation should cover the whole solution domain, the leading term in each subdomain is required. In the entrance flow example for $Re \to \infty$ two expansions are needed, one in the inlet region ($x = O(1)$) and one far downstream ($\tilde{x} := x/Re = O(1)$). The first term of both expansions might serve as a uniformly valid first approximation when they are combined to a composite expansion.

Alternatively one might take the first term of only one of the two (or more) expansions. Of course this one term expansion can be a reasonable approximation only in part of the whole solution domain. It should be the major part leaving small regions of nonuniformity. In the entrance flow example for $Re \to \infty$ we took the first term downstream expansion as an approximation for the whole flow situation. As a consequence the approximation fails in the immediate vicinity of the entrance location (for $|x| \to 0$ asymptotically). In Figure 5 for example major deviation compared to the "exact solution" based on the Navier Stokes equations (5) – (8) only occur for $x \to 0$.

DISCUSSION

In the introduction to this paper three objections have been raised against a "blind computer approach". They now may serve to illustrate the advantages of our ACFD approach:

(1) loss of physical insight; With the ACFD approach the dominating physical effects can be separated. Due to asymptotic considerations certain terms in the equations can be neglected according to their minor physical significance.

(2) very limited results; With the ACFD approach approximate solutions for a large number of similar flow situations are found. This was illustrated for the entrance flow in a channel. Instead of a specific solution for *one* fluid, *one* Reynolds and *one* Prandtl number with the ACFD approach we could find a good approximation for *all* fluids, *all* (high) Reynolds numbers and *one* Prandtl number. Only one explicit solution parameter (the Prandtl number) was left in the equations.

(3) extensive time of computation; With the ACFD approach the mathematical complexity of the equations is reduced considerably. In the entrance flow example instead of a system of elliptic nonlinear and coupled equations only parabolic mostly linear and uncoupled equations had to be solved.

To summarize: We suggest that before a numerical procedure is started to solve a specific problem one always should try to simplify and generalize the problem by means of asymptotic considerations.

REFERENCES

1. Herwig, H., Klemp, K. and Stinnesbeck, J. Laminar Entry Flow in a Pipe or Channel: The Effect of Variable Viscosity Due to Heat Transfer Across the Wall, to be published in Numerical Heat Transfer A, 1990.

2. Van Dyke, M. Perturbation Methods in Fluid Mechanics, The Parabolic Press, Stanford/California, 1975.

3. Van Dyke, M. Entry Flow in a Channel, JFM, Vol. 44, pp. 813–823, 1970.

4. Williams, J.C. Viscous compressible and incompressible flow in slender channels, AIAA J., Vol. 1, pp. 186–195, 1963.

KYOKAI.H - A Heat Transfer Code on EWS

K. Onishi(*), H. Ninomiya(**)
(*) Department of Applied Mathematics,
(**) Department of Applied Physics, Fukuoka University, Jonan-ku, Fukuoka 814-01, Japan

ABSTRACT

The purpose of the present work is to inform of the recent further development of our heat transfer software, named KYOKAI.H, for the finite element analysis of a wide variety of thermal problems in engineering. The code is a complete package, which our pre- and post-processors coordinate with. Engineering Work Stations are target machines for its implementation. However, small Personal Computers are sufficient for the tentative solution of those problems. Included are the equations governing the convective heat transfer, computational schemes with the minimum requirement for the implementation of the code, modular configulation of the software system, and some worked examples as well. The KYOKAI.H is a particularly user-friendly portable code, and it runs on desk-top machines for the numerical modelling of the heat transfer.

INTRODUCTION

Recent development of the micro-computer technology has enabled us to solve a wide variety of heat transfer problems on desk-top computers. Some industrial problems of simultaneous viscous flow and heat transfer with the complex geometry and boundary conditions can be treated nowaday on Engineering Work Stations (EWS) by using the finite element method.

In this article, we shall present a small package of computer programmes, called KYOKAI.H for the numerical modelling of heat transfer problems on EWS. The problems can be defined interactively by the pre-processor. The finite element method is coded in the KYOKAI.H. Calculated results are displayed on a graphics by the post-processor. Their BASIC codes can be duplicated in a standard format on a diskette.

The package was developed as one of the series products, which aimed at the implementation on desk-top computers. This new product is evolved from the computer program KYOKAI.F[1], which has been executed exclusively on super-mini computers. A parallel processing on an EWS network for the fast solution of the flow and heat transfer problems is also considered[2].

GOVERNING EQUATIONS

The basic equations for thermal fluid flow problems involve various physical constants. Based on the *Boussinesq approximation*, we assume that all the coefficients are independent on the temperature except a density of the term which describes the buoyancy F_y per unit mass in the equations of two-dimensional motion:

$$\frac{\partial u}{\partial t} + u\frac{\partial u}{\partial x} + v\frac{\partial u}{\partial y} = -\frac{1}{\rho}\frac{\partial p}{\partial x} + \nu\nabla^2 u , \tag{1}$$

$$\frac{\partial v}{\partial t} + u\frac{\partial v}{\partial x} + v\frac{\partial v}{\partial y} = -\frac{1}{\rho}\frac{\partial p}{\partial y} + \nu\nabla^2 v + F_y , \tag{2}$$

where u, v are the components of flow velocity in the x and y directions, respectively; p is the pressure, ρ id the density, and ν is the kinematic viscosity of the fluid.

We shall derive the expression of F_y, depending on the temperature T. To this end, we consider a fluid at the temperature T_0 and the density ρ_0 in a container. The abscissa x is taken horizontally and the ordinate y is taken in the opposite direction to the gravitational acceleration g. We assume that a lump of the fluid in the container is heated or cooled to the temperature T with its density ρ. Then, the *buoyancy* of the magnitude $(\rho_0 - \rho)g$ (N/m^3) is exerted on the lump of the fluid. If the density difference is small enough, the equation of state of the fluid can be expressed as follows.

$$\frac{\rho}{\rho_0} = 1 - \beta(T - T_0) , \tag{3}$$

where β $(1/K)$ is the *volumetric thermal expansion coefficient.* The buoyancy per a unit mass is therefore given by

$$F_y = (\rho_0 - \rho)g/\rho_0 = \beta g(T - T_0) . \tag{4}$$

Thus, the continuity equation under the Boussinesq approximation takes the form

$$\frac{\partial u}{\partial x} + \frac{\partial v}{\partial y} = 0 . \tag{5}$$

With the buoyancy given by (4), the basic equations expressing the fluid flow in two dimensions are written in terms of the streamfunction ψ and vorticity ω as follows.

$$\nabla^2\psi = -\omega , \tag{6}$$

$$\frac{\partial \omega}{\partial t} + \frac{\partial \psi}{\partial y}\frac{\partial \omega}{\partial x} - \frac{\partial \psi}{\partial x}\frac{\partial \omega}{\partial y} = \nu\nabla^2\omega + g\beta\frac{\partial T}{\partial x} . \tag{7}$$

Thermal energy is transported simultaneously by conduction and convection. When dissipation of the kinematic energy to heat due to viscosity is neglected, the components of the heat flux are given by

$$q_x = -\kappa\frac{\partial T}{\partial x} + u\rho c T , \tag{8}$$

$$q_y = -\kappa\frac{\partial T}{\partial y} + v\rho c T , \tag{9}$$

where c is the specific heat, and κ is the coefficient of heat conduction. From the law of conservation of the thermal energy described as

$$\rho c\frac{\partial T}{\partial t} + \frac{\partial q_x}{\partial x} + \frac{\partial q_y}{\partial y} = 0 , \tag{10}$$

we can obtain the following convective heat conduction equation.

$$\frac{\partial T}{\partial t} + \frac{\partial \psi}{\partial y}\frac{\partial T}{\partial x} - \frac{\partial \psi}{\partial x}\frac{\partial T}{\partial y} = \lambda\nabla^2 T \tag{11}$$

with the thermal conductivity $\lambda = \kappa/(\rho c)$.

FINITE ELEMENT DISCRETISATION

We shall apply the conventional Galerkin finite element method to discretising the equation of streamfunction (6), the vorticity transport equation (7), and the convective heat conduction equation (11) in the flow domain Ω, which is enclosed by the boundary Γ. To this end, let $\delta\psi$, $\delta\omega$ and δT be weighting functions, being arbitrary but $\delta\psi = 0$ on the part of the boundary Γ_ψ where the value ψ is prescribed, $\delta\omega = 0$ on $\Gamma_\omega \cup \Gamma_w$ where the value of ω is prescribed, and $\delta T = 0$ on Γ_T where the value of T is prescribed. We start the finite element formulation with the corresponding weak forms of those equations:

$$\int_\Omega \nabla\delta\psi \cdot \nabla\psi \, d\Omega - \int_\Omega \delta\psi \, \omega \, d\Omega - \int_{\Gamma_s} \delta\psi \, V_s \, d\Gamma = 0 , \tag{12}$$

$$\int_\Omega \delta\omega \frac{\partial\omega}{\partial t} d\Omega + \int_\Omega \delta\omega \left(\frac{\partial\psi}{\partial y}\frac{\partial\omega}{\partial x} - \frac{\partial\psi}{\partial x}\frac{\partial\omega}{\partial y} \right) d\Omega + \int_\Omega \nu \, \nabla\delta\omega \cdot \nabla\omega \, d\Omega$$
$$- \int_\Omega \delta\omega \, g \, \beta \frac{\partial T}{\partial x} d\Omega - \int_{\Gamma_\chi} \nu \, \delta\omega \frac{\partial\omega}{\partial n} d\Gamma = 0 , \tag{13}$$

$$\int_\Omega \delta T \frac{\partial T}{\partial t} d\Omega + \int_\Omega \delta T \left(\frac{\partial\psi}{\partial y}\frac{\partial T}{\partial x} - \frac{\partial\psi}{\partial x}\frac{\partial T}{\partial y} \right) d\Omega + \int_\Omega \lambda \, \nabla\delta T \cdot \nabla T \, d\Omega$$
$$+ \int_{\Gamma_q} Q_B \, \delta T \, d\Gamma + \int_{\Gamma_h} Q_h(T) \, \delta T \, d\Gamma = 0 , \tag{14}$$

where Γ_s is the part of the boundary on which the tangential velocity V_s is prescribed, $Q_B = q_B/(\rho c)$ with the heat flux q_B on the part of the boundary Γ_q, and $Q_h = h(T - T_a)/(\rho c)$ with the heat transfer coefficient h and the ambient temperature T_a on the part of the boundary Γ_h.

The domain Ω is divided into triangular finite elements. Inside each triangle e having its three vertices as the element nodes with their local node numbers 1, 2, 3; the unknown streamfunction, vorticity and temperature are linearly interpolated as follows.

$$\begin{aligned} \psi &= \sum_\alpha \phi_\alpha \psi_\alpha , & \delta\psi &= \sum_\alpha \phi_\alpha \, \delta\psi_\alpha , \\ \omega &= \sum_\alpha \phi_\alpha \omega_\alpha , & \delta\omega &= \sum_\alpha \phi_\alpha \, \delta\omega_\alpha , \\ T &= \sum_\alpha \phi_\alpha T_\alpha , & \delta T &= \sum_\alpha \phi_\alpha \, \delta T_\alpha , \end{aligned} \tag{15}$$

where the notation $\sum_\alpha$ indicates taking sums for all α ($\alpha = 1, 2, 3$), and ϕ_α are linear interpolation functions given by

$$\phi_\alpha = \frac{1}{2\Delta^e} (a_\alpha + b_\alpha x + c_\alpha y) \tag{16}$$

with the area of the triangle Δ^e; $\psi_\alpha, \omega_\alpha, T_\alpha$ are nodal values of the corresponding unknowns; and $\delta\psi_\alpha, \delta\omega_\alpha, \delta T_\alpha$ are their arbitrary nodal variations.

The interpolations (15) are substituted into (12)-(14). From the continuity in the interpolations of ψ, ω, T and from the arbitrariness of $\delta\psi_\alpha, \delta\omega_\alpha, \delta T_\alpha$, the next element equations follow.

$$\sum_{\beta=1}^{3} D^e_{\alpha\beta} \psi_\beta - \sum_{\beta=1}^{3} M^e_{\alpha\beta} \omega_\beta - \Gamma^e_{s\alpha} = 0 , \tag{17}$$

$$\sum_{\beta=1}^{3} M^e_{\alpha\beta}\dot{\omega}_\beta + \sum_{\beta=1}^{3} A^e_{\alpha\beta}\omega_\beta + \nu \sum_{\beta=1}^{3} D^e_{\alpha\beta}\omega_\beta - F^e_\alpha(T) - \Gamma^e_{\chi\alpha} = 0 , \quad (18)$$

$$\sum_{\beta=1}^{3} M^e_{\alpha\beta}\dot{T}_\beta + \sum_{\beta=1}^{3} A^e_{\alpha\beta}T_\beta + \lambda \sum_{\beta=1}^{3} D^e_{\alpha\beta}T_\beta + \Gamma^e_{q\alpha} + \Gamma^e_{h\alpha}(T) = 0 . \quad (19)$$

After assembling all the element equations over the whole domain, we will obtain the total equations in the form:

$$[D]\{\psi\} - [M]\{\omega\} - \{\Gamma_s\} = \{0\} , \quad (20)$$

$$[M]\{\dot{\omega}\} + [A(\psi)]\{\omega\} + \nu[D]\{\omega\} - \{F(T)\} - \{\Gamma_\chi\} = \{0\} , \quad (21)$$

$$[M]\{\dot{T}\} + [A(\psi)]\{T\} + \lambda[D]\{T\} + \{\Gamma_q\} + \{\Gamma_h(T)\} = \{0\} , \quad (22)$$

where $[D]$, $[M]$, $[A]$ denote the total matrices, $\{F\}$ corresponds the buoyancy term; $\{\psi\}, \{\omega\}, \{T\}$ are the nodal unknown column vectors. Other column vectors are due to the respective boundary conditions. One can read (20)-(22) as a nonlinear system of first-order ordinary differential equations for unknown $\{\psi\}, \{\omega\}$ and $\{T\}$.

We shall consider the discretisation of the total equations with respect to time. To this end, we apply the semi-implicit scheme to the initial value problem. The time derivatives of the nodal vorticity and temperature are approximated by the following finite differences.

$$\dot{\omega}_\beta = \frac{d\omega_\beta}{dt} \approx \frac{\omega_\beta^{n+1} - \omega_\beta^n}{\Delta t} , \quad (23)$$

$$\dot{T}_\beta = \frac{dT_\beta}{dt} \approx \frac{T_\beta^{n+1} - T_\beta^n}{\Delta t} , \quad (24)$$

where ω_β^n and T_β^n denote the nodal vorticity and temperature, respectively, at the time level t_n, defined by $t_{n+1} = t_n + \Delta t$ $(n = 0, 1, 2, \ldots)$ with the time increment Δt.

We consider this finite difference approximation at the time level t_{n+1} by replacing the time derivatives with (23) and (24). This results in

$$[D]\{\psi^{n+1}\} = [M]\{\omega_n\} + \{\Gamma_s^n\} , \quad (25)$$

$$\frac{1}{\Delta t}[M]\{\omega^{n+1}\} + \nu[D]\{\omega^{n+1}\}$$
$$= \frac{1}{\Delta t}[M]\{\omega^n\} - [A(\psi^{n+1})]\{\omega^n\} + \{F(T^n)\} + \{\Gamma_\chi^{n+1}\} , \quad (26)$$

$$\frac{1}{\Delta t}[M]\{T^{n+1}\} + \lambda[D]\{T^{n+1}\}$$
$$= \frac{1}{\Delta t}[M]\{T^n\} - [A(\psi^{n+1})]\{T^n\} - \{\Gamma_q^{n+1}\} - \{\Gamma_h(T^n)\} . \quad (27)$$

Initial zero vorticity is assumed for most of the cases. Then, the computed $\{\psi_1\}$ using (25) corresponds the potential flow of the problem. In the equation (27), the radiation term on the boundary Γ_h is evaluated at the time level t_n so that the following nonlinear radiation of the Stefan-Boltzmann type may also be considered.

$$-\kappa\frac{\partial T}{\partial n} = \sigma E(T^4 - T_r^4) \quad \text{on} \quad \Gamma_\sigma , \quad (28)$$

where σ is the Stefan-Boltzmann constant of the boundary surface Γ_σ, and E is given by the expression

$$E = V/(\frac{1}{\varepsilon} + \frac{1}{\varepsilon_r} - 1)$$

with the radiation view factor V, the emissivities ε and ε_r of the surface at the temperature T (K) and of the external radiating source at the temperature T_r (K), respectively.

The artificial kinematic viscosity as well as the artificial thermal conductivity are introduced in the computation: Instead of ν and λ, we use

$$\nu_{ax} = \nu + \frac{1}{2}u^2\,\Delta t \quad , \quad \nu_{ay} = \nu + \frac{1}{2}v^2\,\Delta t \,, \tag{29}$$

$$\lambda_{ax} = \lambda + \frac{1}{2}u^2\,\Delta t \quad , \quad \lambda_{ay} = \lambda + \frac{1}{2}v^2\,\Delta t \,. \tag{30}$$

In terms of the Algol-like statements, the computation proceeds as follows.

Read *topological data.*

Read *material constants and boundary conditions.*

Set initial values.

For $n = 0, 1, 2, \ldots,$ **until** *satisfied,* **do:**

Calculate $\{\psi^{n+1}\}$.
Insert the boundary values ω_{wall} on Γ_w .
Calculate $\{\omega^{n+1}\}$.
Calculate $\{T^{n+1}\}$.

The algorithm is not complete until some stopping criteria are specified for the iteration counter n. When flow duration is given, n runs up to the integer $N = (t_f - t_0)/\Delta t$ with the final time t_f. Otherwise, the iteration may continue until the calculated thermal fluid motion fully develops.

PROGRAM SPECIFICATION

All programs are written in the Quick BASIC, Version 4.5 for IBM PC, IBM-XT, or IBM-AT. However, the programs will run successfully on other IBM-compatible machines. The EGA (Enhanced Graphic Adapter board) and 640 KB memories are required for the implementation.

The total of 33 program files is included; 7 programs for the potential flow, 5 for the incompressible viscous flow, 6 for the natural convection, 4 for the air convective diffusion, 4 for the tidal current, and 7 for the pre- and post-processors. These BASIC programs are duplicated on a diskette in the ASCII form.

The diskette contains the programs for the solution of problems, as well as pre- and post-processing programs. Listed below are the problems, program names, names of sample data files, and short description of the function of each program. The program instructions are also stored in the diskette file, named "README.DOC". The question marks "???" below indicate an extension of the corresponding file names.

Potential flow. Following 7 BASIC programs are included:

PFLOWS.BAS	This program calculates the streamfunction flow.
PFLOWV.BAS	This program calculates the velocity potential flow.
PFLOWU.BAS	This program calculates the ground-water flow.
POTS-D.BAS	creates the data files, named "PS1.???", of the streamfunction flow.
POTV-D.BAS	creates the data files "PV1.???" of the velocity potential flow.
POTU-D.BAS	creates the data files "PU1.???" of the ground-water flow.
PGRF.BAS	displays calculated flow vectors and contours.

Incompressible viscous flow. For the solution of the Navier-Stokes equations, following 5 programs are included:

NAVIER.BAS	calculates the incompressible viscous flow.
NAVC-D.BAS	creates the data files "NC1.???" of the cavity flow in a square box at the Reynolds number $Re = 100$.
NAVC2-D.BAS	creates the data files "NC2.???" of the cavity flow in a square box at $Re = 1000$.
NAVK-D.BAS	creates the data files "NK1.???" of the flow over a half cylinder.
NAVE-D.BAS	creates the data files "NCC.???" of the flow around a cylinder.

Natural convection. This treats of the thermal fluid flow. Following 6 programs are included:

THERMCAL.BAS	calculates the natural convection.
THERMC-D.BAS	creates the data files "TC1.???" of the flow in a box at $Re = 10^3$.
THERMC2D.BAS	creates the data files "TC2.???" of the flow in a box at $Re = 10^4$.
THERMC3D.BAS	creates the data files "TC3.???" of the flow in a box at $Re = 10^5$.
THERMN-D.BAS	creates the data files "TN1.???" of the flow in a cup at $Re = 10^4$.
THERMN2D.BAS	creates the data files "TN2.???" of the flow in a cup at $Re = 10^5$.

Air convective diffusion. Prior to calculation of the convective diffusion, the evaluation of the corresponding flow field using NAVIER.BAS is necessary. Following 4 programs are included:

AIR.BAS	calculates the convective diffusion.
AIRNAV.BAS	creates the data files "DIF1.???" of the smoke diffusion for NAVIER.BAS.
AIRDAT.BAS	creates the data files "DIF1.???" of the smoke diffusion for AIR.BAS.
CONTAIR.BAS	displays the concentration contours.

Tidal current. Following 4 programs are included:

TIDALCAL.BAS	calculates the tidal current in a shallow water.
TIDALDAT.BAS	creates the data files "TID.???" in a sample bay.
TIDVECT.BAS	displays calculated flow vectors at the nodal points.
TIDVECTE.BAS	displays calculated flow vectors at the center of each finite element.

Pre- and post-processors. The following is the list of program names of pre- and post-processors:

4 BASIC programs are included in the pre-processor:

AUT.BAS	generates and renumbers a finite element mesh.
MCHK.BAS	checks nodes and elements of the mesh.
PCHK.BAS	checks the boundary conditions for the potential flow.
CCHK.BAS	checks the boundary conditions for incompressible viscous flow and thermal fluid flow.

3 BASIC programs are included in the post-processor:

ARROW.BAS	displays flow vectors for the NAVIER.BAS and for the THERMCAL.BAS.
CONT.BAS	displays streamlines, vorticity contours and isotherms for NAVIER.BAS and THERCAL.BAS.
CONTC.BAS	displays contours depicted by the CONT.BAS in colours.

Miscellaneous. One data file is included for the database of half-font patterns. The patterns can be displayed by PUT command directly on the graphic screen. This data file is used in all sample programs and in the MCHK.BAS.

FONT.PAT	contains half-font patters in ASCII codes from &H20 to &H7F.

DATA FILE SPECIFICATION

In the following are listed the extensions, indicated by ".???", of data file names together with the contents in the corresponding data files.

NAVIER.BAS The data files; NC1.???, NC2.???, NK1.???, NCC.??? are required.

Input .NO Parameters.
.XY x and y nodal coordinates.
.NOD Node numbers associated with each element.
.WBN Boundary condition for the vorticity $\omega = 0$.
.VOL The other types of boundary conditions for the vorticity.
.FRM Frame node numbers.

Output .ANW Velocity.
.PHI Streamfunction.
.DTI Computing time.

THERMCAL.BAS The data files; TC1.???, TC2.???, TC3.???, TN1.???, TN2.??? are required.

Input .NO Parameters.
.XY x and y nodal coordinates.
.NOD Node numbers associated with each element.
.WBN Boundary condition for the vorticity $\omega = 0$.
.VOL The other types of boundary conditions for the vorticity.
.THE Thermal condition for the program THERMCAL.BAS.
.FRM Frame node numbers.

Output .ANW Velocity.
.PHI Streamfunction.
.DTI Computing time.

AIR.BAS The data files DIF1.??? are required.

Input .NO Parameters.
.XY x and y nodal coordinates.
.NOD Node numbers associated with each element.
.WBN Boundary condition for the vorticity $\omega = 0$.
.VOL The other types of boundary conditions for the vorticity.

.FRM Frame node numbers,

and

.NOA Parameters for the air.

.COA Boundary conditions for the air flow.

Output .ANW Velocity.

.PHI Streamfunction.

.DTI Computing time for the NAVIER.BAS,

and

.COT Concentration.

.TMP The concentration at each time step.

.DTI Computing time for the AIR.BAS.

NUMERICAL EXAMPLES

Bénard cell

We consider natural convection of water in a closed shallow vessel, as shown in Figure 1. The vessel is mildly heated from below, so that the temperature difference between top and bottom is 1 $°C$. With the representative height $H = 0.01\ m$ and the temperature difference $\Delta T = 1\ °C$, the corresponding Rayleigh number is $Ra = 20250$.

Calculated results are shown in Figure 2. We can see the twelve Bénard cells generated in the vessel.

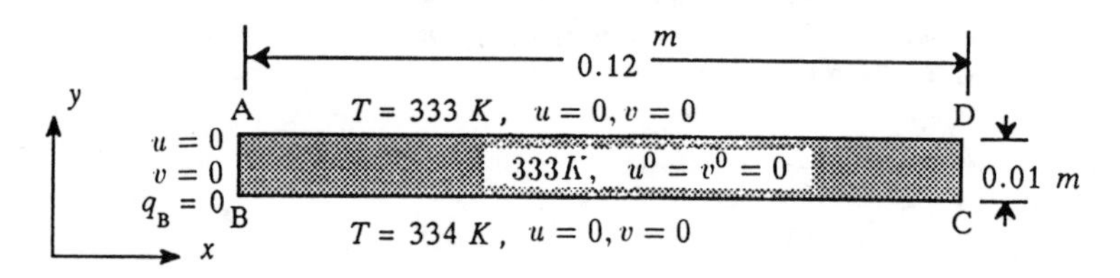

(a) Boundary conditions.

(b) Finite element mesh.

Figure 1: Cross section of a shallow vessel.

Forced thermal convection

We consider vertical forced convection of water between two parallel plates with the distance 0.01 m apart and of 0.3 m long, as shown in Figure 3. The plates are assumed to be adiabatic. Water at a temperature of 288 $K(= 15\ °C)$ is poured into the channel from the top inlet at the velocity 0.02 m/s. Away from the inlet, there are heating stations of 0.02 m long, at a temperature of 353 $K(= 80\ °C)$ on both sides.

Physical constants used are; $\nu = 1.14 \times 10^{-6}\ m^2/s, \lambda = 1.40 \times 10^{-7}\ m^2/s, \beta = 1.5 \times 10^{-4}\ 1/K$ and $g = 9.81\ m/s^2$. With the representative length $L = 0.01\ m$, the representative velocity $U = 0.02\ m/s$ and $\Delta T = 65\ °C$, the corresponding Reynolds and

(a) Isotherms ($333 \leq T \leq 334\ K$).

(b) Streamlines ($-4.47 \times 10^{-4} \leq \psi \leq 4.32 \times 10^{-4}\ m^2/s$).

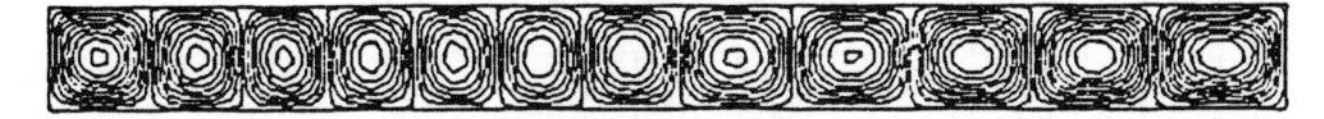

(c) Vorticity contours ($-0.663 \leq \omega \leq 0.674\ 1/s$).

Figure 2: Calculated Bénard convection in quasi-steady state ($Ra = 20250$).

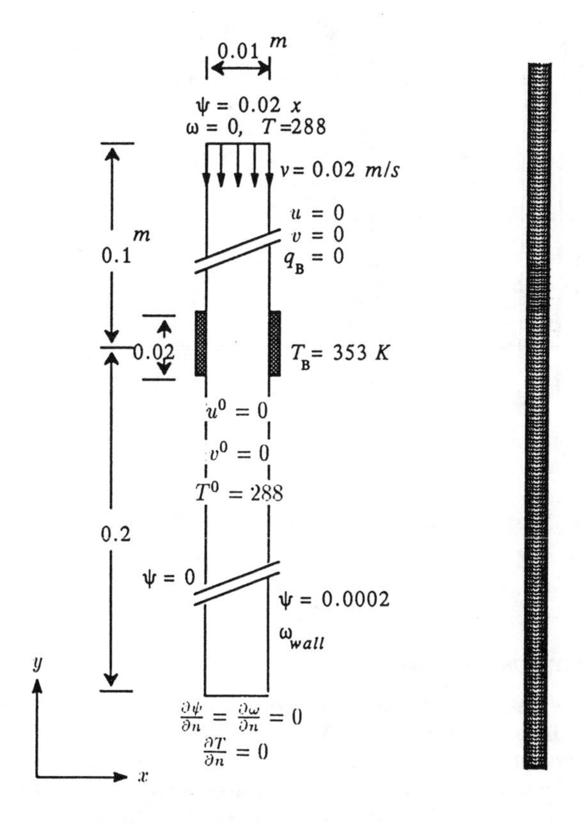

(a) Boundary conditions. (b) Finite element mesh.

Figure 3: Vertical channel between two parallel plates.

Froude numbers are $Re = 175$ and $Fr = 0.418$. The time increment in the computation is $\Delta t = 0.1\ s$.

Calculated results are presented in Figure 4.

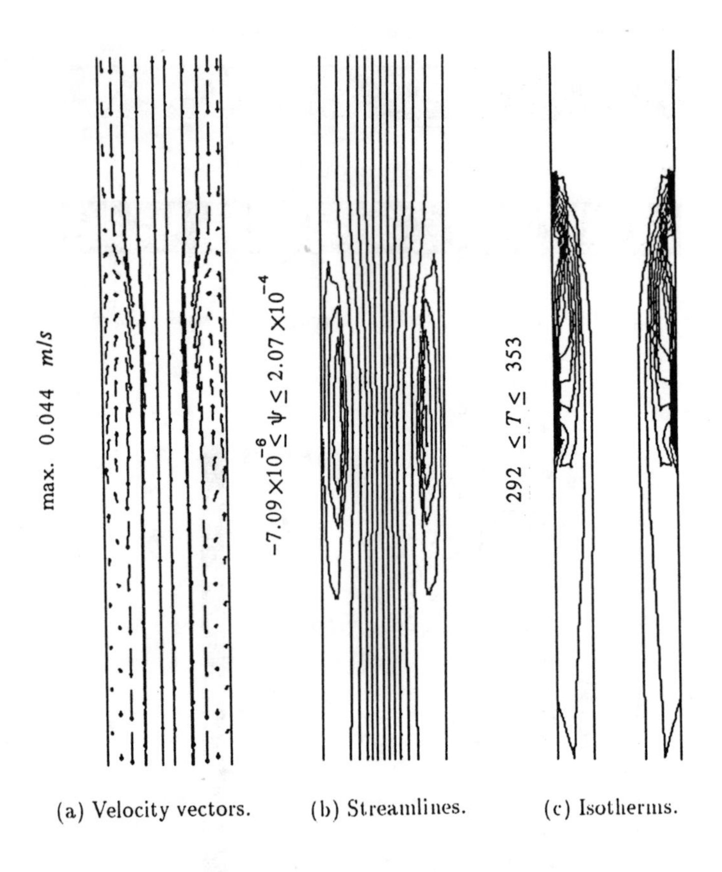

Figure 4: Calculated forced thermal convection near the heating spot ($t = 20\ s$).

Density-dependent viscous flow

We deals with the finite element analysis of the isothermal mass transfer in a viscous fluid. A soluble mass diffuses in the liquid. When the liquid is in motion, the mass is transported simultaneously by the diffusion and convection. The natural convective motion of the fluid is induced by the buoyancy due to the non-uniform density of the fluid under the gravity. It is assumed that the density depends only on the concentration of the mass under consideration. The equation of mass diffusion is coupled to the equations of viscous fluid flow.

If the density difference is small enough, the equation of state of the fluid can be expressed as follows.

$$\frac{\rho}{\rho_0} = 1 + \sigma(C - C_0), \qquad (31)$$

with a proportional constant σ (m^3/kg). The buoyancy per a unit mass is therefore given by

$$F_y = (\rho_0 - \rho) g / \rho_0 = \sigma g (C - C_0) . \tag{32}$$

We consider a closed experimental vessel with its rectangular cross section, as shown in Figure 5. The vessel is divided into two cells by an impervious thin plate. In the right cell is contained the pure water at the initial concentration $C_0 = 0\ kg/m^3$, while in the left cell is contained a solution at the unit initial concentration. We shall calculate the mixing process of the solution after the intermediate plate is removed.

We assume that $\nu = 1.14 \times 10^{-6}\ m^2/s$, $\eta = 0.90 \times 10^{-9}\ m^2/s$ and $\sigma = 0.05\ m^3/kg$. With the representative length $L = 0.02\ m$ and the concentration difference $\Delta C = 1\ kg/m^3$, the Rayleigh number is $Ra = 1.03 \times 10^{11}$. We take $\Delta t = 0.01\ s$. The calculated results are shown in Figure 6.

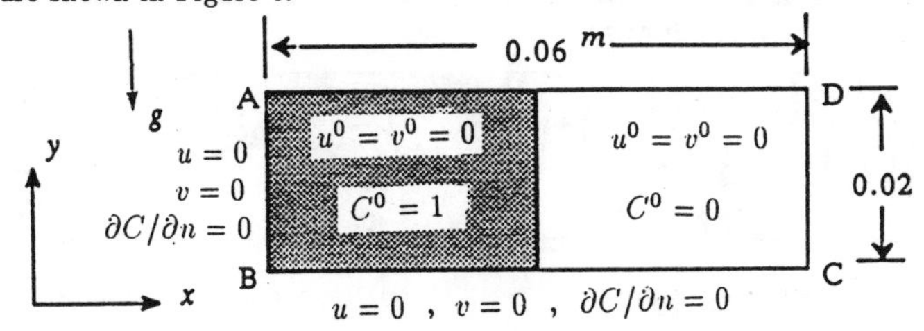

(a) Boundary conditions.

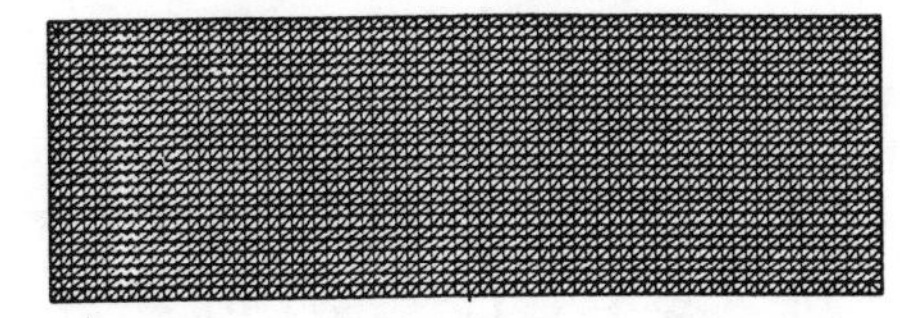

(b) Finite element mesh.

Figure 5: Twin-cell vessel.

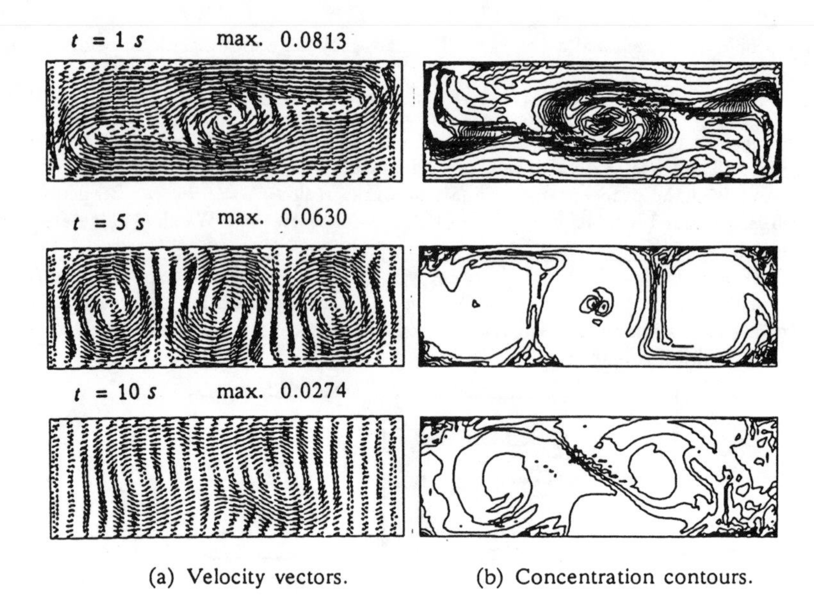

Figure 6: Calculated mixing in aquious solution ($Ra = 1.03 \times 10^{11}$).

REFERENCES

1. Kobayashi, K., Ohura, Y., and Onishi, K. Computer Programme KYOKAI.F for Viscous and Thermal Fluid Flow Problems, in *Boundary Elements IX* (Eds. Brebbia, C. A., Wendland, W. L., and Kuhn, G.), Vol.1, pp.579-592, Proceedings of the 9th Int. Conf. on Boundary Elements, Stuttgart, Germany, 1987. CML Publications, Southampton and Springer-Verlag, Berlin, 1987.

2. Kobayashi, K. and Onishi, K. A Parallel Processing in Viscous Flow Problems, in Proceedings of the 2nd Symposium on Numerical Fluid Dynamics, pp.515-518, Chuo University, Tokyo, 1988 (in Japanese).

An Arc-Length-Based Time Stepping Algorithm to Analyze Thermal Transients

A. Soria, P. Pegon

Applied Mechanics Division, Safety Technology Institute, Commission of the European Communities, Joint Research Centre, I-21020 Ispra (VA), Italy

ABSTRACT

The arc length method is applied to a class of stiff parabolic problems related to heat conduction and radiation. The resulting system of nonlinear equations is solved by iteration not only on the temperature vector, but also on the current time step. The procedure is linked with an a posteriori analysis of the time integration error. In addition, a quasi-Newton method is used to obtain approximate factorizations of the Jacobian matrix during the iterative process.

INTRODUCTION

The solution of the unsteady heat diffusion equation following a standard finite element discretization can lead to a stiff system of equations. This is mainly due to the presence of materials with very different thermal properties, strongly temperature-dependent parameters, and thermal radiation (both across internal cavities and as boundary conditions) [1]. The superposition of these phenomena produces transients in which very different time scales are involved, so that pure explicit methods seem a priori to be penalized. The selection of an adequate time stepping strategy becomes important, specially when large problems are under consideration.

The purpose is to integrate accurately the parabolic equation up to a certain time value, optimizing the time step size at each time, in order to minimize the number of time steps. Several methods have been proposed, in the frame of implicit and mixed implicit-explicit schemes [2], [3]. These methods are based on spectral considerations when using conditionally stable schemes, and on integration error estimation when dealing with unconditionally stable formulas.

In this paper, an arc length method [4] with spherical constraint [5] is combined with the classical Crank-Nicolson formula. This last choice is, however, arbitrary, and the method could be applied also to any implicit time integration scheme. The integration error criterion acts then on the current arc-length rather than on the time step size. The aim of the selected strategy is to reach a uniform time integration error, limiting the number of calculated time steps refused due to accuracy requirements. A smooth variation of the calculated time step is obtained, which contributes to save matrix factorizations during the transient analysis.

AN ARC-LENGTH PARABOLIC FORMULATION

The Crank-Nicolson-Galerkin [6] time integration formula yields the temperature vector at the end of a time step Δt in the form:

$$T^{t+\Delta t} = 2T^{t+\frac{\Delta t}{2}} - T^t \tag{1}$$

where $T^{t+\frac{\Delta t}{2}}$ is the solution of the following system of nonlinear equations, obtained after insertion of the boundary conditions:

$$\left[K^{t+\frac{\Delta t}{2}} + \frac{2}{\Delta t}C^{t+\frac{\Delta t}{2}}\right] T^{t+\frac{\Delta t}{2}} = \left[\frac{2}{\Delta t}C^{t+\frac{\Delta t}{2}} T^t + g^{t+\frac{\Delta t}{2}}\right] \tag{2}$$

which must be solved iteratively. Here, K is the conductance matrix, C is the thermal inertia matrix, g is the thermal load vector, and the superscripts denote the time level. At each iteration level, the temperature vector at time $t + \Delta t/2$ is given by:

$$T_{i+1} = T_i + s_i = T^t + \tau_{i+1} \tag{3}$$

We claim the solution $T^{t+\frac{\Delta t}{2}}$ to lie on a hypersphere of radius l, and centre at T^t, so that the euclidean norm of τ_{i+1} must be constant and equal to l:

$$\tau_{i+1}^T \tau_{i+1} = l^2 \tag{4}$$

If we add the current time step as a variable, and use equation (4) to form with (2) an augmented system of equations, the following nonlinear system on $(T^{t+\frac{\Delta t}{2}}, \Delta t)$ can be solved using an iterative Newton-like method by finding at each iteration level the residual vector:

$$R_i = \left[\frac{2}{\Delta t}C^{t+\frac{\Delta t}{2}} T^t + g^{t+\frac{\Delta t}{2}}\right]_{\Delta t=\Delta t_i} - \left[K^{t+\frac{\Delta t}{2}} + \frac{2}{\Delta t}C^{t+\frac{\Delta t}{2}}\right]_{\Delta t=\Delta t_i} T_i \tag{5}$$

as well as the Jacobian matrix:

$$J_i = -\frac{\partial R_i}{\partial T_i} \tag{6}$$

and solving the linear system of equations:

$$\begin{pmatrix} J_i & -\frac{\partial R_i}{\partial \Delta t_i} \\ a_i^T & e_i \end{pmatrix} \begin{pmatrix} s_i \\ \lambda_i \end{pmatrix} = \begin{pmatrix} R_i \\ l^2 - \tau_i^T \tau_i \end{pmatrix} \tag{7}$$

where λ_i is the correction on the time step:

$$\Delta t_{i+1} = \Delta t_i + \lambda_i \tag{8}$$

and s_i is the correction on the temperature vector given in (3).
The vector a_i and the scalar e_i are formulated as follows:

$$a_i = \frac{\partial(\tau_i^T \tau_i - l^2)}{\partial T_i} = \tau_i \tag{9}$$

$$e_i = \frac{\partial(\tau_i^T \tau_i - l^2)}{\partial \Delta t_i} = 0 \tag{10}$$

The vector $\partial R_i/\partial \Delta t_i$ can be approximated within a time step neglecting the dependence of K, C on the temperature and using the relationship $\partial/\partial \Delta t = \partial/\partial t$, to yield:

$$\frac{\partial R_i}{\partial \Delta t_i} = \frac{2}{\Delta t_i^2}C^{t+\frac{\Delta t}{2}}\tau_i + \dot{g} \tag{11}$$

Note that no term in $\dot{T}$ appears since in the augmented system context T and Δt are independent variables.
The Crisfield indirect approach [5] is used to solve (7), because of the ill-conditioned character of the augmented Jacobian matrix. This method also preserves the structure of the factorized Jacobian matrix and easily allows the switching from a fixed Δt method to a fixed l method. The first equation (7) is used to obtain the following expression:

$$s_i = J_i^{-1} R_i + \lambda_i J_i^{-1} \frac{\partial R_i}{\partial \Delta t_i} = \overline{s_i} + \lambda_i \overline{\overline{s_i}} \tag{12}$$

At each iteration level, one calculates J_i, R_i, r_i, and $\partial R_i / \partial \Delta t$, and the vectors $\overline{s_i}$ and $\overline{\overline{s_i}}$. Equation (12) is then used to solve for λ_i which must exactly verify the condition (4):

$$\left(r_i + \overline{s_i} + \lambda \overline{\overline{s_i}}\right)^T \left(r_i + \overline{s_i} + \lambda \overline{\overline{s_i}}\right) = l^2 \tag{13}$$

Making:

$$v_i = r_i + \overline{s_i} \tag{14}$$

the quadratic equation to be solved is:

$$v_i^T v_i + 2 v_i^T \overline{\overline{s_i}} \lambda_i + \overline{\overline{s_i}}^T \overline{\overline{s_i}} \lambda_i^2 - l^2 = 0 \tag{15}$$

One should choose the λ_i root closest to the linear solution [5]:

$$\lambda^* = \frac{l^2 - v_i^T v_i}{2 v_i^T \overline{\overline{s_i}}} \tag{16}$$

The advantages of the method appear from the definition of the vectors $\overline{s_i}$ and $\overline{\overline{s_i}}$ (12), which require no additional matrix factorizations.

REMARK 1
The application of the arc length method requires two backsubstitutions per iteration. One could calculate $\partial R_i / \partial \Delta t_i$ (and $\overline{\overline{s_i}}$), and keep them constant during the iterative process within a time step. This modified-Newton-like linearization seems to perform well, although no extensive numerical experimentation has been done. Following this approach, sometimes used in structural mechanics, only one additional backsubstitution is needed per time step when comparing the method with a fixed Δt scheme.

REMARK 2
The first iteration represents a particular case due to the index uncoupling in the above formulation. The first guess for T and Δt coming from the preceding time step, the first correction s_1 does not lie in the constraining surface. One could scale the correction onto the hypersphere or just leave the first correction free, and proceed with the algorithm from iteration two onwards. Figure 1 shows both strategies within a modified Newton-Raphson iterative scheme. According to our experience, the latter method seems to have a better performance. The scaling factor $\|s_1\|_2 / l$ is, however, a good predictor of a difficult convergence when its value is far away from unity.

REMARK 3
Special attention must be paid to Dirichlet boundary conditions when finding the vector $\overline{\overline{s_i}}$. Since the solution is exact on nodes with prescribed temperature during the iterative process, the correction on these nodes must be zero. This is necessarily accomplished in the $\overline{s_i}$ part in a fixed Δt scheme, but should not be forgotten in the $\lambda_i \overline{\overline{s_i}}$ part. An additional loop on these boundaries is needed in order to reset the correction to zero.

QUASI-NEWTON ITERATIVE STRATEGY

The calculation and inversion of the Jacobian matrix is in general the most expensive operation when using implicit methods. The LU factorization of the matrix needs a number of basic operations which grows with N^3, N being the number of degrees of freedom of the system, whereas the calculation of the residual vector R_i requires about N^2 operations.

Liu et al. [7] used a quasi-Newton (QN) rank-two formula in the frame of a mixed-time partition algorithm. Their work was based on the Broyden-Fletcher-Goldfarb-Shanno (BFGS) method [9] combined with a line search procedure.

In a recent paper [1], the authors have also pointed out the excellent convergence properties of two quasi-Newton methods for this class of parabolic problems. A comparison between rank-one and rank-two QN formulas was made there, and it was also shown that QN methods are less sensitive to changes in the time step size than the usual modified Newton-Raphson method.

An additional difficulty is present here, since the time step changes within the iterative process itself, so that the residual vector R_i is calculated with a time step different from the time step corresponding to the residual R_{i+1}. This should imply a correction in the recursive formulas used to update the Jacobian matrix. This problem is solved by means of a Taylor polynomial expansion of the residual vector with Δt as independent variable [8] and makes reasonable the use of a two-index formula rather than a three-index formula (e.g. the BFGS formula). It will be pointed out that the QN rank-one Broyden inverse formulation remains unchanged under this new condition, and is then the preferred tool to solve the arc-length equations.

The inverse Broyden approximation to the Jacobian matrix is given in [10] [11]:

$$J_i^{-1} \simeq H_i = \prod_{j=i-1}^{1} (I + w_j s_j^T) H_1 \tag{17}$$

with

$$w_j = \frac{s_j - H_j y_j}{s_j^T H_j y_j} \tag{18}$$

In the above equations, H_1 is the starting factorized Jacobian matrix, I is the identity matrix, and the vectors s_i, y_i are given by:

$$s_i = T_{i+1} - T_i \tag{19}$$

$$y_i = R_i - R_{i+1} \tag{20}$$

The fact of calculating R_i and R_{i+1} at different time levels can be taken into account as follows:

$$T_{i+1} = T_i + \overline{s_i} + \lambda_i \overline{\overline{s_i}} \tag{21}$$

$$y_i = R_i + (\frac{\partial R}{\partial \Delta t})_i \lambda_i - R_{i+1} \tag{22}$$

According to this new definition of y_i, the H_i Broyden matrix verifies:

$$H_{i-1} y_{i-1} = \overline{s_i} + \lambda_i \overline{\overline{s_i}} - H_{i-1} R_i = s_i - H_{i-1} R_i \tag{23}$$

which leads to the same formal scheme used in a fixed Δt context, so that no modification is needed in equation (17) to account for the varying Δt procedure. The complete Broyden iterative method is an extension of the algorithm given by [12], and reads as follows:

1) Calculate R_i and $(\partial R/\partial \Delta t)_i$.

2) To prepare the calculation of $\overline{s_i}$, find
$H_{i-1}R_i = \prod_{j=i-2}^{1}(I + w_j s_j^T)H_1 R_i$.

3) To prepare the calculation of $\overline{\overline{s_i}}$, find
$H_{i-1}(\frac{\partial R}{\partial \Delta t})_i = \prod_{j=i-2}^{1}(I + w_j s_j^T)H_1(\frac{\partial R}{\partial \Delta t})_i$.

4) Find and store w_{i-1}, using equations (18) and (23).

5) Complete the calculation of $\overline{s_i}$:
$\overline{s_i} = (I + w_{i-1}s_{i-1}^T)H_{i-1}R_i$

6) Complete the calculation of $\overline{\overline{s_i}}$:
$\overline{\overline{s_i}} = (I + w_{i-1}s_{i-1}^T)H_{i-1}(\frac{\partial R}{\partial \Delta t})_i$

7) Solve equation (13) for λ_i, then find and store s_i.

8) Update the temperature vector T and the current time step Δt, using equations (3) and (8).

9) Test convergence, and eventually go to step 1.

REMARK 4
If one decides to keep the vector $\partial R/\partial \Delta t$ constant during the iterative process within a time step, in order to save one backsubstitution per iteration, there are two ways to proceed:
a) Find $\overline{\overline{s_i}}$ with the starting vector $\partial R/\partial \Delta t$, and keep both constant.
b) Keep the vector $H_1 \partial R/\partial \Delta t$ unchanged, but update the succesive $\overline{\overline{s_i}}$ vectors according to the preceding QN algorithm.
Since the relative cost of one backsubstitution is the most expensive part of the procedure, the second method would give better results with only two additional scalar products per iteration.

As pointed out above, the aim is to minimize the number of calculations and factorizations of the Jacobian matrix, so that the same matrix is used as H_1 matrix in the preceding algorithm as many time steps as possible, preserving the robustness of the iterative implicit method. Since the time step is continuously changing, it is clear that the convergence becomes harder and harder as the current time step differs from the time step used for the factorization of H_1. The choice of an adequate refactorization strategy is not easy. Depending on the number of nodes, one could prefer to do more factorizations (and less iterations) or vice-versa. The present algorithm calculates a new Jacobian matrix factorization:
a) If the current time step at the beginning of a step calculation verifies:

$$\Delta t \geq \nu \Delta t_\star \;\; or \;\; \Delta t \leq \nu^{-1} \Delta t_\star \tag{24}$$

where $\Delta t_\star$ is the time step used in the last Jacobian calculation, and ν is a constant, given by the analyst ($\nu > 1$). Values ranging from 3 to 10 have been tested with success.
b) If, at any moment, a divergence on s_i or λ_i is detected.
c) Each NSTEPS time steps, NSTEPS being a parameter given by the analyst.

ARC LENGTH SELECTION AND TIME INTEGRATION ACCURACY

The arc length algorithm described in section 2 needs to be combined with an automatic selection of the parameter l from one time step to the next one. This would be equivalent to the time step selection strategy developed for fixed Δt schemes.
The simplest solution would be to proceed with the time integration keeping the radius unchanged. However, according to our experience, the resulting growing rate of the time step is not satisfactory and depends greatly on the choice of the initial parameter l.
From the qualitative point of view, it is desirable that large nonlinearities are taken into account in the arc length selection.
Crisfield [5] proposed to use the number of iterations in a given time step as a measure of nonlinearity, and to drive the arc length according to the heuristic formula:

$$l_{new} = \sqrt{\left(\frac{I_{old}}{I_{opt}}\right)}\, l_{old} \tag{25}$$

where I_{old} is the number of iterations used in the previous time step, and I_{opt} is the desired average number of iterations per time step. However, the number of iterations is not a good indicator of nonlinearity in our case, since I_{old} could be large because the matrix H_1 used is no longer 'fresh'. This implies that the cost of the solution of the nonlinear system of equations should be separated from the actual step length.
The method adopted here is based on the 'a posteriori' error measure suggested by Winget and Hughes [3]. The time step growing rate is now replaced by an equivalent arc length growing rate. The general procedure is outlined in the following.
The iterative process within each time step is carried out until the midpoint residual given by equation (5) is equal to zero within a prescribed tolerance. A time integration error measure is then introduced by computing the corresponding end-of-step residual vector:

$$R^{t+\Delta t} = \left[\frac{2}{\Delta t} C^{t+\Delta t} T^{t} + g^{t+\Delta t}\right] - \left[K^{t+\Delta t} + \frac{2}{\Delta t} C^{t+\Delta t}\right] T^{t+\Delta t} \tag{26}$$

This absolute error is then normalized to obtain the following adimensional time integration criterion:

$$E_{int} = \frac{\|R^{t+\Delta t}\|_2}{\|\left[K^{t+\Delta t} + \frac{2}{\Delta t} C^{t+\Delta t}\right] T^{t+\Delta t}\|_2} \leq \epsilon_{int} \tag{27}$$

where ϵ_{int} is usually prescribed between 10^{-4} and 10^{-2}. In all the calculations presented here, the assumed value was 10^{-3}. It is commonly agreed that the time step should grow slowly, but in case of an unacceptable time integration error it should be reduced more severely. On the other hand, the time step size and the arc length are narrowly correlated. Taking this into account, if equation (26) does not hold at the end of a time step, this time step is refused, both time step and arc length are scaled by a factor η (usually $0.2 \leq \eta \leq 0.5$) and the time step is recalculated.
Note that the selection of ϵ_{int} is not completely free. It should be combined with an effective tolerance in the iterative process to annul the midpoint residual, i.e. this residual should be very small with respect to $R^{t+\Delta t}$. In practice, the convergence criterion used acts on the temperature correction rather than on the residual vector:

$$\frac{\|s_i\|_2}{\|s_0\|_2} \leq \epsilon_{rel} \tag{28}$$

where s_0 is the first correction within a time step and ϵ_{rel} is usually between 10^{-4} and 10^{-2}. However, the following absolute criterion:

$$\|s_i\|_\infty \leq \epsilon_{abs} \tag{29}$$

can be added to equation (28) in order to avoid useless iterations when solving a real case.
If the time step is acceptable, the arc is scaled by a factor ξ greater than unity ($1.1 \leq \xi \leq 1.25$ in practice), and the preceding time step is the first guess for the succesive iteration on Δt.
However, these two rules are too crude to provide an optimal behaviour of the time integration process. It should be desirable to limit the growing rate of the arc when the integration error E_{int} reaches a value close to ϵ_{int}, in order to avoid refused time steps and to stabilize near a fixed value the time integration error. To do this, one should choose a fraction μ of the tolerance ϵ_{int} ($0.25 \leq \mu \leq 0.75$) and apply the following rules:

a) If $E_{int} \leq 0.1\epsilon_{int}$, the arc length growing rate is equal to ξ.

b) If $0.1\epsilon_{int} \leq E_{int} \leq \mu\epsilon_{int}$, the arc length growing rate is linearly interpolated between ξ and 1 in the range $(0.1\epsilon_{int}, \mu\epsilon_{int})$.

c) If $\mu\epsilon_{int} \leq E_{int} \leq \epsilon_{int}$, the arc length is scaled by a factor linearly interpolated between 1 and η in the range $(\mu\epsilon_{int}, \epsilon_{int})$.

This scheme provides the desired feedback effect needed to keep the time integration error E_{int} around the optimal value $\mu\epsilon_{int}$, and a calculated time step is refused only in very severe circumstances (i.e. a sudden change in the boundary conditions or in the heat generation rates).

REMARK 5
The first time step calculation represents a particular case, since one has to specify both the time step guess and the corresponding arc. The correlation between these two parameters is difficult to establish, so that the best solution seems to be to calculate this first step in a fixed Δt mode, by prescribing a reasonable Δt_0. The arc length obtained from this step is then used and modified according to the rules presented above.

REMARK 6
The main numerical problem associated with the arc length algorithm summarized in section 2 comes from the solution of the quadratic equation (15). A negative discriminant in this equation clearly indicates that the algorithm can not find a solution lying on the sphere of radius l. The reasons for this could be:
a) The arc and the time step are no longer well correlated.
b) The transient has reached a quasi-steady state.
The first case is always followed by a divergence on λ_i, and is frequent when sudden changes in the boundary conditions occur. In this case, the best method is to switch the procedure to perform a fixed Δt time step to reset the $l - \Delta t$ relationship.
The second case is more difficult to detect, but it is generally revealed by two indicators: $\|\dot{T}\|$ tends to zero and Δt grows to infinity. In this case, the simulation is stopped.

NUMERICAL EXPERIMENTS

Two practical problems are discussed in this section in order to show the performance of the solution procedure discussed in the present paper. Both of them are solved in terms of absolute temperature.
The first problem involves radiation across internal cavities as well as anisotropic material properties. The analyzed specimen is a piece of the divertor plate of a fusion tokamak

reactor. It is made up of an anisotropic graphite having the following material properties: $\rho = 1800\ Kg/m^3$, $C_p = 578+1.4T\ W/Kg/^OK$, $k_y = 292-0.2T+4.4\ 10^{-5}T^2\ W/m/^OK$, and $k_x = 85 - 0.032T + 5.7\ 10^{-6}T^2\ W/m/^OK$. The graphite armour is cooled by two GlidCop-alloy tubes ($\rho = 8900\ Kg/m^3$, $C_p = 362 + 0.06T\ W/Kg/^OK$, $k = 384 - 0.1T + 2.5\ 10^{-5}T^2\ W/m/^OK$), containing pressurized water at 80 ^OC. The thermal heat exchange coefficient as a function of the tube wall temperature is shown in figure 2. A prescribed constant heat flux ($10MW/m^2$) is applied to the upper side, all the other boundaries being adiabatic. The constant volumetric heat generation rate due to neutron slowing is 10 MW/m^3 in the tube and 5 MW/m^3 in the graphite. At time $t = 0\ s$, a loss of coolant accident is assumed to occur in the left tube. The heat exchange coefficient linearly decreases to zero in 0.1 s. From this point on, the tube is assumed to be void and the only heat transport mechanism in the cavity is the thermal radiation. The assumed value for the GlidCop-alloy emissivity is 0.6.
Figure 3 shows the steady state temperature field. This transient is characterized by two time constants due to the highly anisotropic thermal conductivity. During the first period (up to 0.5 s) the temperature field changes its shape. The temperatures in the faulty side begin to rise to extremely high values in the second period. Figure 4 shows the temperature distribution along the heated line (in mm) at 7 different times.
The problem was solved with the following set of parameters: $\mu = 0.75$, $\epsilon_{int} = 10^{-3}$, $\xi = 1.25$, $\eta = 0.5$, $\nu = 5$, $\Delta t_0 = 10^{-2}s$, $\epsilon_{rel} = 10^{-3}$, and $\epsilon_{abs} = 0.05$. The analized period was about $15s$ in a fixed number of time steps (70).
Three runs were done in order to study the sensivity of the method to the vector $\partial R/\partial \Delta t$ quality (see remark 4). This vector was recalculated and backsubstituted at each iteration level in run A. Runs B and C correspond to methods (a) and (b) in remark 4. The numerical results are summarized in table 1. Although the differences in the CPU time are not relevant (the number of degrees of freedom is not large enough), it appears that one could safely save the calculation and backsubstitution of the vector $\partial R/\partial \Delta t$ provided its succesive updating is done according to the QN formula. When this update is not carried out, the convergence of the method is damaged (all the Jacobian matrix factorizations in run B were triggered by a divergence on λ_i). Figures 5, 6, and 7 show

Table 1: Problem 1. Performance of the methods.

run	A	B	C
iterations	354	348	355
CPU time	33 m 39 s	34 m 13 s	32 m 06 s
factorizations	3	5	3
at steps	1, 19, 46	1, 2, 18, 30, 54	1, 17, 37

the relative error E_{int}, the time step size and the number of Broyden iterations as a function of the time step number ($*$ = run A, $\square$ = run B, $\triangle$ = run C). It should be underlined that the time step size moves along three decades and only three factorizations were required to analyze the whole transient. The algorithm adaptivity is shown in figure 6, which reflects the fact of having two time scales. The relative error quickly reaches the desired value, and the time integration proceeds from this point on with a constant error.
The second problem is taken from [3] with some modification. Figure 8 shows a $1m \times 1m$ square plate made up of a material with $\rho = 1\ Kg/m^3$, $C_p = 50\ W/Kg/^OK$ and $k = 2(1 + 0.005T)\ W/m/^OK$. A constant heat generation per unit volume $Q = 2500\ W/m^3$ is present. The boundary conditions are:
a) The upper side is kept at 273 ^OK.
b) The right side has a constant heat flux $q = 1000\ W/m^2$.

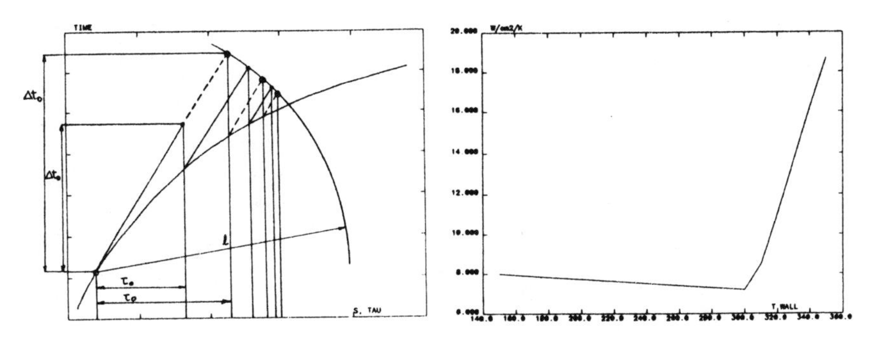

Fig. 1. Arc-length method.

Fig. 2. Problem 1. Heat exchange coefficient.

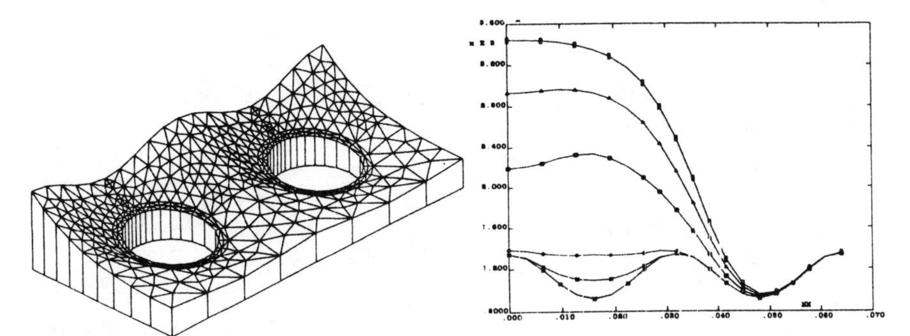

Fig. 3. Problem 1. Steady-state.

Fig. 4. Problem 1. Upper side temperatures.

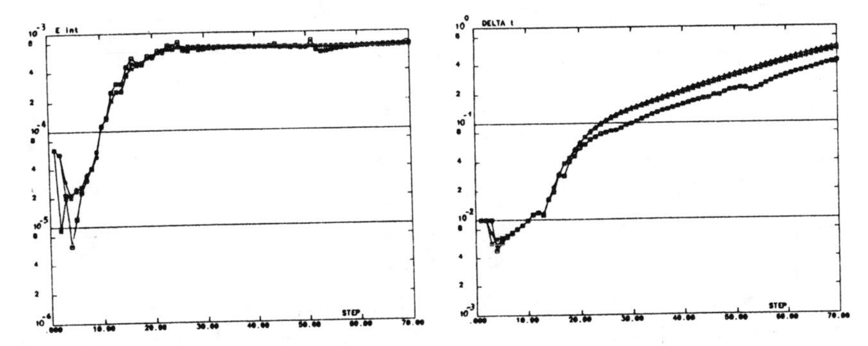

Fig. 5. Problem 1. Relative integration error.

Fig. 6. Problem 1. Time step size.

c) The lower side is adiabatic.
d) The left side has a radiation boundary condition $q = h_{rad}(T^4 - T_{rad}^4)$ with $h_{rad} = 10^{-9}$. The radiation temperature is $T_{rad} = 1473\ ^OK$ if $0\ s \leq t \leq 0.5\ s$ and $T_{rad} = 273\ ^OK$ if $0.5s\ < t$.
The body has an initial uniform temperature of $273\ ^OK$.
The purpose is to analyze how the algorithm behaves traversing the discontinuity in the boundary condition. The problem is solved using 80 time steps, and the following parameters were used: $\mu = 0.25$, $\epsilon_{int} = 10^{-3}$, $\xi = 1.1$, $\eta = 0.2$, $\nu = 5$, $\Delta t_0 = 10^{-3}s$, $\epsilon_{rel} = 10^{-4}$, and $\epsilon_{abs} = 10^{-3}$.

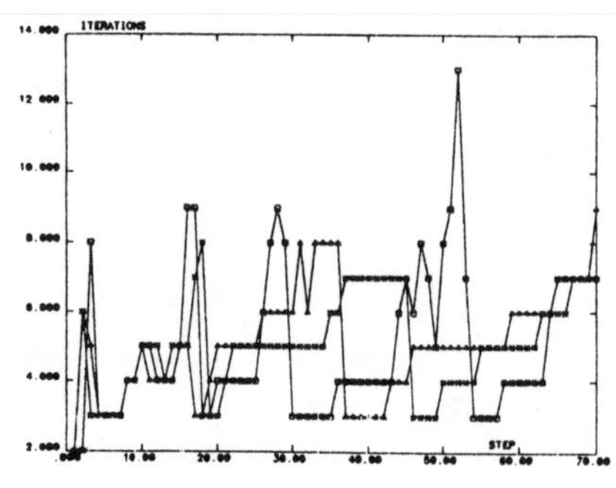

Fig. 7. Problem 1. Number of iterations.

Fig. 8. Problem 2. Finite element mesh.

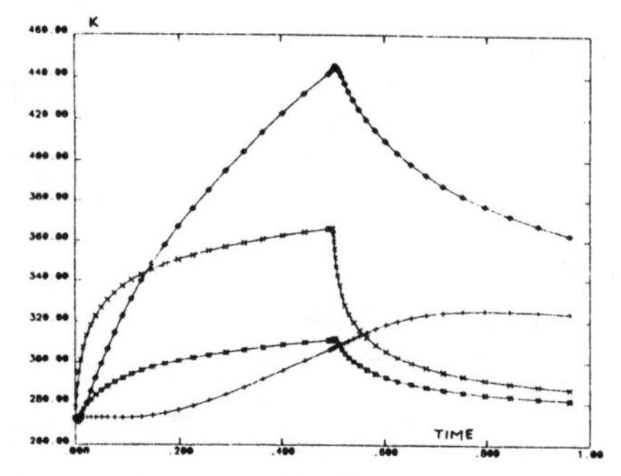

Fig. 9. Problem 2. Nodal temperatures.

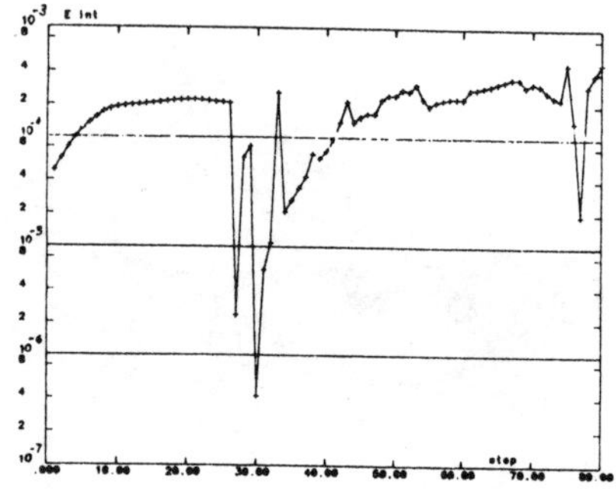

Fig. 10. Problem 2. Relative integration error.

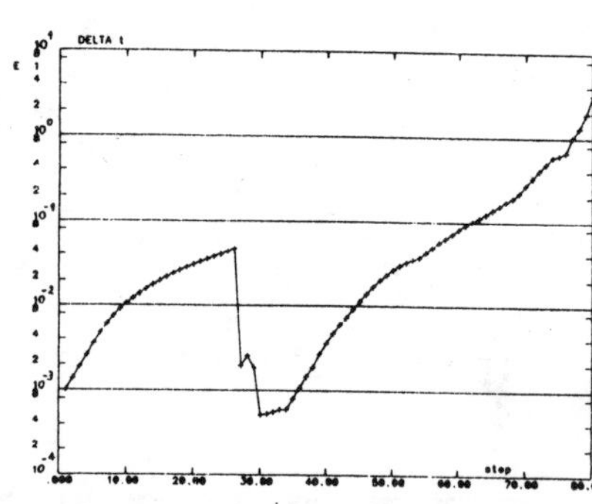

Fig. 11. Problem 2. Time step size.

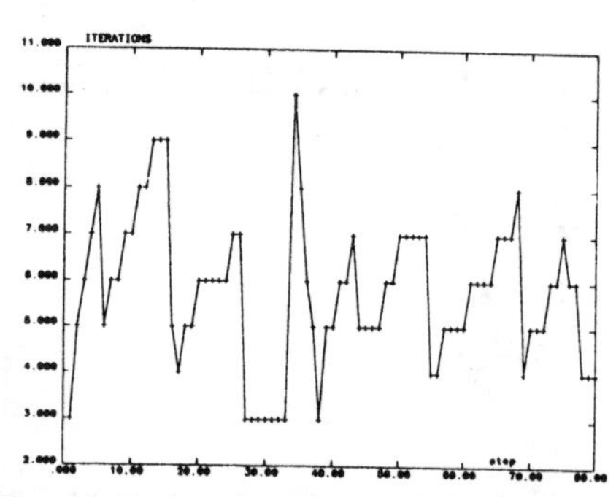

Fig. 12. Problem 2. Number of iterations.

Figure 9 shows the temperature evolution for 4 selected nodes up to 1 s. The period between 0 and 1 s. was analyzed in 58 time steps, and the 80 time steps covered up to about 10 s. Figures 10, 11, and 12 show the relative error E_{int}, the time step size and the number of Broyden iterations needed per time step as a function of the time step number, respectively. The discontinuity at $t = 0.5$ was crossed by recalculating twice the time step 27 and once the time step 30. The Jacobian matrix was factorized 10 times during the analysis (in the steps 1, 6, 16, twice 27, 30, 44, 55, 69 and 78). The total number of iterations was 458, corresponding to an average of 5.7 iterations per time step. Figure 4 shows how the integration error moves towards the value $\mu\epsilon_{int}$ even after the crossing of the discontinuity. The oscilations observed in the last time steps appear as a consequence of the reaching a quasi-steady state.

CONCLUSION

The arc length method has been used as a time stepping algorithm for the parabolic heat diffusion equation. The particular features of the method for this class of problems have been outlined. It has been shown that a properly designed arc length driving strategy can lead to a uniform time integration error during the analysis. A considerable amount of numerical work can be saved if the method is combined with a quasi-Newton approximate factorization of the Jacobian matrix.

ACKNOWLEDGEMENTS

The first author gratefully acknowledges the financial support received from the Commision of the European Communities through a grant.

REFERENCES

[1] A. Soria and P. Pegon, 'On the performances of quasi-Newton iterative methods to solve the nonlinear heat diffusion equation'. Proc. 6th International Conference on Numerical Methods in Thermal Problems, edited by R. W. Lewis and K. Morgan, pp. 1120-1130. Pineridge Press. Swansea (U. K.) (1989).
[2] T. N. Narasimhan, S. P. Neuman and A. L. Edwards, 'Mixed explicit-implicit iterative finite element scheme for diffusion-type problems II: solution strategy and examples', Int. Journal for Num. Meth. in Eng., 11, 325-344 (1977).
[3] J. M. Winget and T. J. R. Hughes, 'Solution algorithms for nonlinear heat conduction analysis employing element-by-element iterative strategies', Comp. Meth. Appl. Mech. Eng., 52, 711-815 (1985).
[4] E. Riks, 'An incremental approach to the solution of snapping and buckling problems', Int. Journal Solids Struct., 15, 529-551 (1979).
[5] M. A. Crisfield, 'An arc-length method including line searches and accelerations', Int. Journal for Num. Meth. in Eng., 19, 1269-1289 (1983).
[6] J. Douglas and T. Dupont, 'Galerkin methods for parabolic equations', SIAM J. Numer. Anal., 7, 575-626 (1970).
[7] W. K. Liu, Y. F. Zhang and T. Belytschko, 'Implementation of mixed-time partition algorithms for nonlinear thermal analysis of structures', Comp. Meth. in Appl. Mech. and Eng., 48, 245-263 (1985).
[8] P. Pegon and P. Guelin, 'Implicit finite strain analysis in convected frames'. Proc. 5th International Symposium on Numerical Methods in Engineering, edited by R. Gruber, J. Periaux and R. P. Shaw, pp. 165-172. Computational Mechanics Publications and Springer Verlag. Lausanne (Switzerland) (1989).
[9] K. J. Bathe and A. P. Cimento, 'Some practical procedures for the solution of nonlinear finite element equations', Comp. Meth. in Appl. Mech. and Eng., 22, 59-85 (1980).
[10] J. E. Dennis and J. J. Moré, 'Quasi-Newton methods, motivation and theory', SIAM Review, 19, 46-89 (1977).
[11] H. Matthies and G. Strang, 'The solution of nonlinear finite element equations', Int. Journal for Num. Meth. in Eng., 14, 1613-1626 (1979).
[12] M. S. Engelman, G. Strang and K. J. Bathe, 'The application of quasi-Newton methods in fluid mechanics', Int. Journal for Num. Meth. in Eng., 17, 707-718 (1981).

Time Integration of Non-Linear Convection-Diffusion Problems

N-E. Wiberg, J. Frykestig

Department of Structural Mechanics, Chalmers University of Technology, S-412 96 Göteborg, Sweden

SUMMARY

The conductive–convective heat transfer equation is characterized by time dependency and non–linearity. The high complexity of the problem from elliptic to strongly hyperbolic puts a great claim on the numerical procedures. They must cover strongly non–linear elliptic problems concerning phase–change and radiation as well as strongly hyperbolic problems such as convective heat flow. The solution to the differential equation can be found by a very general penalty formulation which is a combination of an ordinary Petrov–Galerkin formulation and a Least–Square formulation. Different time–stepping procedures are discussed. Generalized mid–point methods and Rosenbrock type methods, both based on a finite difference approach are compared to space–time finite element procedures such as discontinuous Streamline Upwind Petrov Galerkin formulation. Numerical procedures for non–linearity and time dependency are dealt with. The non–linear problems are solved by Newton–Raphson–iterations. Studies of efficiency are made. A general mesh generator ADMESH is used for creation of general triangular meshes. The paper is concluded with some numerical experiments for 1D– and 2D–problems showing different aspects of the discussed matters.

INTRODUCTION

In the general case heat transfer is of a transient coupled conductive– and convective type. The unified analysis of the transient nonlinear heat transport and storage problems includes a number of nonlinear and time–dependent terms: nonlinear thermal conductivities, convective boundary conditions and sources, coupled convection and conduction, phase change (freezing), and internal heat generation. Because of the complicated time dependency an automatic time–stepping procedure is needed for a "safe" analysis. The physical properties are non–linear. Thus the numerical procedures which are established must range from strongly non–linear elliptic problems such as pure heat conduction to strongly hyperbolic problems, such as pure convective heat flow. The literature is huge.

The differential equation is solved by a Petrov–Galerkin (PG) weak formulation including upwinding described in Reference [1–3]. The most important matters are:

Different time–stepping procedures are discussed. Generalized mid–point methods and Rosenbrock type methods [4], both based on a finite difference approach, are compared to hierarchical space–time finite element procedures [5,6] which may lead to a discontinuous PG method. The nonlinear problems are solved by Newton– and Newton–Raphson–iteration [7] methods. For the solution of the final equation system a direct method is used. Some notes about adaptive procedures in space and time are given using the mesh generator ADMESH [8].

THEORY

An appropriate formulation of the convectice and conductive heat transfer problem is described by the differential equation

$$L\theta - f = 0 \qquad where \qquad L = c(\theta)\frac{\partial}{\partial t} + b(\theta), \tag{1}$$
$$b(\theta) = c_1(\theta)v^t\nabla - \nabla^t k(\theta)\nabla, \qquad f = f(\theta, t) = f_t(t) + f_\theta(\theta)$$

together with the initial condition $\theta_0 = \theta(0)$. In (1) L is the differential operator, ∇ the gradient operator, $\theta = \theta(x,t)$ is the temperature, $x = (x,y,z)$ is the spatial variable, t is the time, v is the fluid velocity, f is the forcing vector where $f_\theta(\theta)$ represent internal heat generation due to phase change and radiation from surface and boundary and finally k, c and c_1 are material parameters.

The differential equation (1) is either cast into a semidiscrete PG formulation in space combined with finite differences or Least–Squares (LS) formulation in time; or into a space–time PG formulation. The field variable θ with nodal values $\bar{\theta}$ is approximated as

$$\theta = \phi_x\bar{\theta}(t) \qquad or \qquad \theta = \bar{\phi}_x\theta(t) = \phi_x \sum_{i=1}^{N} \bar{\phi}_t^i\theta_i \tag{2a,b}$$

where ϕ_x is basis functions in space and ϕ_t the basis functions in time. Using the weighted residual PG weak approach with weighting functions $W = W_x$ or $W_j(x,t) = W_x W_{tj}$, $j = 1, \ldots, N$, where W_x, W_t are weighting functions in space and time respectively, we obtain the equations at a time step $[t_i, t_{i+1}]$ as

$$\int_x W(L\phi_x\bar{\theta}(t) - f)dx = 0 \qquad or \qquad \int_{t_i}^{t_{i+1}}\int_x W_j(L\phi\bar{\theta} - f)dxdt = 0 \tag{3a,b}$$

Different methods can be established. For a PG weak formulation we choose $W = W^G$ and for a LS formulation

$$W = W^L = L\phi = (c\frac{\partial}{\partial t} + v^t c_1\nabla - \nabla^t k\nabla)\phi \qquad convection - diffusion \tag{4}$$

We observe here that the weighting function W^L is dependent on the flow.

Based on the PG–formulation and the LS–formulation defined in (3) and (4) a fairly general formulation is the General Penalty (GP) method defined as

$$a\int\int W^G(L\phi\bar{\theta} - f)dxdt + b\int\int W^L(L\phi\bar{\theta} - f)dxdt = 0, \qquad aI^G + bI^L = 0 \tag{5a,b}$$

where a and b are parameters. Different possibilities of (6) are; $a = 1$ and b small, means a PG formulation plus a LS penalty term; and $b = 1$ and a small, means a LS formulation plus a PG penalty term.

It is interesting to look at some special cases. If $a = 1$ we have the Streamline Upwind PG formulation (SUPG). We can interpret this as using a weighting function

$$W = W^S = W^G + bW^L \approx \quad W^G + b(\frac{\partial}{\partial t} + v^t\nabla)\phi \quad \approx W^G + bv^t\nabla\phi \tag{6a,b,c}$$

where b is an upwinding parameter. Different simplifications of (6a) have appeared in literature, (6b) shows upwinding in space and time according to Johnson [3] and (5c) shows upwinding in space according to Zienkiewicz [1] and Hughes [2].

We also observe that a and b do not need to be constant and thus can be dependent on θ and $\nabla\theta$. This makes it possible to use the PG formulation and the LS–formulation, where they best describe the problem. For example for shock capturing Mizerkami and Hughes [9] have suggested a direction dependent weighting which may be interpreted as $b = b(\theta, \nabla\theta)$.

The basis functions ϕ , may be defined in a hierarchical way so that if ϕ_b denotes the basic basis functions and ϕ_h the hierarchical basis functions. In the SUPG a hierarchical discontinuous ϕ_t can be used. We may write $\theta = \phi_b\theta_b + \phi_h\theta_h$ where $\phi = \phi_x$ or ϕ_t.

FE–PROCEDURES

The time domain is discretized into a number of time–steps with the length $h = t_{i+1} - t_i$. The time discretization is either made by using some finite difference approximation of (3a) or by use of space–time finite elements according to (3b). Finally non–linear solution procedures are discussed.

Semi–discrete FE–procedures

The semi–discrete formulation (3a) gives the matrix equation (7) where C and B are matrices from spatial integration.

$$C(\theta)\frac{d\bar{\theta}}{dt} + B(\theta)\bar{\theta} - f(\theta, t) = 0 \qquad or \qquad C\frac{d\bar{\theta}}{dt} = \bar{H}(\theta, t), \qquad t > 0 \tag{7a,b}$$

A **generalized trapezoidal scheme** to solve (7a), stepwise centered at time $t = t_i + ah$ can be used. Different choices of a gives schemes with different accuracy and stability.

$$\bar{\theta}_{i+1} = \bar{\theta}_i + \bar{\theta}_i^{\Delta}, \qquad \left[C(\theta_a) + haB(\bar{\theta}_a)\right]\,\bar{\theta}_i^{\Delta} + hB(\theta_a)\theta_i - hf_a = 0 \tag{8}$$

Another possibility is to use a **Runge–Kutta method**, called the W(2,4)–method, [4].

$$\bar{\theta}_{i+1} = \bar{\theta}_i + \frac{1}{4}\bar{\theta}_{i1}^{\Delta} + \frac{3}{4}\bar{\theta}_{i2}^{\Delta} \tag{9a}$$

$$W(h, a, Q)\bar{\theta}_{i1}^{\Delta} = h\bar{H}_i = \bar{H}(\bar{\theta}_i, t_i), \qquad W(h, a, Q) = C + haQ, \tag{9b}$$

$$W(h,a,Q)\bar{\theta}_{i2}^{A} = h\bar{H}_i^{*} = h\bar{H}(\bar{\theta}_i + \frac{2}{3}\bar{\theta}_{i1}^{A}, t_i + \frac{2}{3}h) - \frac{4}{3}aQ\bar{\theta}_{i1}^{A}, \qquad a = 1 - 1/\sqrt{2} \tag{9c}$$

$$W(h,a,Q)\bar{\theta}_{i3}^{A} = h\bar{H}(\bar{\theta}_{i+1} + \frac{2}{3}\bar{\theta}_{i+1,1}^{A}, \; t_i + \frac{2}{3}h) + aQ(\frac{2}{3}\bar{\theta}_{i1}^{A} + 6\bar{\theta}_{i2}^{A}) \tag{9d}$$

The matrix Q is an arbitrary matrix but the choice influence the stability and accuracy. We have chosen an approximate value of the Jacobian (20b) with $\partial f/\partial\theta = 0$. A built in local error estimate of third order can be calculated with the extra stages $\bar{\theta}_{i+1,1}^{A}$ and $\bar{\theta}_{i3}^{A}$ satisfying equations (9b) and (9d).

$$\bar{T} = \frac{1}{8}(\bar{\theta}_{i1}^{A} - 5\bar{\theta}_{i2}^{A} + 5\bar{\theta}_{i+1,1}^{A} - \bar{\theta}_{i3}^{A}) \tag{10}$$

The determination of the time step length h_{new} is based on a given error tolerance τ and on the estimated local truncation error $\bar{T}$. The relative local error $\bar{E}$ is calculated from $\bar{T}$ by scaling with a reference level.

$$h_{new} = h_{old} \cdot \left(\tau / \| \bar{E} \|\right)^{1/p+1} \qquad with \qquad \| \bar{x} \| = \left(\frac{1}{N} \sum_{i=1}^{N} x_i^2 \right)^{1/2} \tag{11}$$

where p is the order of the method. The W method is of order 2 independently of the matrix Q. To avoid frequent updating of the matrix W and thereby to increase efficiency, the step size h is halved or doubled instead of using the formula above. The time selecting strategy is based on the following algorithm:

1) if $\| \bar{E} \| > \tau$, the step is rejected and the stepsize is halved.

2) if $R \cdot \tau \leq \| \bar{E} \| \leq \tau$, the step is accepted and $h_{new} = S \cdot 2h_{old}$. R is set to 0.1 – 0.3, and to avoid updating of W, S is given the value 1.0.

3) if $\| \bar{E} \| < R \cdot \tau$, the step is accepted and $h_{new} = 2h_{old}$ if at least M successive steps have occurred since the last change of h. In the current implementation M is set to 3.

The relative local error $\bar{E}$ with components E_i is calculated from $\bar{T}$ with components T_i . The reference level $\theta_{max,i}$ used for scaling, can be chosen as the maximum global solution component or by the numerically greatest solution value so far, encountered for each unknown value.

$$E_i = \begin{cases} T_i/\theta_{max,i} & \text{if } \theta_{max,i} \neq 0 \qquad i = 1, N \\ \tau & \text{if } \theta_{max,i} = 0 \end{cases} \tag{12}$$

A great advantage of the W(2,4) method is that only linear equations have to be solved, even for nonlinear problems.

Space–time finite elements. Hierarchical formulation

The differential equation (1) can also be approximated by space–time finite elements of hierarchical form [5,6]. Using the product form $\phi_x\phi_t$ (2b) it is advantageous to integrate (3b) by a numerical integration in space and a direct integration in time.

Adaptive time integration differing over the space can be obtained by a hierarchical formulation in time of the field variable (2b) for the interval $[t_1, t_2]$

$$\bar{\theta}(t) = \phi_{tb}\bar{\theta}_b + \phi_{th}\bar{\theta}_h = [\phi_1 \quad \phi_2]\bar{\theta}_b + \phi_{th}\bar{\theta}_h, \qquad \bar{\theta}_b = [\bar{\theta}_1, \bar{\theta}_2]^t \tag{13}$$

where $\phi_{tb} = [\phi_1 \quad \phi_2] = [1-\xi, \xi]$ contains the basic basis functions, and ϕ_{th} contains hierarchical functions. Examples of one hierarchical level of functions are $\phi_{h1} = 4(1-\xi)\xi$, $\phi_{h2} = \sin\phi\xi$ or $\phi_{h3} = 1-\xi$, $\xi > 0^+$. If the discontinuous linear function ϕ_{h3} is chosen, the method is identical to the discontinuous PG method in time.

The **Petrow Galerkin formulation** (3b) with weighting functions W_j and replacing θ_2 by the change $\bar{\theta}^{\Delta}$ as the independent variable the following equation is obtained

$$\bar{\theta}_2 = \bar{\theta}_1 + \bar{\theta}^{\Delta}, \qquad (C + a_j hB)\bar{\theta}^{\Delta} + (\beta_j C + \gamma_j hB)\bar{\theta}_h = -hB\bar{\theta}_1 + h\bar{f} \tag{14}$$

with coefficients a_j , β_j , γ_j dependent on the chosen W_j, and C, B and $\bar{f}$ are matrices obtained from the spatial integration, [5]. For the choice $\phi_h = \phi_{h3}$ and $W_2 = \xi$ and $W_3 = 1-\xi$ we obtain the equation

$$\begin{bmatrix} (C+\frac{2}{3}hB) & (-C+\frac{1}{3}hB) \\ (C+\frac{1}{2}hB) & (C+\frac{2}{3}hB) \end{bmatrix} \begin{bmatrix} \bar{\theta}^{\Delta} \\ \bar{\theta}_h \end{bmatrix} = \begin{bmatrix} -hB \\ -hB \end{bmatrix} \bar{\theta}_1 + h \begin{bmatrix} \bar{f} \\ \bar{f} \end{bmatrix} \tag{15}$$

The **least squares formulation** (3b), (4) is suited for purely convective flow, $k=0$. The time approximation (13) with $\phi_h = \phi_{h1}$ gives the equation (16), where C, R and S are obtained from spatial integration

$$\begin{bmatrix} (C+\frac{1}{2}hB+\frac{1}{3}h^2S) & (0+\frac{2}{3}hD+\frac{1}{3}h^2S) \\ (0+\frac{2}{3}hD^t+\frac{1}{3}h^2S) & (\frac{16}{3}C+0+\frac{8}{15}h^2S) \end{bmatrix} \begin{bmatrix} \bar{\theta}^{\Delta} \\ \bar{\theta}_h \end{bmatrix} = \begin{bmatrix} \bar{f}_b \\ \bar{f}_h \end{bmatrix}, \quad \begin{matrix} B = R^t + R \\ D = R^t - R \end{matrix} \tag{16}$$

The PG formulation, the LS formulation, and the GP or SUPG formulation using the weighting function as in (5) or (6), give equations of the same structure as (15) and (16).

The space time formulation using hierarchical basis functions in time makes it possible to apply adaptive procedures. In all formulations we can write the FE–equations as

$$A^i\bar{\theta} = f^i, \qquad i = PG, LS, GP$$

$$\begin{bmatrix} A^i_{bb} & A^i_{bh} \\ A^i_{bb} & A^i_{kh} \end{bmatrix} \begin{bmatrix} \bar{\theta}^{\Delta} \\ \bar{\theta}_h \end{bmatrix} = \begin{bmatrix} \bar{f}_b \\ \bar{f}_h \end{bmatrix}, \tag{17}$$

The adaptive procedure is:

- Solve $A^i_{bb}\bar{\theta}^{\Delta} = f^i_b$, triangular factors A^{i-1}_{bb}
- Estimate errors, select $\bar{\theta}_h$ variables

- Solve (17) as a combination of direct and iterative procedures using starting values $\bar{\theta}^{\Delta}$ and the triangular factors A_{bb}^{i-1} . If A is symmetric PCG–method can be used with the preconditioning matrix.

$$C = \begin{bmatrix} A_{bb} & 0 \\ 0 & diag(A_{hh}^{i}) \end{bmatrix}$$

Non–linear procedures

For nonlinear problems the incremental solutions (9), (14), (15), (16) and (5) for a time–step h can be written as

$$g(\bar{\theta}) = A(\bar{\theta})\bar{\theta} - f(\bar{\theta}) = 0, \qquad \bar{\theta} = \bar{\theta}_{i+1} \tag{18}$$

For the **Newton–Raphson** iteration the incremental solution of (18) is obtained by solving the Jacobian system in

$$\bar{\theta}^{k+1} = \bar{\theta}^{k} + \Delta\bar{\theta}^{k}, \qquad J^{k}\Delta\bar{\theta}^{k} = -g^{k} \tag{19}$$

$$J^{k} = \left(\frac{\partial g}{\partial \theta}\right)_{\theta=\bar{\theta}^{k}}, \qquad \frac{\partial g}{\partial \theta} = \frac{\partial A}{\partial \theta}\theta + A - \frac{\partial f}{\partial \theta}, \qquad g^{k} = g(\bar{\theta}^{k}) \tag{20}$$

where k is the iteration number. The new updated vector is $\bar{\theta}^{k+1}$, $\Delta\bar{\theta}^{k}$ is the increment of the vector $\bar{\theta}^{k}$, J^{k} is the Jacobian matrix, and g^{k} is the residual vector. An example of the true Jacobian and an approximation matrix for use in (9) with $\partial f/\partial\theta = 0$ becomes, if C is constant

$$J^{k} = \left[C + \frac{\partial C}{\partial \theta}\theta + ah\left(B + \frac{\partial B}{\partial \theta}\theta - \frac{\partial f}{\partial \theta} \right) \right]_{\theta=\bar{\theta}^{k}}, \tag{21}$$

$$\approx \left[C + \frac{\partial C}{\partial \theta}\theta + ah\left(B + \frac{\partial B}{\partial \theta}\theta \right) \right]_{\theta=\bar{\theta}^{k}} \tag{22}$$

The iterations are completed by the convergence criterion $\| g^{k} \| < \epsilon$. Using the modified Newton–Raphson technique, J is updated only once per time–step or per a few time–steps in order to reduce computational cost.

COMPUTER IMPLEMENTATION

The numerical procedures above are implemented by use of the SITU–FE–programming system to the specific program SITU–FLOW–H [10].

The subprogram ADMESH, [8], generates triangular meshes for general two–dimensional domains based on simple input data. The technique of advancing front grid generation is adopted in the program. We assume that a domain consists of areas which could be multi–connected. An area is defined by lines composing the boundary of the area as well as interval lines, along which triangles align. A line is defined by points through which the line passes. The triangulation is based on a given nodal density along boundary and internal lines. After the triangulation of the domain, a smoothing can be applied. One can require the coordinates of the nodes on certain internal lines to remain unchanged during smoothing. The recent development towards adaptive finite element methods makes the demands on mesh generators very high. ADMESH has also the possibility of adaptive remeshing and h–refinement.

NUMERICAL EXAMPLES

Time stepping procedures for 1D linear problems

The equation (1) with linear properties is solved for a 1D problem with the initial value $\theta = 0$ and the surface $x = 0$ is kept at constant value $\theta = \theta_0$ for $t > 0$ and x=L is kept at $\theta = 0$. To study the influence of the time–step length convection–conduction problems with Peclet number $P_e = v\Delta x/k$, $P_e = 2.5$, is calculated. A conventional generalized trapezoidal rule is compared with a hierarchical formulation, $\phi_h = \phi_{h3}$, according to (15) with $W_2 = \xi$, and $W_3 = 1 - \xi$ and the W–method. A linear element in space is used. The 1D domain of length L=1 is divided into equal elements of length Δx. Good behavior, with increased time–step length, of the hierarchical procedure, equivalent to discontinuous Galerkin in time, is demonstrated in Figure 1. The time–step and the total elapsed time are denoted Δt and T.

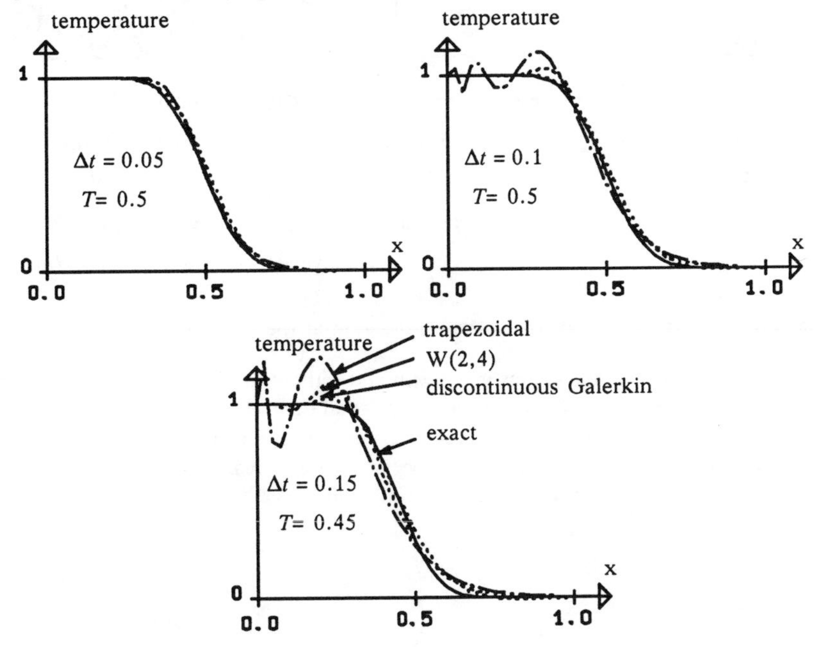

Figure 1. Convection–conduction, $P_e = 2.5$, different time–step lengths.

Heat Storage – Overall Behavior

The heat storage system shown in Figure 2 consists of a number of 10–m pipes emplaced in soil. The pipes are installed so that an equal energy input per unit volume can be obtained within a cylinder with a diameter of 12 m and a height of 10 m. By insulating the upper surface, energy losses can be diminished. Figure 2 shows a heat storage system that has an insulating layer, 0.5 m thick, of expanded clay with a radius of 20 m on the surface. In order to show the importance of the insulation on the temperature distribution, the behavior of the storage system was simulated for a period of four years. The volume source $f_c(t)$ during this period is given in Figure 3a. The variation of air–temperature $\theta_s(t)$ is given in Figure 3b. Figure 2 also shows the material data and the chosen mesh for the finite element analysis. At the upper surface an induced boundary was assumed.

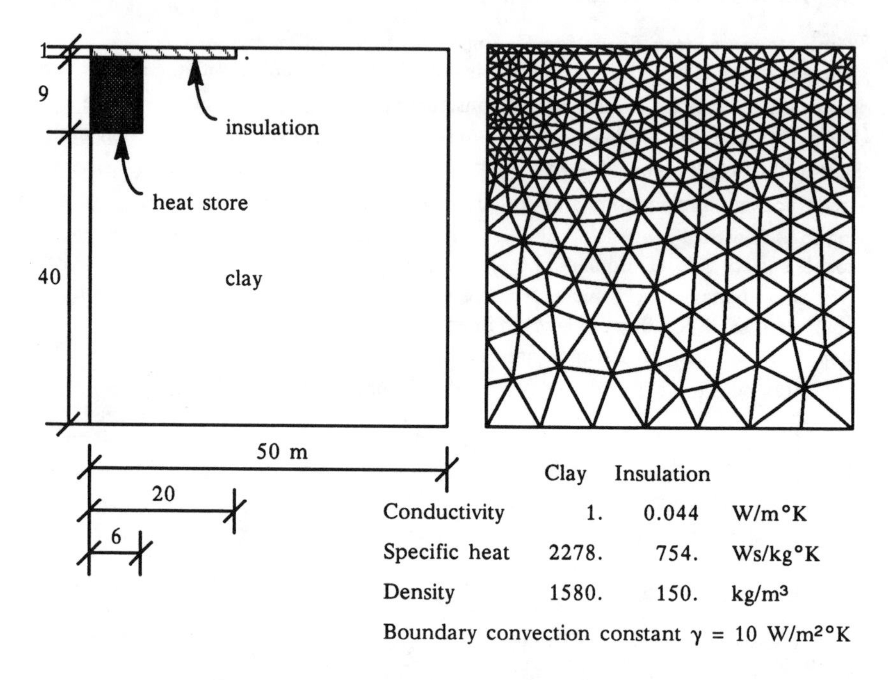

	Clay	Insulation	
Conductivity	1.	0.044	W/m°K
Specific heat	2278.	754.	Ws/kg°K
Density	1580.	150.	kg/m^3

Boundary convection constant γ = 10 $W/m^2{}^\circ K$

Figure 2. Energy storage system with insulation at the surface.

Comparison of simulations made with and without insulation clearly shows that the storage system is more effective with insulation than without it, see Figure 4. The seasonal effects were as expected, very local around the recharge area. The energy loss through the soil was a fairly slow process. Figure 4 clearly shows that the energy losses were much less in the case with insulation.

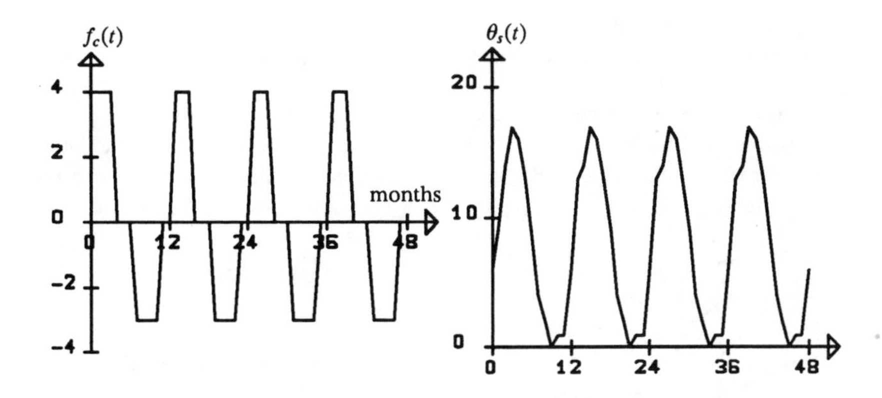

Figure 3a. Volume source (W/m^3). 3b. Outer temperature–variation $\theta_s(t)$

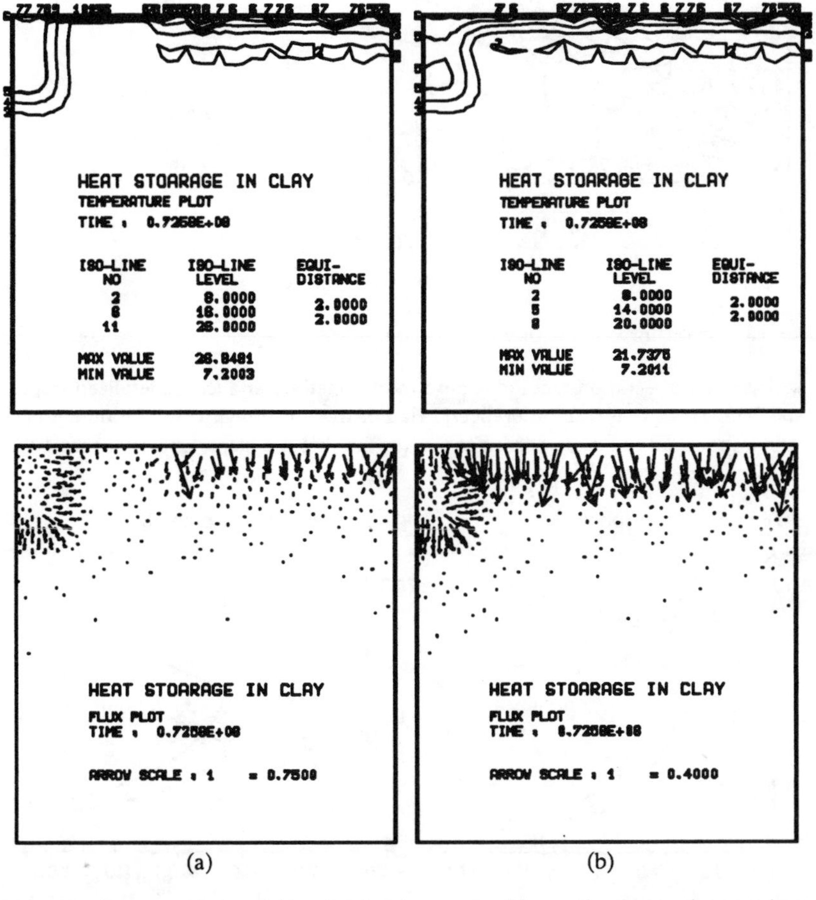

Figure 4. Temperature and flux distributions at t = 28 months during the transient heat storage process (a) with insulation, (b) without insulation.

Nonlinear thermal analysis

In order to study the time integration in a non–linear case we study a quarter of a circular plate with the radius 5.0 with insulated sides. The mesh of the problem is shown in Figure 5. A source at the center of the plate is a bilinear function of time which rises to a value 20.0 at t = 20.0 and declines to zero for $t \geq 40.0$. The initial temperature for all nodes is 9.0. For the linear case the thermal conductivity is chosen as $k = 0.04$ and for the nonlinear case is given by $k = 0.04(1.0 + 0.01\theta)$.

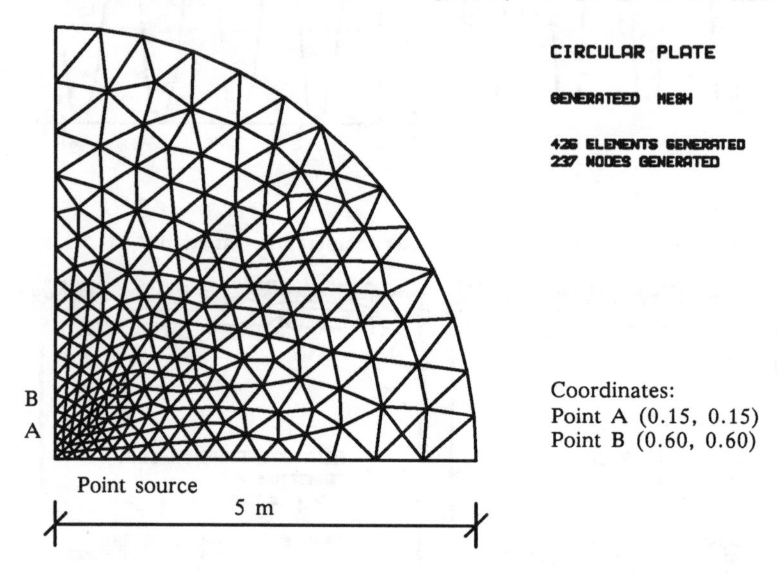

Figure 5. Mesh for quarter of circular plate.

The used integration method is here the W–method and the generalized trapezoidal. The example is from reference [11]. For the non–linear case, the time–steps for the trapezoidal rule were chosen to 0.5 for the first two steps, and after that 1.0.

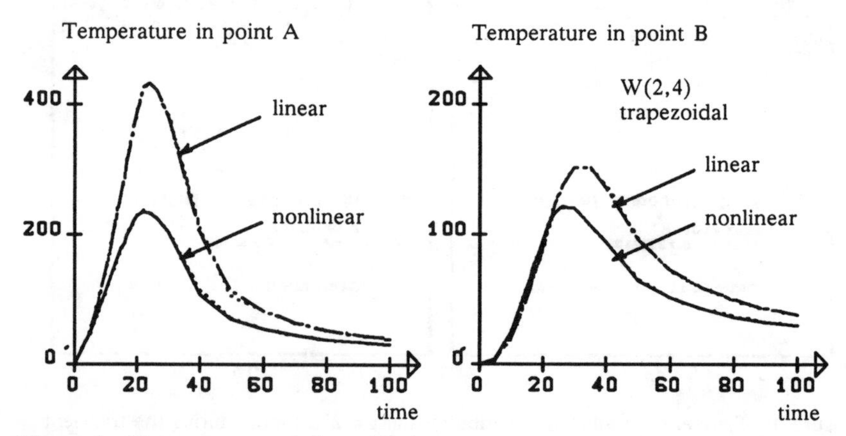

Figure 6. Temperature variation with time.

The automatic procedure in W(2,4) gave the time-step 0.1 in the beginning, it increased up to 0.8 at the time 6 and up to 25 in the end of the studied period. Similar time-steps were obtained for the linear case. The two methods are compared in the points A and B, see Figure 6, for both the linear and the non-linear case. For the non-linear calculations using the trapezoidal rule true Newton was used. In each time-step 2-3 iterations were used and if the tangent contributions in the Jacobian matrix are excluded the number of iterations increased to 3-4 for an ϵ=0.01. The W(2,4) method has a build in property to manage moderately nonlinear problems without non-linear iterations. This is demonstrated in Figure 6.

REFERENCES

1. HEINRICH, J.C., HUYAKORN, P.S., ZIENKIEWICZ, O.C. and MITCHELL, A.R. - An Upwind Finite Element Scheme for Two-Dimensional Convective Transport Equation, Int. J. of Num. Meth. in Eng., 11, , 131-143 (1977).

2. BROOKES, A.N. and HUGHES, T.J.R. - Streamline Upwind/PG Formulations for Convection Dominated Flows with Particular Emphasis on the Incompressible Navier-Stokes Equations, Comp. Meth. in Appl. Mech. and Eng., 32, 199-259 (1982).

3. JOHNSON, C., NÄVERT, U. and PITKÄRANTA, J. - Finite Element Methods for Linear Hyperbolic Problems, Comp. Meth. in Appl. Mech. and Eng., 45, 285-312 (1984).

4. RUNESSON, K., WIBERG, N.-E. and WOLFBRANDT, A. - 'A New Rosenbrock-Type Method Applied to a Finite Element Discretized Nonlinear Diffusion-Convection Equation', MAFELAP 78, Academic Press, London 1979, pp. 279-288.

5. WIBERG, N.-E. - Adaptive and Hierarchical Weighted Residual and LS Time Integration of Convection, Diffusion Problems, Comm. in Appl. Num. Meth., 4, 499-507 (1988).

6. WIBERG, N.-E. and FRYKESTIG, J. - 'Adaptive and Hierarchical Time Integration of Convection-Diffusion Problems', MAFELAP 87, Academic Press. London 1988, pp. 533-540.

7. HOGGE, M.A. - Secant Versus Tangent Methods in Non-Linear Heat Transfer Analysis, Int. J. of Num. Meth. in Eng., 16, 51-64 (1980).

8. JIN, H. and WIBERG, N.-E. - Two-dimensional mesh generation, Adaptive remeshing and refinement, Int. J. of Num. Meth. in Eng. In press (1990).

9. Mizerkami, A. and Hughes, T.J.R. - A Petrov-Galerkin Finite Element Method for Convetion-Dominated Flows - an Accurate Upwinding Technique for Satisfying the Maximum Principle., Computer Methods in Applied Mechanics and Engineering, Vol. 50, pp. 181-193, (1985).

10. Frykestig, H. - SITU-FLOW-H - A computer program for the finite element analysis of plane and axisymmetric heat transfer problems, Report 89:9, Department of Structural Mechanics, Chalmers University of Technology, Göteborg (1989).

11. SMOLINSKI, P. - Explicit multitime step integration for non-linear thermal analysis of structures, Comp. and Struct., 26, 439-444 (1987).

Thermal Template for the Prediction of the Thermal Resistance of a Pressed Contact

R.F. Babus'Haq, H.E. George, P.W. O'Callaghan, S.D. Probert

Department of Applied Energy, School of Mechanical Engineering, Cranfield Institute of Technology, Bedford MK43 0A1, England

ABSTRACT

The heat flowing across the interface between two solids, which are pressed together, encounters a relatively high thermal-resistance in the region of the interface. This arises because in many engineering situations, the true contact area is only a small fraction (< 1%) of the nominal contact area, and so the actual heat-flow path lengths are considerably increased. Also contaminants or oxides on the surfaces inhibit the heat flows. Despite the many pertinent theoretical and experimental investigations, there exists no fully comprehensive theory to predict, with reasonable accuracy, the contact resistance even between nominally-flat engineering uncontaminated surfaces. In the present investigation, the topographies of two dissimilar machined surfaces are measured and the true area of contact as well as its distribution over the nominal contact area is predicted. A mathematical model, using the adaptive and hierarchical finite-difference technique, is employed to predict the thermal resistance of the pressed contact under different mechanical loadings which will aid design engineers.

INTRODUCTION

Due to microminiaturization and the high-density concentration of electronic components, heat dissipation in semi-conductor devices has become increasingly important. In the near future, chip power dissipation is expected to increase from 20 W/cm^2 to about 100 W/cm^2. Nevertheless, the maximum junction temperatures are required to remain below 125 °C, because the active lifetime of the package deteriorates with increasing temperatures[1]. The useful lifetime of such devices is an exponential function of the junction temperature and decreases by a factor of approximately two for every 10 °C rise in temperature above the maximum recommended level which is normally $\ngtr$ 40 °C[2].

Thus the thermal performance of electronic equipment is often crucially affected by the thermal resistance between various components of the system, i.e. the higher the resistance between the electronic component and the ambient environment, the higher the temperature of the component. When pressed contacts are involved, considerable uncertainty in the thermal performance predictions for the assembly occurs[3]. In order to estimate values for thermal resistance of each pressed contact, several requirements have to be met: (i) an adequate specification of each of the surfaces in contact; (ii) an accurate prediction of the micro-contact area distribution resulting when the two surfaces are pressed together under isothermal room conditions; (iii) a knowledge of how the contacting members will distort thermoelastically when heat is passed across the joint; (iv) estimates of the macroscopic thermal resistances arising from the distortions; and (v) estimates of the microscopic thermal resistances arising at the actual contact bridges. Unfortunately, there is as yet no satisfactory theory which will predict thermal contact resistance for all types of engineering applications, nor have experimental studies yielded completely reliable empirical correlations. Thus the contact resistance problem still plagues engineers in the electronic industry[4]. Nevertheless, several different methods have been used to reduce the contact resistance, these include improving the surface finish, using a high-conductivity fluid at the interface, increasing the contact pressure, as well as inserting high conductivity foils or plating the surfaces with relatively soft materials, etc.[3-7].

It is the aim of the present investigation to: (i) identify the topographies of two dissimilar machined surfaces, one of duralumin and the other of stainless steel; (ii) predict the true area of contact at the pressed interface between the two surfaces; and (iii) employ a mathematical model, using a finite-difference technique, to describe the behaviour of these contacts under different mechanical loadings. The thermal resistance will then be predicted.

SURFACE TOPOGRAPHY

The textures of engineering surfaces dictate their cosmetic appearances, the thermal and electrical resistances of contacts between such surfaces when pressed together, as well as their tribological behaviours. Thus industry has equipped itself with stylus-based profilometers and recently optical reflectometers, electron microscopes, image analysers, laser-scanning and fibre-optic devices to obtain sufficiently-accurate surface-finish data. In 1976, a USA national study concerning the impact of surface roughness measurements pointed out that the metal-working industry by itself spent ~ £37 million per year making surface roughness measurements on products whose market value exceeded ~ £27 billion[8]. Many parameters have been used or proposed for characterising qualitatively the distribution of surface asperities, which may be obtained from a profile trace. However, a single profile will not describe comprehensively the topographical features of a surface. So topographical measurements for the whole considered area must be recorded to obtain an adequate description.

An accurate profile-tracing instrument was used to provide three-dimensional assessments of the two considered surfaces and the quantification of their deformations. It incorporates an automatically-controlled, three-dimensional relocating stage, as well as micro-computer-based data-handling and processing facilities[9]. The surface topographies of the specimens are presented as isometric and contour plots. The isometric view shown in Fig. 1 is made up from 128 parallel traces at 25.01 μm increments, each trace containing 128 samples taken at 38.65 μm increments along the machined surface. The various colours of the parallel traces correspond to the contour-height scale indicated. Thus, they aid in identifying the magnitudes of the surface characteristics and hence, show clearly any waviness present, which is usually greater the larger the sampling intervals employed. The common surface parameters were computed by averaging the profile data obtained from the parallel traces over the surface.

AREA OF CONTACT

When two surfaces (of different materials and roughnesses) are pressed together, true contact occurs only where the surface-asperity peaks intersect. Depending upon the materials' properties and the mean direction of heat flow, the macroscopic contact usually takes the form of either a disc or an annulus. Conduction of the heat is then constrained to occur via the cluster (in the macroscopic contact region) of microscopic asperity bridges formed where the micro-asperity tips meet and deform plastically[10]. An important geometric parameter, which controls the rate of heat transfer through the contacting spots, is the ratio of actual to apparent areas of contact. This ratio is determined by the non-dimensional contact pressure, which is defined as the ratio of the applied pressure to the contact microhardness. The relative contact pressure also influences the effective thickness of the layers of gas entrapped in the interface voids and thus directly affects the rate of conduction[5]. Between low emissivity surfaces at or near room temperature, the rate of radiation exchange is relatively small compared with the conduction heat tranasfer[10].

The total true-contact area may be estimated by considering the geometrical configuration which ensues when an equivalent rough surface, having a mean effective roughness equal to the root mean square of the roughnesses is brought into contact with an imaginary, perfectly-flat surface[11]. The distance between the parallel mean-planes of the two surfaces, which are pressed together, is known as the mean-plane separation. This parameter varies in magnitude from three times the mean roughness, if the surfaces are just touching under no-load condition, to zero if the mean planes coincide exactly[6]. The latter condition would be unlikely to occur, even under relatively high mechanical loadings for soft surfaces pressed together. The slope of each surface terrain varies continuously across the surface, according to the distribution of valleys and peaks. Most surfaces generated by normal machining procedures exhibit mean absolute surface slopes of between 2 and 10°. In the absence of more detailed information, a value of 6° would suffice for many estimations[10].

Because the microcontact conductance is proportional to the product of the number of microcontacts and their mean size, contacts involving surfaces having lower asperity flank slopes would in general present higher resistances to interfacial heat flows. This is because the heat would then be constricted to flow by conduction through fewer true contact spots, even though all the other physical and applied conditions are the same for both contacts.

Any misalignment of the surface under consideration, during its scanning, will introduce an apparent low order of form for the surface. However, such apparent states of waviness must be removed from the descriptions of both surfaces before considering them as an interfacial assembly. This can be done by subtracting an appropriate two-dimensional polynomial curve describing the apparent form of waviness of the face of each surface (see Fig. 2). After thereby extracting the underlying form from the three-dimensional images of the two surfaces (see Fig. 3), their behaviour when placed in contact under a specific loading can then be assessed. The magnitude of the macroscopic contact areas may be predicted from a knowledge of the contact patterns, and the pressure distributions enable the plastically deformed microcontact configurations to be quantified.

FINITE-DIFFERENCE TECHNIQUE

In this instance, the numerical analysis requires the replacement of the differential equation, which represents the continuous temperature-distribution behaviour in both space and time, with a finite-difference equation which can yield results only at discrete locations within the considered system. Finite-difference methods are usually adopted for thermal systems with discontinuities that can be described adequately with relatively large grid-sizes. The classical technique is to partition the medium into a regularly-spaced two-dimensional grid of nodes, each having a unit depth. It is assumed that all the heat transfers occur across rectangular faces in the plane of the node. The solution of the general finite-difference equation, for the heat transfers across the medium, can be obtained via direct methods (such as the conventional matrix-inversion method) or by iterative techniques (such as the Gauss-Seidel method or the successive over-relaxation method).

The adaptive and hierarchical steady-state finite-difference technique, employed in the current investigation, requires one to impose a mesh on the medium and then sub-divide it into several groups[12]. Average thermal conductivities, as well as the respective temperatures for each node, are then relaxed until a steady-state is attained. These average temperatures are used as priming values for the groups. Subsequently, each of the groups is relaxed individually until a steady-state condition is attained once again. Following this, the average temperatures are employed to up-date the group temperatures.

This process of calculating the temperatures across the group and fine mesh is repeated iteratively until a full steady-state solution is obtained. This achieves a hierarchical division of the thermal problem. Refinements can be achieved (e.g. computing time requirements reduced) by the correct choice of the processing hierarchy.

THERMAL ANALYSIS AND PREDICTIONS

By a suitable choice of mesh, the corresponding degree of accuracy can be achieved from finite-difference techniques. Nevertheless, the accuracy of the final solution will be limited by the quality of the input data. If a mathematical model describing the thermal behaviour has been produced during the development of an electronic system, it should be updated reflecting any design refinements that have occurred or the better estimates of the temperature/power dissipation that have subsequently become available.

A fast-action thermal analysis template has been developed in the present investigation. It is a facility for the computer-aided analyses of thermal systems. A database, which contains information for over 2000 base materials and composities, is included within this template. Once the schematic representations of the two contacting surfaces (namely duralumin and stainless steel in the current investigation) has been introduced, the software proceeds automatically to provide square or triangular meshing. This produces connectivities and thermal-conductance arrays from each element to adjacent elements in the "north, south, east and west" directions (see Fig. 4). Facilities to specify boundary heat- transfer coefficients are included.

The program then proceeds to obtain the adaptive and hierarchical finite-difference solution in order to predict the temperature arrays, and hence local heat fluxes. The final temperature-difference distributions, over the whole pressed contact, can be displayed as three-dimensional isometric projections.

The dimensionless interfacial thermal-conductance, C, of the pressed contact was predicted for several applied loadings from:

$$C = (\dot{Q}\ \sigma)/(\Delta T\ A_a\ m\ k_s) \tag{1}$$

where A_a is the nominal area of contact, (m^2); k_s is the harmonic mean conductivity, (W/mK); m is the effective absolute surface slope, (rad); $\dot{Q}$ is the power dissipation, (W); σ is the effective root-mean-square surface roughness, (m); and ΔT is the temperature difference between the contacting surfaces, (K)[13].

For the equivalent (i.e. theoretically simplified version of the) interface formed between the two pressed surfaces (namely duralumin and stainless steel), i.e. with one surface perfectly flat, the values for the other surface are have to be compounded: the RMS roughness is then 1.96 μm, the compound surface slope being 0.065 rad., and the harmonic mean conductivity 78.22 W/mK[13].

The resulting predictions were compared with those obtained from the correlation developed by Yovanovich[13,14] for rough surfaces (see Fig. 5):

$$C = 1.25\ (A_r/A_a)^{0.95} \tag{2}$$

where A_r is the real area of contact, m^2.

CONCLUSIONS

The topographies of two dissimilar surfaces were measured and the true area of contact (when they were pressed together) was predicted for different applied mechanical loadings. An easy-to-use thermal template was developed for the prediction of the thermal resistance of such contacts. The present investigation demonstrated the power of the adaptive and hierarchical finite-difference technique employed.

For the sizes of the surfaces chosen (a limitation imposed because of the micro-computer's capacity), the program predicted successfully the dimensionless interfacial thermal-conductance of the pressed contact. Moreover, the trend observed is in broad agreement with the previous correlated results. However, as expected, higher thermal conductaces were anticipated. This indicates that the program, when used to its maximum potential, would provide solutions of high level of realiability.

Experimental measurements for typical pressed contacts are needed, so that the quality of the predictions made either via the Yovanovich correlation[13,14] or by the present template mathematical model can be accurately assessed.

REFERENCES

1. Kadambi, V. and Abuaf, N., Numerical Thermal Analysis of Power Chip Packages, Heat Transfer in Electronic Equipment, (Eds. Oktay, S. and Moffat, R.J.), HTD-Vol. 48, pp. 77-84, Amer. Soc. Mech. Engrs., New York, 1985.

2. Peterson, G.P. and Fletcher, L.S., Thermal Contact Resistance of Silicon Chip Bonding Materials, Cooling Technology for Electronic Equipment, (Ed. Aung, W.), pp. 523-533, Hemisphere Publishing Corporation, New York, 1988.

3. Pinto, E.J. and Mikic, B.B., Novel Design Concept for Reduction of Thermal Contact Resistance, Cooling Technology for Electronic Equipment, (Ed. Aung, W.), pp. 547-559, Hemisphere Publishing Corporation, New York, 1988.

4. Chou, S.F. and Jou, S.K., The Effect of Metal Powder on Thermal Contact Resistance, Cooling Technology for Electronic Equipment, (Ed. Aung, W.), pp. 513-522, Hemisphere Publishing Corporation, New York, 1988.

5. Song, S. and Yovanovich, M.M., Relative Contact Pressure: Dependence on Surface Roughness and Vickers Microhardness, J. Thermophysics, Vol. 2, pp. 43-47, 1988.

6. O'Callaghan, P.W. Babus'Haq, R.F. and Probert, S.D., Surface Topography Assessment: Precursor to the Prediction of Pressed-Contact Behaviour, Trans. Inst. Meas. Control, Vol. 10, pp. 207-217, 1988.

7. Babus'Haq, R.F., O'Callaghan, P.W. and Probert, S.D., Reducing the Thermal Resistance of a Pressed Contact by Employing an Interfacial Filler, Engineering Science Preprints-26, (Ed. Koh, S.L.), Paper No. ESP26.89009, Soc. Engg. Sci., Baltimore, 1989.

8. Jones, T.S., Rosen, M., Gaynor, E.S. and Hsieh, T.M., Ultrasonic Measurement of Surface Roughness for Machined Surfaces, Advances in Manufacturing Systems Integration and Processes, (Ed. Dornfeld, D.A.), pp. 93-100, Soc. Manuf. Engrs., Michigan, 1989.

9. O'Callaghan, P.W., Babus'Haq, R.F., Probert, S.D. and Evans, G.N., Three-Dimensional Surface Topography Assessments Using a Stylus/Computer System, Int. J. Comp. Appl. Techn., Vol. 2, pp. 101-107, 1989.

10. O'Callaghan, P.W., Babus'Haq, R.F. and Probert, S.D., Predictions of Contact Parameters for Thermally-Distorted Pressed Joints, 24th. AIAA Thermophysics Conf., Buffalo, Paper No. 89-1659, 1989.

11. O'Callaghan, P.W. and Probert, S.D., Prediction and Measurement of True Areas of Contact Between Solids, Wear, Vol. 120, pp. 29-49, 1987.

12. George, H.E., Babus'Haq, R.F., O'Callaghan, P.W. and Probert, S.D., Adaptive and Hierarchical Steady-State Solutions of Non-Homogeneous 2-D Thermal Problems, Numerical Methods in Thermal Problems, (Eds. Lewis R.W. and Morgan, K.), Vol. 6, pp. 1600-1609, Pineridge Press Ltd., Swansea, 1989.

13. Yovanovich, M.M., Thermal Contact Correlations, Spacecraft Radiative Transfer and Temperature Control, (Ed. Horton, T.E.), Vol. 83, pp. 83-95, AIAA, New York, 1982.

14. Yovanovich, M.M. and Nho, K., Experimental Investigation of Heat Flow Rate and Direction on Contact Resistance of Ground/Lapped Stainless Steel Interfaces, 24th. AIAA Thermophysics Conf., Buffalo, Paper No. 89-1657, 1989.

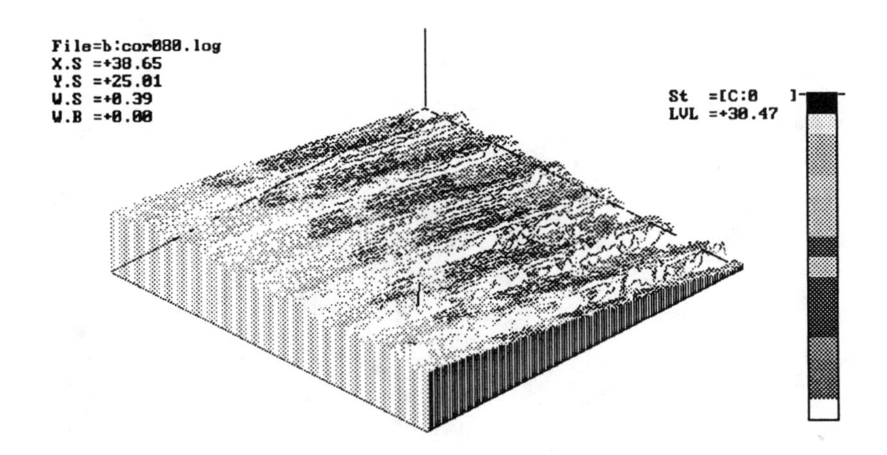

Fig. 1. Isometric plot of the stainless-steel specimen.

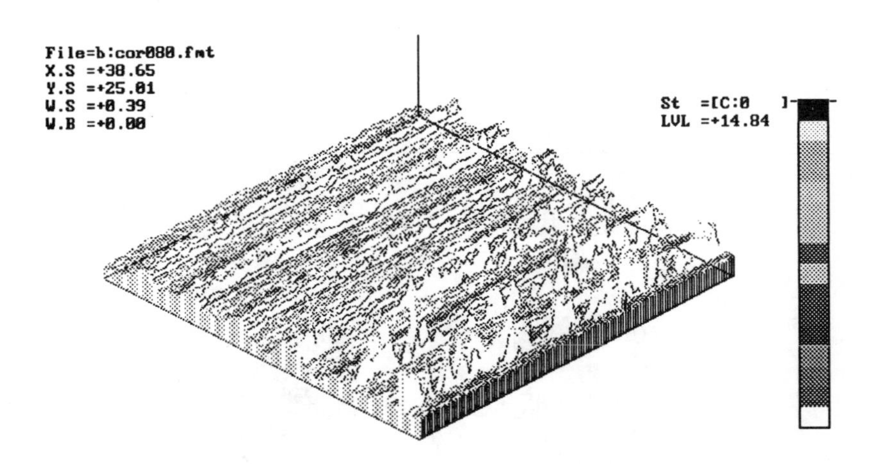

Fig. 2. The same surface as was used in Fig. 1. after extracting its underlying form.

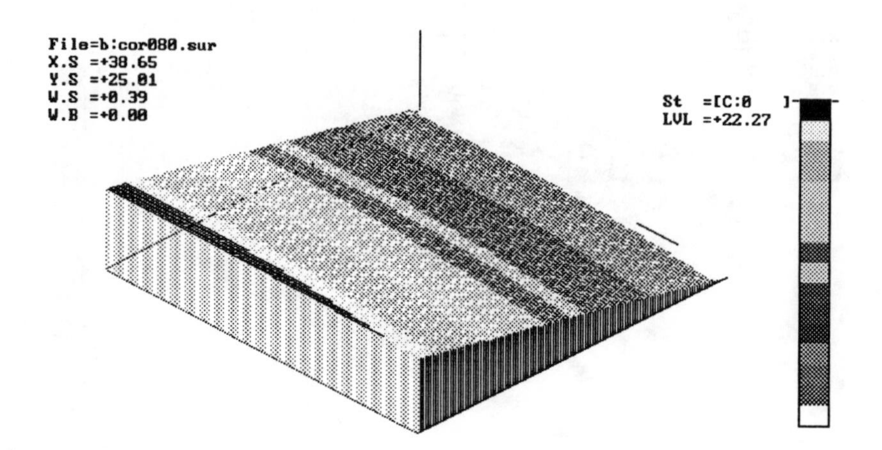

Fig. 3. The underlying form.

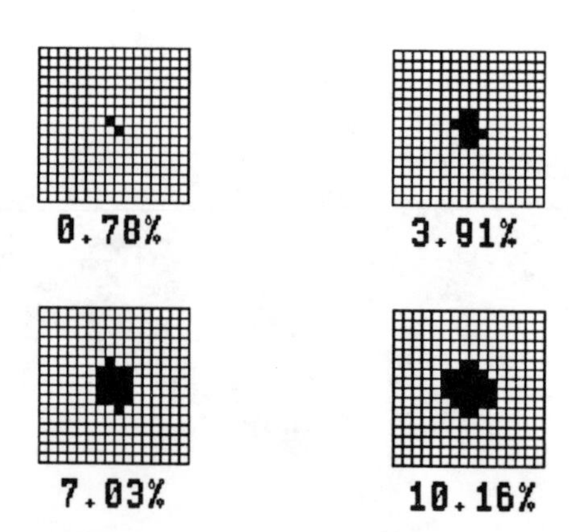

Fig. 4. Schematic representations of the true areas of contact.

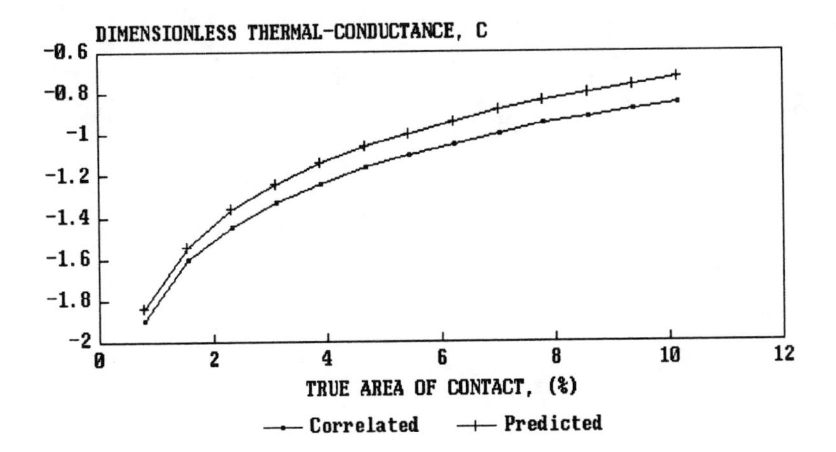

Fig. 5. Comparison between the correlated and the predicted conductances.

A New Method for Adaptive Grid Generation for Fluid Mechanics

G. Krieger, W. Braga

Mechanical Engineering Department, Pontifical Catholic University, 22453, Rio de Janeiro, RJ, Brazil

ABSTRACT

In this paper we present a new adaptive criteria to generate moving grids, specially fitted for fluid mechanics problems. Instead of using first and/or second order derivatives of the involved dependent variables as in Dwyer et al [3], we propose to generate adaptive grids based on the source term of the governing equation, which indicates the most important balance between convective and diffusion effects. Numerical results for some test problems taken from the literature are also presented, to allow comparison with other methods.

INTRODUCTION

One of the most severe limitations to a more wide spread use of computer simulation to study three-dimensional physical situations is certainly the need for an adequate specification of the grid points number and location to describe them accurately. Even considering only regular geometries, which do not have the additional requirement for an accurate description of a complicate boundary, the number of grid points may become so drastically large that the computer memory available becomes insufficient. On the other hand, if the grid is carefully selected to match the requirements at a specific physical situation, say at a certain temperature level, there is no clear indication that it will keep its good characteristics as the condition changes. In fact, the experience indicates that it will not.

The problem becomes more involved in situations such as those described by the Navier-Stokes equations. It is well established that the accuracy of the numerical solution is strongly dependent on several factors, the most important probably being the grid selected. Even assuming discretizing schemes with second order truncation errors and simple Dirichlet boundary conditions, the adequate description of localized effects, such as boundary layers and discontinuities (shocks), require the use of sufficiently fine meshes. For these strongly convective flows, the situation becomes even more severe due to the undesired appearance of numerical diffusion effects or the formation of wiggles resulting from the most common discretization of the inertia terms, the upwind and the central diference schemes, respectively. Up to date, the most effective solution for these problems is the use of fine grids. Besides the computer memory limit, the methods available to solve the large system of algebraic equations resulting from these fine grids have usually poor convergence rates, precluding any detailed investigation of the physical situation due to the involved costs.

Clearly, the moving grid proposition, capable of "following" the flow characteristics is very interesting. In fact, the user finds out, through a literature survey, that adaptive grids are strongly recommended to increase solution accuracy, which is achieved by the automatic move of the mesh points location, from smooth regions towards other regions found to be critical according to some physical criterion (e.g. see Thompson [1] or Eiseman [2] who, amongst others, made recently an extensive review of this subject). In some other situations, a number of additional points is introduced in these critical regions to obtain the same results.

The main objective of this paper is the discussion of the possible physical criteria to be used and the study of their most convenient implementation. Through a series of simple model problems with known exact solutions, some of the available criteria including the one proposed in the present paper and specially suitable for strongly convective flows are analysed and their results studied for error assessment. The proposed criterion is in fact a extension of the one proposed by Dwyer et al [3]. Initially, simple 1-D and 2-D problems described by a generalized Burgers' equation are studied. However, recognizing that the criterion for grid placement must rely on a dependent variable and that in most cases of interest more than one is available, all exhibiting steep gradients, it is also briefly investigated the use of several different criteria, one for each governing equation. For that, a more complicated case involving a system of two linear convection-diffusion equations, roughly similar to Navier Stokes problems, is analysed.

BACKGROUND

As mentioned, the basic idea for the development of an adaptive grid method is the reduction on the node intervals according to some property of the dependent variable. Dwyer et al [3] developed a method for moving the grid based on a transformation of coordinates. Through a mapping function to be introduced, they moved from a moving non-orthogonal grid in the physical space (x,y) to a fixed orthogonal grid in the computational space (ξ,η). As proposed, this mapping function depends on the first and/or second derivatives of one of the problem dependent variables. For example:

$$\xi (x,y) = \frac{\int_0^x W \, dx}{\int_0^{x_{max}} W \, dx} \tag{1}$$

where ξ is the general coordinate, W is the weighting function and x the arc length in the physical space. As a consequence of a more commonly used fixed increment adopted in the computational space, equation (1) implies that:

$$W_i \ \Delta x_i = \text{Constant} \tag{2}$$

Clearly, if the weighting function is held constant, an uniform spacing is achieved in the physical plane. However, to increase the accuracy of the solution to be obtained, W must be large in the regions requiring small spacing, possibly due to a large solution variation, resulting on the equidistribution process mentioned by Eiseman [2]. Dwyer et al [3] proposed that the weighting function should include the first and the second derivatives of the variable, resulting in:

$$W = 1 + b_1 \left| \frac{\partial \phi}{\partial x} \right| + b_2 \left| \frac{\partial^2 \phi}{\partial x^2} \right| \tag{3}$$

where b_1 and b_2 are the adjustable constants used for the optimization of the grid distribution, that is, to regularize the mapping transformation in those regions where the exact solution Φ is nearly zero. For the case $b_1 = 0 = b_2$, a uniform distribution of points is obtained, as mentioned. Naturally, this is not the only weighting function. For instance, Verwer et al [4] recently introduced a different form :

$$W = \alpha + \sqrt{|| \phi_{xx} ||} \tag{4}$$

where $|| \phi_{xx} ||$ may represent a norm, for instance, the weighted Euclidean norm (e.g. Dahlquist & Björck [5]), for the case of a system of equations, and clearly $\alpha = 1 / b$, the adjustable constant, is exactly similar to the procedure proposed by Dwyer et al [3]. As it may be observed, the difficulty of using b (or α) values relates to the eventually large range they may assume. Small b values should indicate smooth grid whereas larger values should increase the adaptive effect. However, the first and/or the second derivatives may eventually display very large variations in some regions, typically boundary layers, thus making it more difficult to select convenient b's value. This was investigated by Lee et al [6], who introduced user's selected parameters R_1 and R_2, limited in range, which are related to b_1 and b_2 through the expressions:

$$R_1 = \frac{b_1 \int_0^{x\ max} \left| \frac{\partial \Phi}{\partial x} \right| dx}{\int_0^{x\ max} W\ dx} \tag{5}$$

and

$$R_2 = \frac{b_2 \int_0^{x\ max} \left| \frac{\partial^2 \Phi}{\partial x^2} \right| dx}{\int_0^{x\ max} W\ dx} \tag{6}$$

With this, the constants b_1 and b_2 are normalized so that their products with the first and the second order derivatives are set in the same order of unity. In their work, Lee et al [6] used R_1 and R_2 = 0.2. In many situations, however, the resulting grid was found to have strong skewness that tend to deteriorate the results, therefore reducing the good advantages of moving grids. This led Jeng & Liou [7] to develop a smoothing strategy, based in the weight average of the neighboring gradients, with good results.

PROPOSED METHOD

Lee et al [6] made it clear that the relative importance of the terms in the monitor function should be maintained, that is, the combined value of R_1 and R_2 should not be large. However, during the initial studies done in [8], it was observed that better results were obtained provided only one of them was used. This is somewhat confirmed by Verwer et al [4] and Jeng & Liou [7], who, respectively, considered only the second order derivative and the first order derivative.

The objective of the present work is to simulate computational fluid mechanics situations described by Navier Stokes equations, which may be represented by the one-dimensional form of a generalized Burgers' equation :

$$Re \frac{d\ \Phi}{d\ x} = \frac{d^2\ \Phi}{d\ x^2} + So(x) \tag{7}$$

where Re is the Reynolds number, defined by $\bar{u} / \nu$, assuming an unit length and So(x) may represent the pressure gradient, for instance. As it is now fairly known, the dominant aspect of strongly convective flows is not the gradient nor the curvature of the involved dependent variable themselves, but the relationship between convective and diffusion effects, which is implicitly defined in the source term. Consequently, it is proposed the use of the source term as the weighting function. Furthermore, recognizing that in a system of equations involving several dependent variables, say u,v,and T, as typically found on CFD situations, not all governing equations display source terms, the proposition may involve the use of different physical criteria, each one being the most suitable for each governing equation and therefore generating different grids. This is currently under investigation [8].

In the next sections, some of these criteria will be applied to several 1-D and 2-D governing equations to study their efficiency. In these situations, equation (7) or its 2-D counterpart will be used.

DISCRETIZING SCHEMES AND MODEL PROBLEMS DESCRIPTIONS

Before attempting to investigate the most convenient form of the monitor function and its implementation, a brief description of the numerical scheme used for the discretization of the governing equation will be made. As mentioned in the introduction, most of the available schemes introduce spurious numerical effects. As a literature search indicates, these efffects are significantly reduced provided the cell Reynolds number, defined by $Re_c = \bar{u}\, h / \nu$, is a small number.

Recently, an adaptive class of schemes that depends directly on the cell Reynolds number, was introduced in [9]. One of the schemes, as was shown, is a simplified version of the exponential one proposed by Allen & Southwell [10] and Spalding [11]. Although this is not the most effective one proposed, it involves a simple tri-diagonal relation involving neighbouring nodes so, it was used. Consequently, in the present study, the first order derivative was discretized as:

$$\frac{d\Phi}{dx} = \left(\frac{1-\alpha}{2}\right)\Phi_{i+1} + \alpha\,\Phi_i - \left(\frac{1+\alpha}{2}\right)\Phi_{i-1} + \frac{\alpha\, h^3}{2}\,\frac{d^2\Phi}{d\,x^2} + \;.. \tag{8}$$

where $\alpha = 1 - 2 / Re_c$ if $|Re_c| > 2$ and $\alpha = 0$, if $|Re_c| < 2$. It should be noted that for large values of α, the upwind scheme is obtained and $\alpha = 0$ reproduces the central difference scheme. In any case, the values for α are selected in order to assure that the matrix of coefficients is dominant diagonally at all times, what indicates a good convergence rate for the iterative procedure. The discretization of the diffusion terms is made through the standard central scheme.

As may be observed, the adaptive effect comes in play as small values for α have smaller truncation errors and. as it is easily seen. this depends on the local grid. In other words, through the use of moving grids, it is proposed not only to increasy the accuracy through better resolutions of localized effects such as gradients but also through a better description of the inertia terms of the governing equations. As it will be indicated, this adaptive scheme coupled to a most efficient monitor function seems to be most suitable for the present purposes.

To examine the efficacy of the proposed monitor function in handling strongly convective effects, several test problems were studied, all based on the 1-D or 2-D generalized Burgers' equations. These problems were selected to indicate clearly the desired effects. which were achieved through the variation of the Reynolds number, Re. Unless stated, the results are presented in terms of the percentage average error in the domain, ε_{av}, defined as :

$$\varepsilon_{av} = \frac{100}{\text{no of unknowns}} \sum \left| 1 - \frac{\Phi_{num}}{\Phi_{exa}} \right| \tag{9}$$

where Φ_{num} and Φ_{exa} indicate the numerical solution and the exact profile. All cases studied involve Dirichlet boundary conditions, with the selected number of points being indicated in the corresponding tables or figures. The source term is obtained through direct substitution of the exact profile in equation (7) and the boundary conditions are readily obtained.

In all the present tests, the weighting function was given by an equation similar to equation (3) above, reproduced here for completeness:

$$W = 1 + b_1 \left| \frac{\partial \phi}{\partial x} \right| + b_2 \left| \frac{\partial^2 \phi}{\partial x^2} \right| + b_3 \left| So \right| \tag{10}$$

where b_1 and b_2 are given by equations (5) and (6), respectively, and b_3 is given by a similar equation, however involving the source term instead. Furthermore, the results were always obtained using only one type of monitor function at a time. No combination of these effects proved to be equally important nor clearly indicated results possible to be generalized. Consequently, this option is not recommended.

The moving grid is estimated through an algebraic system of equations based on equation (2), which is solved after some iterations of the governing equations, corresponding to the static method mentioned by Verwer et al [4]. This was found to be the most convenient in the present study.

Problem 1
The first test made involved a simple 1-D problem with a known exact solution given by $\Phi_{exa} = 1 + x + x^2 + \exp[-Re(1-x)] + \sin(\pi x)$. This problem was used by Prakash [12], for instance. The boundary conditions are $\Phi(x=0) = 1 + \exp(-Re)$ and $\Phi(x=1) = 4$.

Problem 2
This problem combines an exact profile given by $\Phi_{exa} = (B x - A)\exp[-C x^2] + A$ with the standard Burgers profile given by $[\exp(Re\, x - 1)] / [\exp(Re) - 1]$. The objective of this test is to obtain a steep boundary layer at the downstream wall. The first profile indicates a function basically similar to the one used by Leonard [13], for example.

Problem 3
In this problem, we investigate the behavior of the proposed monitor function for a 2-D situation. The test problem is given by

$$Re\left(u\frac{\partial \Phi}{\partial x} + v\frac{\partial \Phi}{\partial y}\right) = \frac{\partial^2 \Phi}{\partial x^2} + \frac{\partial^2 \Phi}{\partial y^2} + So(x,y) \tag{11}$$

where $\Phi_{exa} = u = 2x^2 y$ and $v = -2 x y^2$. This test was introduced by Shih [14] to help debug computer programs.

Problem 4
This last test was conducted to investigate the behavior of the proposed monitor function to help solving the solution of a system of 2-D equations, loosely related to the Navier-Stokes equations. The test problem is given by

$$Re\left(Vel_1\frac{\partial \Phi_1}{\partial x} + Vel_2\frac{\partial \Phi_1}{\partial y}\right) = \frac{\partial^2 \Phi_1}{\partial x^2} + \frac{\partial^2 \Phi_1}{\partial y^2} + So(x,y) \tag{12}$$

$$Re\left(Vel_3\frac{\partial \Phi_2}{\partial x} + Vel_4\frac{\partial \Phi_2}{\partial y}\right) = \frac{\partial^2 \Phi_2}{\partial x^2} + \frac{\partial^2 \Phi_2}{\partial y^2} + S_1(x,y) \tag{13}$$

where $\Phi_{1exa} = 2x^2 y$ and $\Phi_{2exa} = -2 x y^2$. Vel_1, Vel_2, Vel_3

and Vel_4 are selected accordingly to the test to be studied. This test is an extension of problem 3.

RESULTS: COMPARISON WITH OTHER MONITOR FUNCTIONS

Figure 1 indicates, at Re = 500, the dependence of the average error on the user-selected R factor, which will weight the grid, for test 1. At this Re number, the balance between convective (or inertia) and diffusion mechanisms is important and this is clearly indicated in the results. Although no reduction in the error is observed for the moving criterion based on both the first or the second order derivative, the criterion based on the combined mechanism is effective for this purpose. The investigation stoped at a R factor equal to 1 for several reasons. It was observed that increasing the R factor generally results in a much higher number of intermediate grids, resulting in poor convergence rates. Also, it was observed that in some cases, large values for R induce oscillation in the iterative procedure, sometimes resulting in a non-convergent solution. Some ways of overcoming these limitations are currently being investigated and they will be reported in [8]. Results for even higher Reynolds number were basically similar.

It is instructive to understand the basic reasons that led the better monitoring characteristics for the source term. Figure 2 indicates the normalized values for the function, first and second derivatives and the source term based on problem 1. It should be observed that at Re = 1000, the boundary layer characteristic of this problem is clearly displayed. As it is known, the boundary layer implies in the existence of very large second derivatives close to the downstream walls.

According to equation (2), if the monitor function is based only on the second derivative, it will indicate a very small grid in the boundary layer region, which is most convenient. However, outside this region, the second derivative profile drops sharply, indicating now a large grid. Unfortunately, this is bad and seems to deteriorate drastically the average errors. As may be observed, the same occurs to the first derivative. On the other hand, the adaptive grid obtained based on the source term works fine.

It may be argued if the type of function used in problem 1 is indicative of the type of profile to be expected in fluid mechanics situations. This certainly is a guess but it should be pointed out that it was used in [9], in [12] and in [13] to test discretization schemes for strongly convective flows. Futhermore, the same conclusions are obtained with the other cases tested, as indicated in figure 3, for instance, for problem 2 at Re = 1000, and in the following tables.

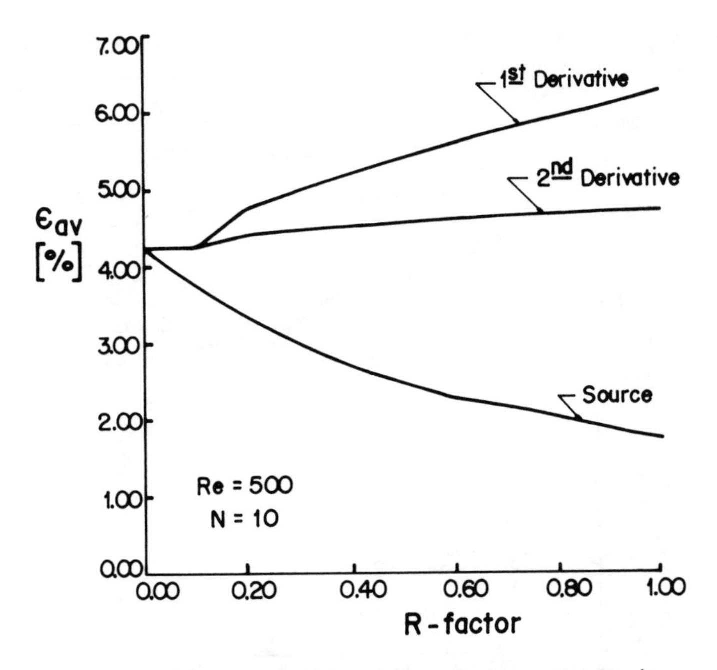

Fig. 1 - Average Error Dependence on the Monitor Function and R-factor. Test 1

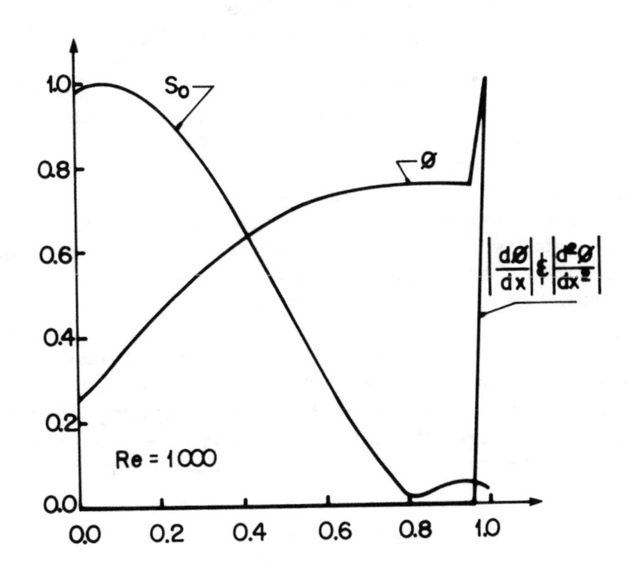

Fig.2 - Normalized Exact Monitor Functions Test 1

Before proceeding to more complex situations, the advantage of using the adaptive scheme defined by equation (8), if

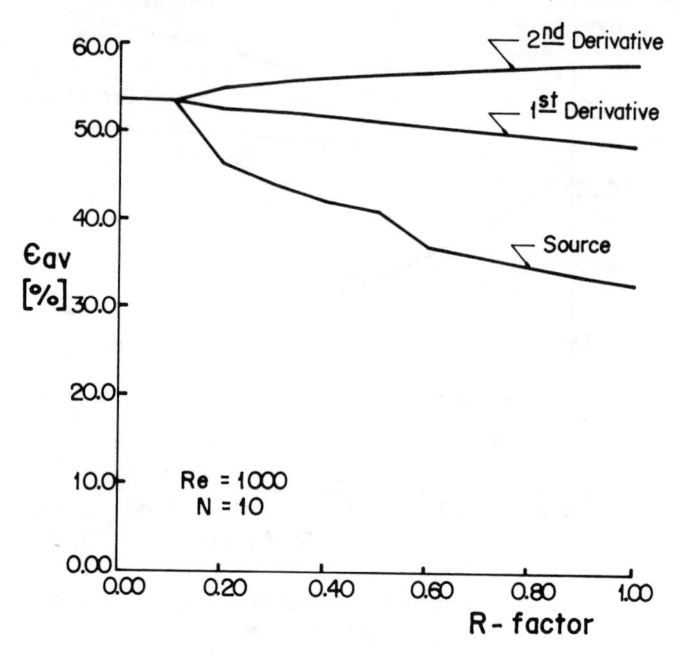

Fig. 3 - Average Error Dependence on the Monitor Function and R - factor. Test 2

compared to the upwind scheme, will be analysed briefly. The analysis made in problem 1 was repeated with this last scheme and the comparative results appear on table I. The test was conducted at Re = 100, using 10 points and the chosen monitor function was the source term. As may be observed, the adaptive scheme reduces even more the average errors and, as it is easily implemented on existing codes, its use is recommended provided only three point adaptive schemes are to be used (see, in any case, the discussion made in [9]).

Table II below indicates some results for the test problem 3 at Re = 1000, using several grid points. Also indicated in this table are the cossine of the angle between grid lines. As indicated by Maliska et al [15], this angle indicates the nonorthogonality of the grid and it is defined as:

$$\text{Cos } \theta = \frac{\beta}{\alpha \quad \gamma}$$

where α, β and γ are the components of the metric tensor. Clearly, if cos θ = 0, the grid lines are orthogonal and if cos

$\theta = \pm 1$, the Jacobian of the transformation reduces to zero. As it is observed, with the increase in the number of grid points and the also the user specified R-factor, the nonorthogonality increases significantly, what is not desired. It may be observed that the results obtained for the proposed method are better than for the other, specially at the smaller grids. Also, increasing the r-factor results in better results, at least up to the point where oscillations appear, as previously mentioned.

Table I : Influence of Discretizing Scheme on the Average Errors (%). Problem 1. Re = 100 N = 10 points. Monitor : Source Term

r_3 - factor	Adaptive Scheme	Upwind Scheme
0.0 (uniform grid)	3.53	4.05
0.2	2.79	3.17
0.4	2.26	2.53
0.6	1.92	2.08
0.8	1.68	1.87
1.0	1.46	1.61

Table II : Averave Errors (%) for Adaptive Grids Problem 3. Re = 1000, 11x11 grid points

Monitor Function	R-factor	Error	no of grids	Cos θ
Uniform	0	10.2	1	0
First Deriv	0.2	13.1	4	0.74
First Deriv	0.4	14.4	5	0.86
Source	0.2	6.2	7	0.74
Source	0.4	5.9	18	0.75

Finally, some of the results obtained for test problem 4 will be shown, always using R-factor = 0.2 and 11x11 points. The results for higher R-factors were neglected as oscillations ocurred, disturbing the convergent process. Three cases will be reported here: a linear case defined with $Vel_1 = \Phi_{1exa}$, $Vel_2 = \Phi_2$, $Vel_3 = \Phi_1$ and $Vel_4 = \Phi_{2exa}$ and two non linear cases, with $Vel_1 = \Phi_1 = Vel_3$ and $Vel_2 = \Phi_2 = Vel_4$. One of these non linear situations was defined using a known source term and the other was chosen in such a way that the source terms, So and S_1, depend on the both the unknowns, Φ_1 and Φ_2. The results are indicated on table III, below:

Table III : Average Errors (%) for Adaptive Grids Problem 4. Re = 1000, 11x11 grid points

Monitor Function	Errors Φ_1	Φ_2	no of grids	Cos θ
Linear Case				
Uniform	12.4	4.9	1	0
First Deriv	10.5	6.7	4	0.74
Source	5.4	8.6	7	0.74
Non Linear Case with known source terms				
Uniform	24.1	140.3	1	0
First Deriv	31.1	112.1	4	0.74
Source	18.1	65.5	7	0.74
Non Linear Case with unknown source terms				
Uniform	11.3	36.2	1	0
First Deriv	9.1	34.9	4	0.74
Source	14.3	6.9	8	0.78

CONCLUSIONS

In this paper, we investigate a new formulation suitable to fluid mechanics problems to generate adaptive grid. Through the analysis of the balance between convective and diffusive effects, displayed on the source terms, the formulation seems to indicate better results, at least if compared with others in several examples taken from the literature. At this stage, a more complex fluid mechanics problem is being analysed, to proceed with the investigation.

References

1. Thompson J.F., A Survey of Dynamically Adaptive Grids in the Numerical Solution of Partial Differential Equations , Appl. Num. Math., 1, pp. 3-28, 1985.

2. Eiseman P.R., Adaptive Grid Generation, Comp. Methods Appl. Mech. Eng., Vol. 64, pp 321-376, 1982.

3. H.A. Dwyer, R.J. Kee & B.R. Sanders, Adaptive Grid Method for Problems in Fluid Mechanics and Heat Transfer, AIAA Journal, Vol. 18, no. 10, pp 1205-1212, 1980.

4. J.G. Verwer, J.G. Blom & J.M. Sanz-Serna, An Adaptive Moving Grid Method for One-Dimensional Systems of Partial Differential Equations, J. of Comp. Physics, vol 82, pp 454-486, 1989.

5. G. Dahlquist & A. Björck, Numerical Methods, Prentice Hall, New Jersey, 1974.

6. D. Lee, W. Shyy & Y.M. Tswei, Local and Global Methods in Adpative Grid for Highly Convective Flow Computation, Proceedings, Computational Techniques and Apllications: CTAC-87, edited by J. Noye & C. Fletcher, pp. 401-410, Sydney, Australia, North-Holland, 1987.

7. Y.N. Jeng & S.C. Liou, Modified Multiple One-Dimenional Adaptive Grid Method, Numerical Heat Transfer, Part B, Vol. 15, pp. 241-247, 1989.

8. G. Krieger, Filho, M. Sc. thesis, Mechanical Engineering Department, PUC/Rio, Rio de Janeiro, Brazil, in portuguese, 1990.

9. W. Braga, On the use of some weighted upstream schemes for strongly convective flows, to appear on Numerical Heat Transfer, Part B, 1990.

10. D.N. de G. Allen and R.V. Southwell, Relaxation Methods Applied to Determine the Motion in Two Dimensions of Viscous Fluid Past a Fixed Cylinder, Q.J. Mech Appl. Math., Vol. 8, pp. 129-145, 1955.

11. D. B. Spalding, A Novel Finite-Difference Formulation for Differential Expressions Involving Both First and Second Derivatives, Intern. J. for Num. Meth. in Eng. Vol. 4, pp. 551-559, 1972.

12. C. Prakash, Application of the Locally Analytic Differencing Scheme to some Test Problems for the Convection-Diffusion equation, Numerical Heat Transfer, vol. 7, pp. 165-182, 1984.

13. B.P. Leonard, A Survey of the Finite Differences of Opinion on Numerical Muddling of the Incomprehensible Defective Confusion Equation, in T.J.R. Hughes (ed.), Finite Element Methods for Convection Dominated Flows, p. 1, ASME, New York, 1979.

14. T.M. Shih, A Procedure to Debug Computer Programs, Intern. J. for Num Meths in Eng., Vol 21, pp 1027-1037, 1985.

15. C.R. Maliska & A.F.C. Silva, Local Effects of Highly Nonorthogonal Grids in the Solution of Heat Transfer Problems in Cusped Corners, Proceedings of the First International Conference on Numerical Grid Generation in Computational Fluid Dynamics, Landshut, West Germany, 1986.

Formalisation of Field Problem Specification for Two-Dimensional Finite Element Analysis

S. Das, S.K. Saha

Mechanical and Engineering Department, Jadavpur University, Calcutta, India 700 032

INTRODUCTION

Finite Element Analysis has achieved quite rigour, and formalism in this respect is expected. With a view to automate the analysis procedure, the extent of formalism that can be achieved needs investigation. This paper attempts in that direction taking steady state heat conduction problems as example, confining itself to two dimensional Finite Element Analysis.

A REVIEW OF DIFFERENTIAL EQUATION IN 2-D HEAT CONDUCTION

The most general heat conduction equation may be given as the energy principle with fluid motion(3)

$$\rho \frac{D}{Dt}(e + \tfrac{1}{2}v^2) = -\nabla \cdot \bar{q} + \nabla \cdot (\overleftrightarrow{T} \cdot \bar{v}) + \rho\bar{g} \cdot \bar{v} + \Phi$$

which after substitution of the heat conduction equation

$$\bar{q} = -\overleftrightarrow{k} \cdot \nabla\theta$$

becomes

$$\rho \frac{D}{Dt}(e + \frac{v^2}{2}) = \nabla \cdot (\overleftrightarrow{k} \cdot \nabla\theta) + \nabla \cdot (\overleftrightarrow{T} \cdot \bar{v}) + \rho\bar{g} \cdot \bar{v} + \Phi$$

Different cases will arise with

a) Steady state condition $\frac{D}{Dt}$ term = 0,

b) No fluid motion $\bar{v} = \bar{0}$ & $v^2 = 0$,

c) Noenergy generation $\Phi = 0$, etc.

Expansion of ∇ will depend on the orthonormal basis (Cartesian, Cylindrical, or Spherical) chosen.

Three types of boundary conditions may arise on some

boundary S of the domain in steady state heat conduction problems(2)

i) Temperature specified $\Theta = \Theta_S$ on S
ii) Heat flux specified $-\hat{n} \cdot \overleftrightarrow{k} \cdot \nabla\Theta = q_S$ on S
iii) Convection condition $-\hat{n} \cdot \overleftrightarrow{k} \cdot \nabla\Theta = h(\Theta_S - \Theta_\infty)$ on S.

EXISTENCE OF VARIATIONAL BASIS OF A SYSTEM OF PARTIAL DIFFERENTIAL EQUATIONS(PDE's)

Following Finlayson's(1) treatment of Frechet derivatives, firstly an mth order PDE is taken with a single dependent variable u in N-dimensional space as shown below

$$N(u) = f(x_1;\ x_2;\ ..\ ;\ x_N;\ u;\ u_{,1};\ ..\ ;\ u_{,NN}\ ..\ m\ \text{times}) = 0.$$

Let us define expanded form of a partial differential expression(PDEX) (like N(u)) to be one where there is no differential operator operating on a product or sum of two or more operators or functions.

The enumeration scheme for sorting derivatives of u in order is shown with an example considering N = 2 and m=3 (i.e. a 2-dimensional space & the PDE is of 3rd order)

$$u;\ u_{,1};\ u_{,2};\ u_{,11};\ u_{,12};\ u_{,22};\ u_{,111};\ u_{,112};\ u_{,122};\ u_{,222}$$

where the partial derivative

$$u_{,i_1 i_2 \,..\, i_p} = \frac{\partial}{\partial x_{i_1}}\left(\frac{\partial}{\partial x_{i_2}}\ ..\ \left(\frac{\partial u}{\partial x_{i_p}}\right)\ ..\ \right).$$

It may be noted that the derivatives of u are always in increasing order. This is done because we have assumed that the nature of u is such that

$$u_{,i_1 i_2} = u_{,i_2 i_1}$$

and so on. The following algorithm may now be presented.

<u>Algorithm 1.1: To find the adjoint of Frechet derivative $N_u'(\phi)$ of N(u)</u>

1. Convert the mth order PDE N(u) in expanded form and let it be $f(x_1;\ ..\ x_N;\ u;\ u_{,1};\ ..\ )$.
2. Define the matrix A_{pq} with p,q = 1 to M(m) where

$$A_{pq} = 0 \quad \text{if } p > q$$
$$= (-1)^{G(q)} * B(p,q) \quad \text{if } q \geq p.$$

In this

$G(x) = i$ such that $M(i) \geqslant x > M(i-1)$, x varying from 1 to $M(m)$

$$M(i) = \sum_{j=-1}^{i-1} S(j) \quad \text{if } i \geqslant 0$$
$$= 0 \quad \text{if } i < 0, \text{ i varying from } -1 \text{ to } m.$$

$$S(i) = \sum_{r_i=1}^{N} \sum_{r_{i-1}=1}^{r_i} \cdots \sum_{r_2=1}^{r_3} \sum_{r_1=1}^{r_2} r_1 \quad \text{if } i > 0$$
$$= N \quad \text{if } i = 0$$
$$= 1 \quad \text{if } i = -1, \text{ i varying from } -1 \text{ to } m-1.$$

3. The adjoint of the Frechet derivative has now the form

$$\tilde{N}'_u(\emptyset) = \sum_{p=1}^{M(m)} \emptyset,_{C(p)} \sum_{q=p}^{M(m)} A_{pq} * \left(\frac{\partial f}{\partial u,_{C(q)}} \right),_{D(p,q)}.$$

<u>Algorithm 1.2: To find the coefficient B(p,q) in the algorithm 1.1</u>

1. Get the two sequences of digits (each digit may vary from 1 to N) in ascending order C_1 and C_2 as

$$C_1 = C(p) \quad \& \quad C_2 = C(q).$$

2. If all the digits in C_1 are not contained in C_2, $B(p,q)=0$ and return.

3. If C_1 and C_2 both contain only zero's then $B(p,q)=1$ and return.

4. Find $D(p,q)$ using algorithm 1.4.

5. Make an array $U(r)$, r=1 to N, such that
$U(r)$ = number of repetitions of digit r in sequence C_1.
Make an array $V(r)$, r=1 to N, such that
$V(r)$ = number of repetitions of digit r in sequence $D(p,q)$.

$$B(p,q) = \prod_{r=1}^{N} {}^{U(r)+V(r)}C_{V(r)}.$$

<u>Algorithm 1.3: To find the sequence C(i)</u>
The algorithm is presented through an example because the implementation is through an ascending enumeration procedure.

If N = 3 then C(1)=0, C(2)= 1, C(3)=2, C(4)=3, C(5)=11, C(6)=12, C(7)= 13, C(8)=22, C(9)=23, C(10)=33 and so on.

<u>Algorithm 1.4: To find the sequence D(p,q)</u>
D(p,q) is the ordered set difference C(q) - C(p). For exam-

ple, if C(q) = 1133 and C(p) = 13 then D(p,q)= 13.

Algorithm 1.5: To find the Fréchet derivative $N'_u(\emptyset)$ of N(u)

1. Convert the PDEX N(u) to expanded form

$$N(u) = f(x_1; \;..\; ; x_N; u; u_{,1} \;..\;).$$

2. $$N'_u(\emptyset) = \sum_{p=1}^{M(m)} \emptyset_{,C(p)} \frac{\partial f}{\partial u_{,C(p)}} .$$

Algorithm 1.6: To find the existence of variational basis of the PDE N(u) = 0 of order m

1. Convert the PDEX N(u) to the expanded form as shown in algorithm 1.5 step 1.
2. If m is odd, the variational basis does not exist.Return.
3. For p=1 to M(m) - S(m-1), check validity of the equations

$$\sum_{q=p}^{M(m)} A_{pq} * \left(\frac{\partial f}{\partial u_{,C(q)}} \right)_{,D(p,q)} = \frac{\partial f}{\partial u_{,C(p)}} \qquad (1)$$

which are obtained by equating the coefficients of $\emptyset_{,C(p)}$ in the equation

$$N'_u(\emptyset) = N^{*\prime}_u(\emptyset).$$

If all the equations (1) are satisfied, then the variational basis exists, otherwise not.

Algorithm 1.7: To find the existence of variational basis of a system of PDE's of k dependent variables in expanded form

$$\begin{aligned} &N_1(u_1, u_2, \;..\;, u_k) = 0 && u_1 \text{ major} \\ &N_2(u_1, \;..\; \;..\;, u_k) = 0 && u_2 \text{ major} \\ &\quad\vdots && \quad\vdots \\ &N_k(u_1, \;..\; \;..\;, u_k) = 0 && u_k \text{ major}. \end{aligned}$$

Here u_i major means that u_i has the highest order and number of terms in the PDE.

For variational basis to exist

$$\int_V \lfloor \emptyset_1, \emptyset_2, \;..\; \emptyset_k \rfloor \begin{bmatrix} N'_{1u_1} & N'_{1u_2} & \cdots & N'_{1u_k} \\ N'_{2u_1} & N'_{2u_2} & \cdots & N'_{2u_k} \\ & \vdots & & \\ N'_{ku_1} & N'_{ku_2} & \cdots & N'_{ku_k} \end{bmatrix} \begin{Bmatrix} \psi_1 \\ \psi_2 \\ \vdots \\ \psi_k \end{Bmatrix} dV =$$

$$\int_V \lfloor \psi_1, \psi_2, \;..\; \psi_k \rfloor \begin{bmatrix} N'_{1u_1} & N'_{1u_2} & \cdots & N'_{1u_k} \\ \vdots & & \vdots & \\ N'_{ku_1} & \cdots & \cdots & N'_{ku_k} \end{bmatrix} \begin{Bmatrix} \emptyset_1 \\ \vdots \\ \emptyset_k \end{Bmatrix} dV$$

which implies satisfaction of the following equalities :

$$N'_{iu_j} = \tilde{N}'_{ju_i} \quad , \quad \text{for } i,j = 1,2, \ldots ,k.$$

The existence of variational basis produces a favourable weak form that can utilize the Ritz type Finite Element solution procedure.

FORMULATION OF ESSENTIAL BOUNDARY CONDITIONS(EBC) AND THE CORRESPONDING NATURAL BOUNDARY CONDITIONS(NBC) FOR EVEN ORDER(2m) LINEAR PDE's IN N=DIMENSIONAL SPACE

Algorithm 2.1: To find EBC's and NBC's of a PDE N(u)=0 of order 2m

1. Express N(u) in expanded form, so that the PDE becomes

$$a_0 + a_1 u + a_2 u_{,1} + a_3 u_{,2} + \ldots = 0.$$

2. The following boundary terms are obtained after integrating by parts the equation (so that continuity requirement of u is halved):

$$\int_V vN(u)\, dV = 0$$

where v is the test function or variation of u, i.e. δu.

$$\sum_{p=0}^{m-1} \sum_{k_p=1}^{N} \ldots \sum_{k_1=k_2}^{N} (-1)^p v_{,k_p \ldots k_1} \sum_{r_p=1}^{m-p} \sum_{r_{p-1}=r_p+1}^{m-p+1} \ldots \sum_{r_1=r_2+1}^{m-1}$$

$$\sum_{w=2r_1+2}^{2m} \sum_{j \in Int(p,w,k,r)} \sum_{q=r_1}^{S(w)} (-1)^{q-p} n_{i_{q+1}(j)} \, G(a_j)_{,i_1(j) \ldots i_q(j) - i_{r_1}(j) \ldots i_{r_p}(j)}$$

$$* \; u_{,i_{q+2}(j) \ldots i_w(j)}$$

with $r_1 = 0$ for $p = 0$.
With this the following EBC's and NBC's can be formulated.

3. For each p varying from 0 to m-1, there will be $T_2(p)$ EBC's and NBC's in the following way:

For each EBC $(-1)^p \, u_{,k_p \ldots k_1}$, where $k_p, \ldots, k_1 = 1,2, \ldots, N$ and $k_p \ldots k_1$ is an increasing sequence, the corresponding NBC will be

$$\sum_{r_p=1}^{m-p} \sum_{r_{p-1}=r_p+1}^{m-p+1} \ldots \sum_{r_1=r_2+1}^{m-1} \sum_{w=2r_1+2}^{2m} \sum_{j \in Int(p,w,k,r)} \sum_{q=r_1}^{S(w)} (-1)^{q-p} n_{i_{q+1}(j)} \; *$$

$$G(a_j),_{i_1(j) \,..\, i_q(j)} \quad * \quad u,_{i_{q+2}(j) \,..\, i_w(j)} \\ -i_{r_1}(j) \,..\, i_{r_p}(j)$$

with $r_1 = 0$ if $p = 0$.
Here

$$\text{Int}(p,w,k,r) = \bigcap_{l=1 \text{ to } p} \{St(w,k_l,r_l) \text{ to } St(w,k_l,r_l) + Sp(w,k_l,r_l) - 1\}$$

k and r are arrays with p components containing $k_p \,..\, k_1$ and $r_p \,..\, r_1$ respectively.

St(p,i,x) is a set of starting points

$$= (\,..\, (T_1(p) + \sum_{t=1}^{i-1} \sum_{r=0}^{t-1} (-1)^r \, {}^{t-1}C_r \, T_2(p-s_t-r))_{s_{i-1}} = x \text{ dto } s_{i-2} \,..\,)_{s_1} = x \text{ dto } 1.$$

dto means counting downto.

Sp(p,i,x) is a span number

$$= \sum_{s=0}^{i=0} (-1)^s \, {}^{i-1}C_s \, T_2(p-x-s)$$

$$T_2(h) = {}^{N+h-1}C_h$$

$$T_1(h) = 1 \text{ for } h=0$$

$$= 1 + \sum_{i=1}^{h-1} T_2(i) \quad \text{for } h > 0$$

$$S(h) = h \text{ div } 2 - 1$$

The function $i_t(j)$ returns the tth digit in the jth sequence. Thus, for example, if N=3 then

j	sequence
1	no sequence
2	1
3	2
4	3

j	sequence
5	11
6	12
7	13
8	22
9	23
10	33
....	

and $i_2(7) = 3$, $i_1(9) = 2$, and so on.

$n_{i_{q+1}(j)}$ is the component of the normal to the boundary surface of the $x_{i_{q+1}(j)}$ axis

The function $G(a_j)$ is defined as follows:

$$G(a_j) = a_j \quad \text{if } h(j) > m$$

$$= a_j + (-1)^{h(j)} \sum_{\substack{j1 \in \\ Int(h(j),\, 2h(j), k1, r1)}} a_{j1'i_1(j1)\,..\,i_{h(j)}(j1)}$$

$$+ (-1)^{h(j)-1} \sum_{\substack{j2 \in \\ Int(h(j), 2h(j)-1, k2, r2)}} a_{j2'i_1(j2)\,..\,i_{h(j)-1}(j2)}$$

where $k1_l = i_l(j)$ and $r1_l = l + h(j)$, $l = 1$ to $h(j)$

and $k2_l = i_l(j)$ and $r2_l = l + h(j) - 1$, $l = 1$ to $h(j)$

for $h(j) \leq m$.

$i_1(j) \,..\, i_q(j) - i_{r_1}(j) \,..\, i_{r_p}(j)$ is the ordered set difference resulting the subsequence when $i_{r_1}(j) \,..\, i_{r_p}(j)$ is set subtracted from $i_1(j) \,..\, i_q(j)$.

For example, with N=3, q=2, j=7, p=1, $r_1=1$

$$i_1(7)i_2(7) - i_1(7) = 13 - 1 = 3.$$

Lastly,

$$h(j) = t \text{ such that } T_1(t) \leq j < T_1(t) + T_2(t).$$

Algorithm 2.2: To find EBC's and NBC's corresponding to a system of even order linear PDE's involving w dependent variables

1. Convert the w linear PDE's to the expanded form as shown in algorithm 1.7.
2. For i = 1 to w do steps 3 through 4.
3. For j = 1 to w do the following:
   Transfer v_i (test function for u_i) on $N_i(u_1, .. u_w)$ to find the boundary terms on u_j and form EBC's on u_i and NBC's on u_j using algorithm 2.1.
4. Add the NBC's from step 3 for same EBC on u_i. Thus the Nbc corresponding to the EBC on u_i is formed.

DERIVATION OF SHAPE FUNCTIONS FOR TWO DIMENSIONAL ELEMENTS WITH SINGLE DEPENDENT VARIABLE

Generalisation of shape functions for triangular and rectangular master elements is investigated. For C^0 continuous elements, generalisation has been achieved with Lagrange family and Serendipity family of interpolation functions.

Area coordinate system(4) has proven fruitful in this respect (Fig. 1). For C^n continuous elements, the generalisation is yet to be found out. But an attempt is made to provide non-conforming shape functions for n-noded rectangular elements.

Algorithm 3.1: To find a general expression for shape functions for triangular and rectangular master elements with single dependent variable and C^0 continuity.

Lagrange Family:

Shape function for node i

$$N_i = \prod_{k=1}^{sides} \prod_{k=0}^{P_k(i)-1} \frac{L_k - L_k(j)}{L_k(P_k(i)) - L_k(j)},$$

where sides = 3 for triangular element and = 4 for rectangular element.

$P_k(i)$ is a projection operator returning the index of the line parallel to L_k and passing through the node i. The index obviously depends on the number of equidistant parallel lines from $L_k = 0$ to $L_k = 1$.

Serendipity Family:

As is known, Serendipity family of interpolation functions consider nodes only on the element boundaries. The shape functions given below generalise rectangular elements, but for triangular elements, internal nodes come into play for sides above cubic.

For non-corner nodes i on $L_p = 0$ and (n+1) nodes on L_p (numbered 0,1,2, ... ,n)

$$N_i = \prod_{\substack{j=0 \\ j \neq P_{k(p)}(i)}}^{n} \frac{L_{k(p)} - L_{k(p)}(j)}{L_{k(p)}(P_{k(p)}(i)) - L_{k(p)}(j)} * L_{k(p+1)}$$

where k(p) = (p+1) mod sides.

For corner nodes i on $L_p = 0$ with (n+1) nodes, and on $L_{k(p)}$ with (m+1) nodes

$$N_i = \prod_{j=1}^{sides} L_{k(p+j)} - \sum_{j=1}^{n-1} (1-j/n) N_{i(L_p=0, L_{k(p)}=L_{k(p)}(j))} - \sum_{j=1}^{m-1} (1-j/m) N_{i(L_p=L_p(j), L_{k(p)}=0)}$$

Algorithm 3.2: To find non-conforming shape functions for rectangular master elements with C^n continuity and single dependent variable

In this algorithm normalised coordinate system (x_1, x_2) has been used. For an element with n nodes dof's/node are the EBC's evaluated by algorithm 2.1. Let us consider a 2m order linear PDE. Here N=dimension of the system=2. The following steps may be followed:

1. No. of dof's/node

$$= \sum_{p=0}^{m-1} T_2(p) = m(m-1)/2.$$

2. There will be m(m-1)/2 shape functions associated with each node i. Let us denote them by

$$N_i^j(x_1, x_2), \quad j=1 \,..\, m(m-1)/2.$$

3. Each shape function N_i^j may be represented by a series of n x m(m-1)/2 terms each of the form

$a_r \, x_1^{k(r)} \, x_2^{l(r)}$, r = 1 .. n x m(m-1)/2, where k(r) and l(r) are so chosen that a symmetric pattern is marked on the Pascal's triangle. In this connection the following rules may be used

Let b(n,m) = n x m(m-1)/2
If b(n,m) $\in$ {x: x=4i, i=1,2,3, ..} then a Serendipity combination may be chosen.
If b(n,m) $\in$ {x: x=4+5i, i=0,1,2, ..} then a Lagrangian combination may be chosen.
If b(n,m) $\in$ {x: x=i(i+1)/2, i=1,2, ..} then a complete (i-1)th degree polynomial may be exploited.

4. Define a matrix A_{pq} such that

$$A_{pq} = \left[(-1)^{j(p)} \left(x_1^{k(q)} x_2^{l(q)} \right)_{,C(j(p))} \right] \Big|_{\text{at node } i(p)}$$

where $i(p) = (p-1) \text{ div } n + 1$
$j(p) = (p-1) \text{ div } n$
and C(j(p)) is the sequence of digits found out by the algorithm 1.3.
This matrix A_{pq} is same for all the shape functions.

5. Define vector B_p, p = 1 to b(n,m) for a given N_i^j in the following way :

$$B_p = 1 \text{ for } j = j(p) \text{ and } i = i(p)$$
$$= 0 \text{ otherwise.}$$

6. Solve for the coefficient vector a_p considering A_{pq} and B_p by Gauss elimination method.

7. The shape function

$$N_i^j(x_1, x_2) = \sum_{p-1}^{b(n,m)} a_p \, x_1^{k(p)} \, x_2^{l(p)}.$$

It may be noted that the foregoing algorithm exploits the idea of Hermitian polynomials.

APPROACH TO AUTOMATION THROUGH A NEWLY DEFINED LANGUAGE

It has been shown above that quite a lot of information is contained in the field PDE('s), which may be utilized very well in checking the variational basis existence, formulating EBC's and NBC's, and formulating shape functions. As almost 70% of the whole job in Finite Element Analysis is covered by preprocessing, and about 10% by postprocessing, it is worth trying to automate these portions so that less skilled expertise is needed and mistakes can be avoided. Following is a programme written in a language newly defined, compilation of which may produce part of preprocessor and postprocessor - the extent to be evaluated after implementation. The problem is stated first.

The problem

Consider the steady heat conduction in a two dimensional domain V, enclosed by lines AB, BC, CD, DE, EF, FG, GH, and HA (Fig. 2). The governing equation is given by

$$-k\left(\frac{\partial^2 T}{\partial x^2} + \frac{\partial^2 T}{\partial y^2} \right) = 0 \text{ in V},$$

where k is the conductivity of the material of the domain (considered isotropic). The boundary conditions are as follows:

S1 = AB specified heat flux, $q(y)$
S2 = BC specified temperature, $T_o(x)$
S3 = CD convective boundary, T_∞
S4 = DEFGHA insulatedboundary, $\frac{\partial T}{\partial n} = 0$.

To find temperature distribution throughout the domain.

The programme

```
specify EXMP;
axes
   XYZ;  #expansion of V#
data
   k,a,b,a1,b1,h,TINF: scalar;
   Q,TO: function;
var
   T: scalar;  #dependent variable#
field   #field PDE#
   -k * del(T) = 0;
domain  #any geometrical modeller commands may be used #
        #N=2 found #
   two=(0.0,b) for one=(0.0,(a-a1)/2)+((a+a1)/2,a);
```

```
   two=(0.0,b-b1) for one=((a-a1)/2,(a+a1)/2);
   S1=(one=0.0 and two=(0.0,b));
   S2=(one=(0.0,a) and two=0.0);
   S3=(one=a and two=(0.0,b));
   S5=(one=(0.0,(a-a1)/2) and two=b);
   S6=(one=(a-a1)/2 and two=(b-b1,b));
   S7=(one=((a-a1)/2,(a+a1)/2) and two=b-b1);
   S8=(one=(a+a1)/2 and two=(b-b1,b));
   S9=(one=((a+a1)/2,a) and two=b);
   S4=S5+S6+S7+S8+S9;
object  #data definition, some extra data are given #
        #for example #
   k=50; a=40; b=30; a1=8; b1=10; h=100; TINF=25;
boundary  #imposition of boundary conditions #
          #cross checking with EBC's & NBC's #
   -normal dot (k * del(T)) = Q(two) on S1;
   T= TO(one) on S2;
   -normal dot (k * del(T)) = h * (T-TINF) on S3;
   -normal dot (k * del(T)) = 0 on S4;
output   #for postprocessor #
   T at nodes;
end.
```

The bold faced keywords are self explanatory. Here **one** means the first coordinate axis, x, and **two** means the se-coordinate axis y.

LIST OF SYMBOLS

$\frac{D}{Dt}$	Material derivative
ρ	Density
e	Internal energy
v	Magnitude of the velocity vector v
$\bar{q}$	Heat flux vector
$\overleftrightarrow{T}$	Stress tensor
$\bar{g}$	Gravity vector
Φ	Heat generation rate
$\overleftrightarrow{k}$	Conductivity tensor
θ	Temperature
T	Temperature
$\hat{n}$	Unit normal to surface
div	Integer division operator
mod	Remainder on division operator

REFERENCES

1. Finlayson,B.A., The Method of Weighted Residuals and Variational Principles, Academic Press, New York, 1972.
2. Reddy,J.N., An Introduction to Finite Element Method, McGraw-Hill Book Co., Singapore, 1985.
3. Whitaker,S., Fundamental Principles of Heat Transfer, Pergamon Press Inc., 1977.

4. Zienkiewicz, O.C., The Finite Element Method, Third edn. McGraw-Hill Book Co.(UK) Ltd., 1977.

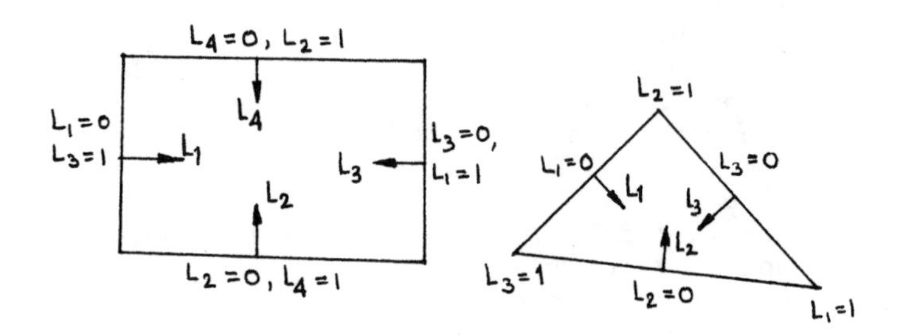

FIG.1 AREA COORDINATE SYSTEM

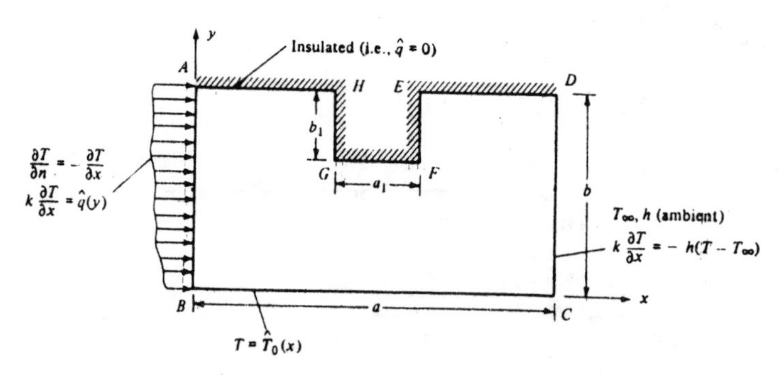

FIG. 2 The Problem

On the Boundedness of Convective Approximation

S. He

Simon Engineering Laboratories, The University of Manchester, Manchester M13 9PL, England

ABSTRACT

The property of boundedness of differencing schemes for conservation equations are discussed and new opinions presented. A methodology is then proposed guaranteeing boundedness of a scheme, particularly for schemes of higher order. A new scheme for boundedness called Quasi-QUICK is derived based on the idea. Satisfactory results have been achieved for test problems including a cavity driven flow and pure convective problems and have been compared with solutions from some well known schemes.

1. INTRODUCTION

A general tensorial form of the equations describing the conservation of any quantity ϕ, can be expressed as:

$$U_j \frac{\partial \phi}{\partial x_j} = \frac{\partial}{\partial x_j} \left(\mu \frac{\partial \phi}{\partial x_j}\right) + S \tag{1}$$

In the discretization of the above expression, the approximation of the convective terms have been found to be most significant. Considerable efforts have been made by many workers in order to achieve accuracy, stability and boundedness or a compromise of them in approximating the terms. Upwind differencing (Gentry 1966) was a great success in representing convective transport, which removed the stability problems found with the central differencing and was once considered to open the way to making numerical simulations of flow phenomena at indefinitely high Reynolds numbers. However, it soon became clear that numerical diffusion imparted increasingly important errors to the solution as the inter-nodal Reynolds numbers were raised, unless the numerical mesh could be aligned fairly close with the fluids path lines. Indeed, physical diffusion could be far outweighed by false diffusion in many flow patterns. There has therefore been considerable effort aimed at devising formulations which do not display errors due to false diffusion of standard upwind differencing while retaining its desirable stability properties. Many schemes have been built up with high accuracy and reasonable stability and some of them, such as QUICK and Skew-upwind differencing, have been used with great success in many applications. However, such schemes including QUICK and Skew-upwind still suffer from so-called unboundedness in applications with sharp changes of the transferred quantities, this could be seen in the original papers where they were first proposed (Raithby 1976, Leonard 1979) and in later application papers (Leschziner 1980, Huang 1985). In fact, as pointed out later in this presentation, none of the normal high order schemes could guarantee boundedness for all sorts of problems. In some calculations, the property of boundedness is strictly required so upwind differencing has to be employed instead of higher order schemes, as a result "artificial diffusion" is introduced and accuracy may be greatly reduced.

In this presentation, the property of boundedness of differencing schemes will be discussed and new opinions presented. A methodology is then proposed guaranteeing boundedness, particularly for higher order schemes. A new scheme of boundedness called Quasi-QUICK is derived based on the idea. Test problems including cavity driven flow and pure convective problems are presented and compared with some well known schemes.

2. ANALYSIS

A general expression for the finite difference discretization of the conservation equations can be written as:

$$AP\ \phi_P = AE\ \phi_E + AW\ \phi_W + AN\ \phi_N + AS\ \phi_S + S, \quad (2)$$

where ϕ_E, ϕ_W, ϕ_N, ϕ_S are the ϕ values at the points located to the east, west, north and south of point P (see Fig 1), AP = AE + AW + AN + AS and S is the source term and the contribution of the other neighboring points. Different methods in approximating the convective terms in the conservation equations result in different coefficients and a different source term in the above expression. "Unboundedness", which is used to describe nonphysical oscillatory behavior of a numerical solution of a certain differencing scheme to the above problem can be expressed as:

$$\phi_p > \text{Max}\{\ \phi_E,\ \phi_W,\ \phi_N,\ \phi_S\ \}$$

or $$\phi_p < \text{Min}\{\ \phi_E,\ \phi_W,\ \phi_N,\ \phi_S\ \}. \quad (3)$$

This reveals the fact that provided that there is no source for the quantity ϕ, the value of ϕ at the centre should not be higher than the highest value and lower than the lowest one of the points around it. Absolutely bounded differencing schemes can be further defined as those for which the bounded solutions are always guaranteed whatever problem the scheme is used for. Based on the above definition, the following statement can be proved,

Absolute boundedness condition: A scheme is absolutely bounded when and only when the corresponding coefficient matrix is diagonally dominant, i.e.,

$$ABS(AP) \geq ABS(AE) + ABS(AW) + ABS(AN) + ABS(AS) \quad (4)$$

Alternatively it can be stated as: none of the coefficients in equation 2 is negative.

The two statements are identical simply because AP = AE + AW + AN + AS . This condition is well known to be a sufficient one for guaranteeing bounded solutions to the present kind of problem, and if absolute boundedness is required it becomes a necessary condition as well. This can be demonstrated by an extreme situation. Consider a very simple problem with five points, four boundary points and one internal. The value of the dependent variable ϕ at the internal point is then expressed by its values at surrounding points as

$$\phi_i = (\ AE^*\phi_E + AW^*\phi_W + AN^*\phi_N + AS^*\phi_S)/(AE + AW + AN + AS). \quad (5)$$

If the diagonally dominant condition is not satisfied, one or more of the four coefficients must be negative. Here we assume AE is negative. Then, from the above expression, a negative ϕi will be expected provided that $\phi_E, \phi_W, \phi_N, \phi_S \geq 0$ and $\phi_E >$ ABS(AW * ϕ_W + AN * ϕ_N + AS * ϕ_S) / AE); undershooting then occurs. Similar results can be obtained if it is assumed that more than one coefficient is negative. Therefore all the coefficients must be non-negative when absolute boundedness is demanded.

However, it is obvious that diagonal dominance is not a practical criterion for assessing the boundedness of a scheme in its development. In fact it is widely known that most of the successful higher order schemes do not satisfy the condition, yet they give excellent results in many applications. Although schemes like QUICK and upwind-skew produce overshooting and undershooting for pure convective problems (Huang 1985, Leschziner 1980), they are still widely applauded and work well in a great number of applications. The reasons for this are simply that either boundedness of the solution is not strictly required in the application, or that the scheme is bounded for the specific problem. Indeed, boundedness is a matter that has to be inspected carefully for each individual application. In some cases, overshooting or undershooting are not serious offenses, instead accuracy is most important. In this sort of problem, some small nonphysical oscillations may help accelerate convergence of the solution. Problems for which boundedness is essential should still be considered individually, since generally speaking the boundedness property of a scheme is problem-dependent, i.e., a scheme could be bounded when it is used with some problems but produces over- or undershooting with others as mentioned in the earlier part of this section. Problems with lower real diffusion and/or with sharp changes of the values of dependent variables are more sensitive to unboundedness. Some problems or even some parts of the field of a problem need more attention than others in achieving boundedness.

As a closure of the discussion in this section, the author would like to suggest the following practical criterion for the development of a universally bounded scheme,

Practical boundedness criterion: A higher order scheme could

be built-up guaranteeing universal boundedness for all problems by taking into account information provided by the specific problem in the scheme formulation, i.e., the coefficient matrix of the scheme includes the local values of the dependent variables.

3. QUASI-QUICK

One of the applications of the practical boundedness criterion is to modify an existing higher order scheme to be universally bounded. This can be achieved by either introducing artificial-diffusion terms controlled by local values of the dependent variable into a high order scheme, or by combining a higher order scheme with an absolutely bounded lower order scheme, say upwind differencing, by weighting factors which are determined by the local values of the dependent variables. As an example, Quasi-QUICK based on the latter approach will be described below.

Upwind differencing is introduced, with weighting factor β, to modify the normal QUICK scheme to guarantee boundedness, i.e.,

$$Ai = (1-\beta)\ Aui + \beta * Aqi, \qquad i = E, W, N, S \qquad (6)$$

where Aqi and Aui are coefficients used in QUICK and Upwind respectively. $\beta \in [0,1]$, is the weighting factor and should be determined by the local values of dependent variables in order to satisfy the conditions for boundedness. The policy of Quasi-QUICK is to use QUICK as a main differencing method and introduce upwind as a portion of the new scheme but minimize the use of it to that really necessary for boundedness. A scheme built-up this way will possess most of the advantages of the ordinary QUICK, including high accuracy, as it is mainly QUICK except at some "sharp" changing points, which standard QUICK could not simulate well. On the other hand, the property of boundedness of the solution can always be guaranteed since the upwind scheme satisfies the absolute boundedness conditions and in extreme situations Quick is completed replaced by upwind differencing.

In order to obtain an expression for β, Ai in Eq 3 is

substituted by $Ai = (1-\beta)\, Aui + \beta * Aqi$, so that,

$$\phi min \le \frac{\sum_{E,W,N,S} (\beta Aq_i + (1-\beta)) Au_i) \phi_i}{\sum_{E,W,N,S} (\beta Aq_i + (1-\beta) Au_i)} \le \phi max, \tag{7}$$

After rearrangement, βi is found to be

$$\beta \le \sum Aqi(\phi i-\phi min)/\sum [(Aui-Aqi)(\phi i-\phi min)] \tag{8}$$

and

$$\beta \le \sum Aqi(\phi max-\phi i)/\sum [(Aui-Aqi)(\phi max-\phi i)] \tag{9}$$

A practical difficulty is that at the nth iteration, ϕi^n are needed in the calculation for β^n, but they are unknown. This can be solved by limiting the movement of values of the dependent variable by $D\phi$ in each iteration, i.e., $\phi^{n-1} - D\phi \le \phi i^n \le \phi i^{n-1} + D\phi$. Equations 8 and 9 then become:

$$\beta \le \sum Aqi((\phi i-D\phi)-\phi min)/\sum [(Aui-Aqi)((\phi i+D\phi)-\phi min)] = \beta 1 \tag{10}$$

$$\beta \le \sum Aqi(\phi max-(\phi i+D\phi))/\sum [(Aui-Aqi)(\phi max-(\phi i-D\phi))] = \beta 2 \tag{11}$$

And finally the expression for determining the value of β is

$$\beta = \max\{ 0, \min\{ \beta 1, \beta 2, 1 \}. \tag{12}$$

4. TEST PROBLEMS

Problem 1. Convective transport of a scalar step.

A scalar step-shaped ϕ-distribution is convected across a square solution domain by a uniform stream inclined to the numerical mesh, as shown in Fig 2(a). Molecular diffusion is assumed to be zero. The problem appears to be exceedingly simple, yet has been often used to examine the relative performance of different numerical approximations to convection due to its extremely sharp gradient in ϕ. A comparison of $\phi(y)$ along x=0.5 obtained with upwind, QUICK and Quasi-QUICK at three different flow to grid inclinations with mesh 30×30 are shown in Fig 2. It can be seen that the upwind approximation results in a very large diffusion ϕ profile for all the three values of ϑ. This is clearly due to the influence of artificial diffusion, which is the most serious at the maximum flow angle ϑ = 45. On the other hand, the step gradient is fairly well preserved by QUICK approximation at all flow angles. However unboundedness (both undershooting and overshooting) is produced. In contrast to the above, both accuracy and boundedness are achieved by Quasi-QUICK. The sharp gradient in ϕ is fairly well preserved and boundedness is guaranteed.

Problem 2. Wall-driven square-cavity flow.
The wall-driven cavity flow has been examined by many workers and accurate numerical data are readily available for this case. The numerical solution presented by (Schreiber, 1983) with a 180×180 mesh is chosen to be the "accurate solution" of the problem. Solutions obtained by upwind, Power-law and QUASI-QUICk, together with those from (Schreiber, 1983) for $Re=UwL/v = 1000$ with a 50×50 mesh are plotted in Fig 3. It is clear that the velocity profile obtained by Quasi-QUICK is much better than those obtained by the upwind and Power-law and is fairly close to the "accurate" solution.

5. CONCLUSION

It has been shown in the presentation that the property of diagonal dominance of the coefficient matrix of a scheme is a sufficient and necessary condition for guaranteeing that the scheme is absolutely bounded. This is obviously too stringent a requirement for the development of differencing schemes of higher orders. In fact, since boundedness is problem dependent, the assessment of the property of boundedness of a scheme should normally by carried out when it is applied to certain problems. On the other hand, a universally bounded scheme can always be developed by taking account of information provided by the problems in scheme development, i.e., by including the local values of the dependent variables in the coefficient matrix of the scheme. Quasi-QUICK, proposed in the presentation, produces very good solutions for the two test problems in respects of both its accuracy and boundedness.

6. REFERENCE

1. Gentry, R. A., Martin, R. E. and Daly, B. J.

An Eulerian differencing method for unsteady compressible flow problems, J. Comput. Phys. 1 (1966) 87.

2. Huang, P. G., Launder, B. E. and Leschziner, M. A.
Discretization of nonlinear convection processes: a broad-range comparison of four schemes, Computer Methods in Applied Mechanics and Engineering 48 (1985), 1-24.

3. Leonard, B. P.
A stable and accurate convective modelling procedure based on quadratic upstream interpolation, Computer Methods in Applied Mechanics and Engineering 48 (1979) 1 - 24.

4. Leschziner, M. A.
Practical evaluation of three finite difference scheme for the computation of steady-state recirculating flows, Computer Methods in Applied Mechanics and Engineering 423 (1983), 293-312.

5. Raithby, G. D.
Skew upstream differencing schemes for problems involving fluid flow, Computer Methods in Applied Mechanics and Engineering 9 (1976) 153 - 164.

6. Schreiber, R. and Keller, H. B.
Driven cavity flows by efficient numerical techniques. J. Comput. Phys. 49 (1983) 310.

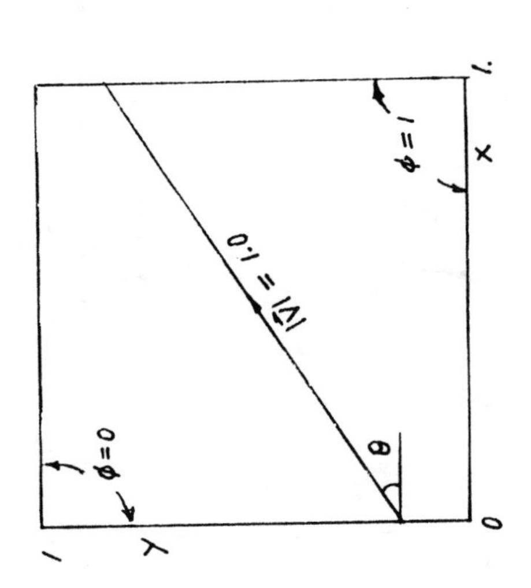

Fig. 2 (a). Convective transport of scalar.

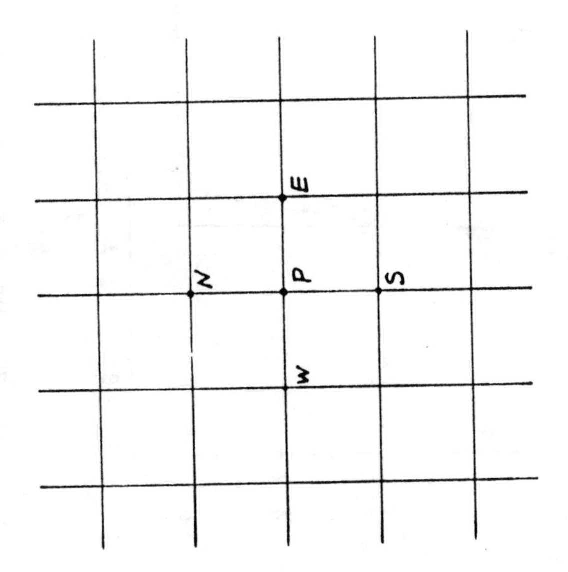

Fig. 1. Computational nodal.

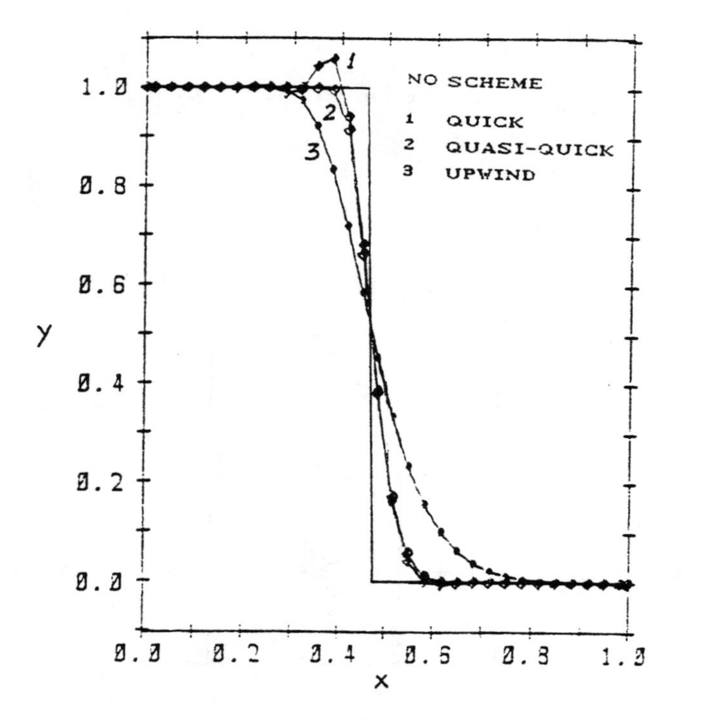

Fig. 2 (b). Convective transport of scalar (ϑ=25°).

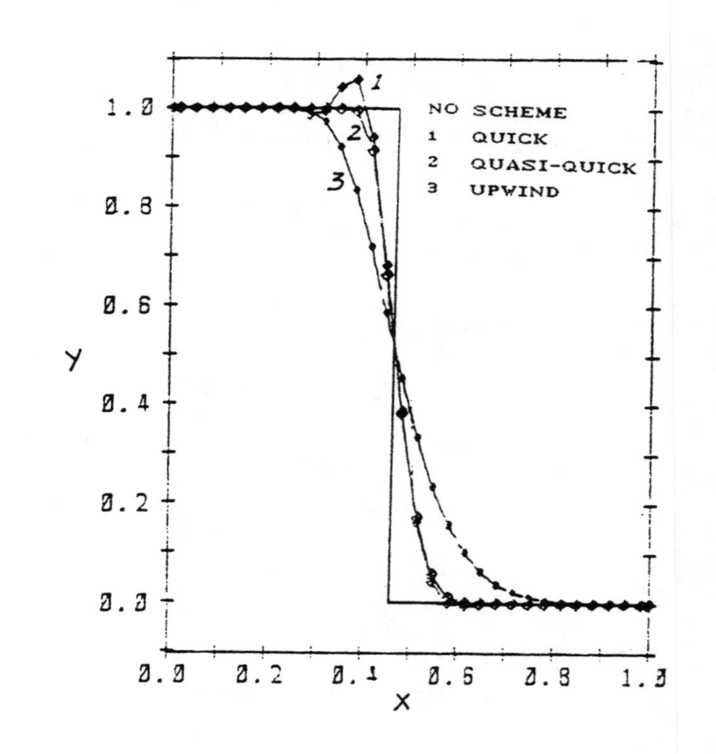

Fig. 2 (c). Convective transport of scalar (ϑ=35°).

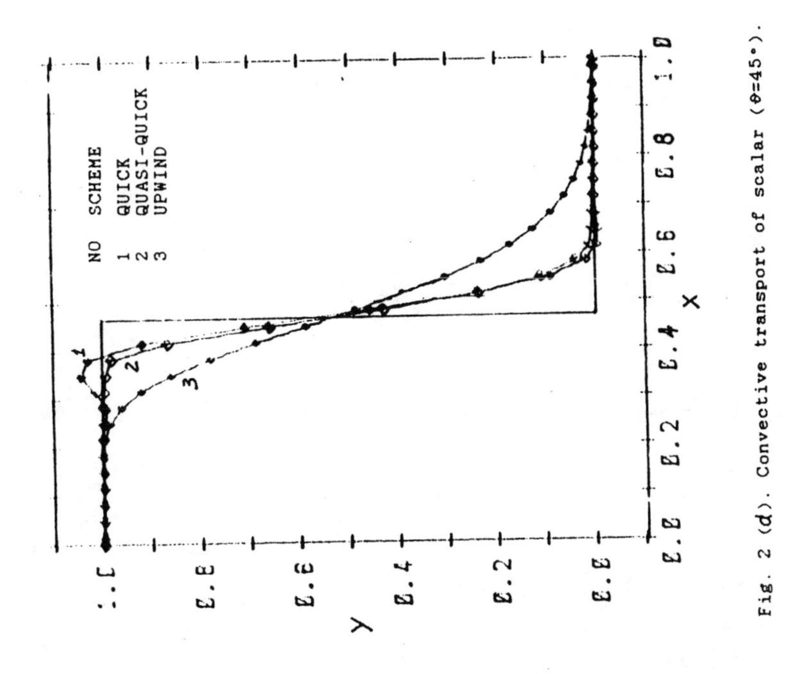

Fig. 2 (d). Convective transport of scalar (θ=45°).

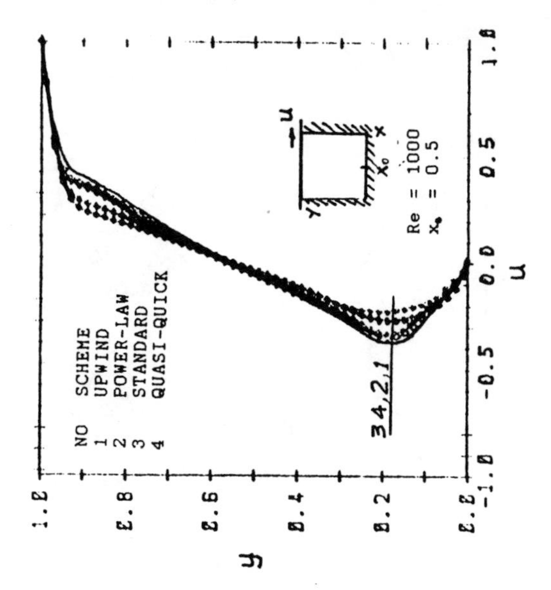

Fig. 3. Flow in a wall-driven cavity.